Electromagnetic Sensing and Its Applications

Electromagnetic Sensing and Its Applications

Guest Editors

Wuliang Yin
Mingyang Lu
Ruochen Huang

MDPI

Basel • Beijing • Wuhan • Barcelona • Belgrade • Novi Sad • Cluj • Manchester

Guest Editors
Wuliang Yin
Department of Electrical and Electronic Engineering
University of Manchester
Manchester
UK

Mingyang Lu
CNDE
Iowa State University
Ames
USA

Ruochen Huang
College of Electrical Engineering and Automation
Fuzhou University
Fuzhou
China

Editorial Office
MDPI AG
Grosspeteranlage 5
4052 Basel, Switzerland

This is a reprint of the Special Issue, published open access by the journal *Sensors* (ISSN 1424-8220), freely accessible at: https://www.mdpi.com/journal/sensors/special_issues/0WPS1Z60B6.

For citation purposes, cite each article independently as indicated on the article page online and as indicated below:

Lastname, A.A.; Lastname, B.B. Article Title. *Journal Name* **Year**, *Volume Number*, Page Range.

ISBN 978-3-7258-6714-1 (Hbk)
ISBN 978-3-7258-6715-8 (PDF)
https://doi.org/10.3390/books978-3-7258-6715-8

Contents

About the Editors

Wuliang Yin

Wuliang Yin is a Professor of Electromagnetic Sensing Systems and Instrumentation at the University of Manchester, Manchester, U.K. He is a world-leading expert in both theoretical and applied electromagnetic research. He developed new EM formulations and innovative instrumentation for materials characterization and defect inspection. He has authored one book and more than 450 papers and has been granted more than twenty patents in electromagnetic (EM) sensing and imaging. He received multiple awards for his work, including the Williams Award from The Institute of Materials, Minerals, and Mining in 2014 and 2015, the Best Paper Award from the IEEE I2MTC Conference in 2016, the Best Application Award from the IEEE Instrumentation and Measurement Society in 2020, an IEEE Gold Medal as the most productive reviewer in the U.K. for *IEEE Trans.* Instrumentation and Measurement in 2020, and the IEEE Instrumentation and Measurement Society Graduate Fellowship as a Supervisor in 2018, 2022, 2022, 2024, and 2025. He recently received the William Gardner Award from the BINDT for his work on sensor-driven autonomous robotics.

Mingyang Lu

Mingyang Lu received the B.S. (2014) and Ph.D. (2018) degree from the School of Electrical and Electronic Engineering, University of Manchester. He is currently an Adjunct Assistant Professor in the Department of Electrical and Computer Engineering, Iowa State University; He is also a Scientist and Principal Investigator (PI) of eddy current projects and in charge of 3 electromagnetic eddy current labs at the Center for Nondestructive Evaluation, Iowa State University. He was a researcher at the University of Manchester. His research interests include electromagnetic (EM) testing, eddy current techniques, and instrumentation and measurement. He is an Associate Editor of *IEEE Open Journal of Instrumentation and Measurement* and *Research in Nondestructive Evaluation*. He was a recipient of the *IEEE Instrumentation and Measurement* Outstanding Reviewer Award in 2020 and 2021. He was an Associate Editor of *IEEE TIM* (2021-2023).

Ruochen Huang

Ruochen Huang received the B.S. (2018) degree from the School of Electrical and Electronic Engineering, University of Manchester and North China Power Electric University; and the Ph.D. (2021) degree from the School of Electrical and Electronic Engineering, University of Manchester, Manchester, U.K. She is currently an Associate Professor with the College of Electrical Engineering and Automation, Fuzhou University. She was selected as a Qishan Scholar (Overseas Project) of Fuzhou University in 2022 and has published more than 40 peer-reviewed articles. Her main research interests include finite-element method (FEM) modeling software packages for electromagnetic (EM) simulations, EM inversion algorithms, and model-based and data-driven fault diagnosis.

Preface

We would like to dedicate this Reprint to the late Professor Huaxiang Wang, who devoted his life to the field of measurement and influenced many younger generations of researchers, including myself, who benefited immensely from his mentorship during my MSc studies and later in my professional life.

Wuliang Yin, Mingyang Lu, and Ruochen Huang
Guest Editors

sensors MDPI

Editorial

Electromagnetic Sensing and Its Applications

Wuliang Yin [1,*], Mingyang Lu [2,*] and Ruochen Huang [3]

[1] Department of Electrical and Electronic Engineering, University of Manchester, Manchester M60 1QD, UK
[2] Center for Nondestructive Evaluation, Iowa State University, Ames, IA 50011, USA
[3] College of Electrical Engineering and Automation, Fuzhou University, Fuzhou 350108, China; ruochen_huang@fzu.edu.cn
* Correspondence: wuliang.yin@manchester.ac.uk (W.Y.); mingylu@iastate.edu (M.L.)

Electromagnetic sensing offers the ability to interact with the physical world beyond our five senses. The technique is used to interrogate the integrity and properties of materials in many industrial settings.

This Special Issue collected a small but diverse set of studies, covering applications ranging from measuring flow velocity [1], thickness [2,3], defects [4,5], steel microstructure [3], and magnetic fields [6,7] to evaluating shielding effectiveness [8], imaging of uniaxial objects [9] and generating leaky waves [10]. The reported frequency range spans from typical eddy current frequencies (below MHz) to radio frequency and up to microwave frequencies. The industrial sectors involved include the rail industry, the steel industry, etc., using multiphase flow measurements for magnetic measurement in synchrotrons. New techniques have been proposed such as data processing techniques using convolutional neural networks [9] and array antenna designs [10].

These issues show the versatility of the technique and the vibrant research activities carried out in our community.

Despite these achievements, important gaps remain. Many sensing techniques still face challenges in scaling from controlled laboratory environments to complex industrial settings, where noise, variability in material properties, and harsh operating conditions can limit accuracy and reliability. Furthermore, while individual applications have demonstrated success, there is a need for unified frameworks that connect electromagnetic sensing data with broader digital manufacturing and predictive maintenance systems. The field must also more deeply explore multiphysics interactions, especially in contexts such as multiphase flow measurement and fracture aperture monitoring, where electromagnetic signals must be interpreted alongside mechanical and fluid dynamics.

This Special Issue has contributed to bridging some of these gaps by presenting diverse approaches—from novel sensor designs to advanced data processing methods—that demonstrate both the versatility and adaptability of electromagnetic sensing. By showcasing applications across industries, the collected studies highlight how the community is actively pushing the boundaries of what can be measured and monitored.

Looking forward, several research directions stand out as promising:

- Integration with AI and machine learning: Beyond CNN-based imaging, future studies should explore adaptive algorithms capable of real-time decision-making in noisy industrial environments;
- Scalability and robustness: Developing sensors and systems that maintain accuracy under variable operating conditions remains a critical challenge;
- Multiphysics modeling: Coupling electromagnetic sensing with mechanical, thermal, and fluid models will enhance interpretation and predictive power;

- Miniaturization and portability: Compact, low-cost sensors could expand applications into new domains, including field inspections and consumer technologies;
- Standardization and interoperability: Establishing common protocols for data acquisition and interpretation will accelerate adoption across industries.

By addressing these areas, the community can ensure that electromagnetic sensing continues to evolve as a cornerstone technology for industrial diagnostics, materials characterization, and beyond. The progress documented here is a testament to vibrant research activity, and it sets the stage for the next wave of breakthroughs that will further extend our ability to interact with and understand the physical world.

Overall, we would like to thank all authors who have made great efforts to present their studies in their best way for this Special Issue, as well as the reviewers, all of whom have provided critical and constructive comments that contributed to the successful completion of this Special Issue.

Finally, I would like to dedicate this Special Issue to the late Professor Huaxiang Wang, who have devoted his life to the field of measurement and influenced many younger generations of researchers, including myself, as I benefited immensely from his mentorship during my MSc studies and later in my professional life.

Conflicts of Interest: The authors declare no conflict of interest.

References

1. Krauter, N.; Stefani, F. Simultaneous Measurement of Flow Velocity and Electrical Conductivity of a Liquid Metal Using an Eddy Current Flow Meter in Combination with a Look-Up-Table Method. *Sensors* **2023**, *23*, 9018. [CrossRef] [PubMed]
2. Shi, Y.; Tian, S.; Jiang, J.; Lei, T.; Wang, S.; Lin, X.; Xu, K. Thickness Measurements with EMAT Based on Fuzzy Logic. *Sensors* **2024**, *24*, 4066. [CrossRef] [PubMed]
3. Wilson, J.W.; Jolfaei, M.A.; Zhou, L.; Slater, C.; Davis, C.; Peyton, A.J. Development of an In-Situ Multifrequency Electromagnetic Sensor for Real-Time Microstructure Monitoring in a Continuous Annealing Furnace. *Sensors* **2025**, *25*, 5158. [CrossRef] [PubMed]
4. Shao, Y.; Xia, Z.; Ding, Y.; Crocker, B.; Saunders, S.; Bai, X.; Peyton, A.; Conniffe, D.; Yin, W. Eddy Current Sensor Array for Electromagnetic Sensing and Crack Reconstruction with High Lift-Off in Railway Tracks. *Sensors* **2024**, *24*, 4216. [CrossRef] [PubMed]
5. Talebkeikhah, M.; Moradi, A.; Lecampion, B. Measurement of Hydraulic Fracture Aperture by Electromagnetic Induction. *Sensors* **2024**, *24*, 6660. [CrossRef] [PubMed]
6. Vella Wallbank, J.; Buzio, M.; Parrella, A.; Petrone, C.; Sammut, N. Pulsed-Mode Magnetic Field Measurements with a Single Stretched Wire System. *Sensors* **2024**, *24*, 4610. [CrossRef] [PubMed]
7. Azaro, R.; Franchelli, R.; Gandolfo, A. Performance Evaluation and Calibration of Electromagnetic Field (EMF) Area Monitors Using a Multi-Wire Transverse Electromagnetic (MWTEM) Transmission Line. *Sensors* **2025**, *25*, 2920. [CrossRef] [PubMed]
8. Cardillo, E.; Carcione, F.L.; Ferro, L.; Piperopoulos, E.; Mastronardo, E.; Scandurra, G.; Ciofi, C. Development of a Simple Setup to Measure Shielding Effectiveness at Microwave Frequencies. *Sensors* **2024**, *24*, 3741. [CrossRef] [PubMed]
9. Chiu, C.-C.; Chiang, J.-S.; Chen, P.-H.; Jiang, H. Convolutional Neural Network-Based Electromagnetic Imaging of Uniaxial Objects in a Half-Space. *Sensors* **2025**, *25*, 1713. [CrossRef] [PubMed]
10. Calcaterra, A.; Simeoni, P.; Migliore, M.D.; Frezza, F. Leaky Wave Generation Through a Phased-Patch Array. *Sensors* **2025**, *25*, 2754. [CrossRef] [PubMed]

https://doi.org/10.3390/s26020574

sensors

MDPI

Article

Simultaneous Measurement of Flow Velocity and Electrical Conductivity of a Liquid Metal Using an Eddy Current Flow Meter in Combination with a Look-Up-Table Method

Nico Krauter and Frank Stefani *

Helmholtz-Zentrum Dresden-Rossendorf, Bautzner Landstraße 400, 01328 Dresden, Germany; n.krauter@hzdr.de
* Correspondence: f.stefani@hzdr.de

Abstract: The Eddy Current Flow Meter (ECFM) is a commonly employed inductive sensor for assessing the local flow rate or flow velocity of liquid metals with temperatures up to 700 °C. One limitation of the ECFM lies in its dependency on the magnetic Reynolds number for measured voltage signals. These signals are influenced not only by the flow velocity but also by the electrical conductivity of the liquid metal. In scenarios where temperature fluctuations are significant, leading to corresponding variations in electrical conductivity, it becomes imperative to calibrate the ECFM while concurrently monitoring temperature to discern the respective impacts of flow velocity and electrical conductivity on the acquired signals. This paper introduces a novel approach that enables the concurrent measurement of electrical conductivity and flow velocity, even in the absence of precise knowledge of the liquid metal's conductivity or temperature. This method employs a Look-Up-Table methodology. The feasibility of this measurement technique is substantiated through numerical simulations and further validated through experiments conducted on the liquid metal alloy GaInSn at room temperature.

Keywords: eddy current; inductive flow measurement; look-up-table; liquid metal

1. Introduction

The measurement of flow velocities in liquid metal is generally difficult due to their opacity and in many cases further aggravated by their high temperature and chemical reactivity. In some sense, however, these drawbacks are compensated by the high electrical conductivity of liquid metals which allows for the application of inductive flow measurement techniques. These include the Contactless Inductive Flow Tomography (CIFT) for the global reconstruction of entire two- or three-dimensional flow fields [1], and quite a few techniques for the measurement of local velocities, including the phase-shift sensor [2], the magnetic distortion probe [3], Lorentz force velocimetry [4–8], and Eddy Current Flow Meters (ECFM) [9–18]. The latter sensors are widely used in liquid metal cooled fast breeder reactors (LMFBR) as part of the reactor safety instrumentation to monitor the coolant flow velocity. Usually they consist of three solenoid coils: one primary coil that generates an alternating magnetic field and two passive secondary coils above and below the primary coil, which detect flow induced changes in the primary field [11] (see Figure 1). These changes are caused by flow induced eddy currents, which lead to an asymmetry of the eddy current distribution around the sensor and thus also to different induced voltages within the secondary coils. By measuring the difference between the magnitude and/or phase [19] of the secondary coil voltages, the flow velocity of the liquid metal can be inferred.

A significant drawback of the ECFM is that the output signal depends not only on the fluid velocity but also on the electrical conductivity, which in turn depends on the temperature of the liquid metal [20]. Typically, the coolant temperature of LMFBR is not constant, so that an extensive calibration of the ECFM is necessary for different temperatures. For small changes in the electrical conductivity of the liquid metal a temperature compensation can be performed [21], but this method only works for a certain excitation frequency. The Transient Eddy Current Flow Meter (TECFM) [22,23] overcomes the temperature-dependence problem by tracking the position of an impressed eddy current system which is moving with the same velocity as the conductive fluid. Yet, there are a couple of applications, in particular in LMFBR's, in which a complementary determination of the temperature would also be desirable.

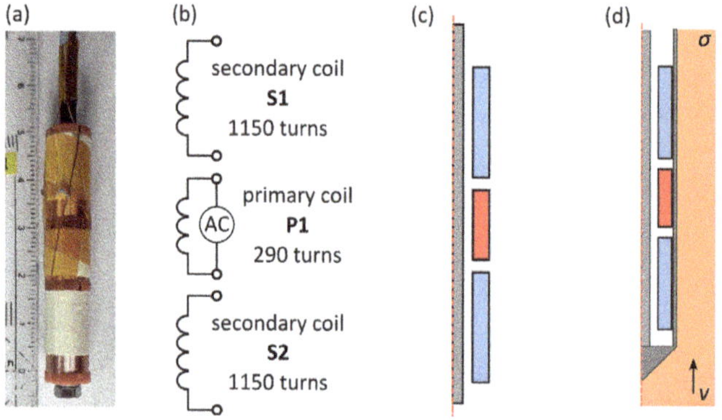

Figure 1. (**a**) Photograph, (**b**) coil overview with specifications and (**c**,**d**) simulation models of the ECFM.

In this paper we present a new method for the simultaneous determination of flow velocity v and electrical conductivity σ of the liquid metal when using an ECFM. This technique relies on the creation of a so-called Look-Up-Table (LuT) which contains a large number of output voltages for different parameter combinations of v, σ and excitation frequencies f that were obtained in advance by numerical simulation of the sensor and its surroundings. By calculating the deviations between a given measurement result (with unknown v and σ) and each value in the LuT, a certain parameter combination of velocity and conductivity can be identified for which the deviation between measurement and simulation is minimal. Here, the method of least squares is used to represent this deviation in the form of a mean squared error (MSE). The parameter combination with the lowest MSE is most likely the unknown parameter combination, assuming that the numerical model depicts reality with sufficient accuracy. Up to three excitation frequencies are used in order to increase the accuracy of this method. For the level detection of liquid magnesium in the titanium production process the LuT method has already yielded promising results [24,25]. Here we will apply a similar technique to the ECFM, demonstrating how to simultaneously obtain v and σ by a detailed analysis of the numerical simulations. We will also present first results of measurements that were performed in the liquid metal alloy GaInSn.

2. Materials and Methods

2.1. Sensor and Measurement Setup

The ECFM sensor which is used for the experiments has a total height of 50 mm and an outer diameter of 10.6 mm. The secondary coils S1 and S2 each have a height of 16 mm and 1150 turns of enamelled copper wire with a diameter of 0.15 mm. The primary coil P1 has a height of 10 mm and 290 turns of enamelled copper wire with a diameter of 0.25 mm. The distance between two neighbouring coils is 2 mm. All coils are wound on a PVC coil holder which is screwed to a stainless steel threaded rod with a diameter of 3 mm (see Figure 1a,b). The primary coil is supplied with an alternating current with an amplitude of 280 mA and frequencies f of 500 Hz, 1 kHz, and 2 kHz. This ECFM is most sensitive to changes of the flow velocity in this frequency range. The coil voltages are measured with a Lock-In-Amplifier (LIA) which has the advantage that voltages that do not have a certain frequency, in this case the excitation frequency f, are filtered out and thus electromagnetic disturbances with a broad frequency spectrum have a strongly reduced impact on the measurement results. This enables highly accurate measurements of the coil voltages. The LIA also provides the amplitude and frequency of the excitation current, which is supplied to the primary coil via a power amplifier that ensures a constant current amplitude.

The experiments are performed in a liquid metal loop (see Figure 2) that uses the liquid metal alloy GaInSn, which has an electrical conductivity of 3.3 MS/m [26] and is liquid at room temperature. The chemical composition of the alloy is 67% gallium, 20.5% indium and 12.5% tin. The flow velocity of the liquid metal can be adjusted with a permanent-magnet pump that is connected to an electric motor. Velocities up to ±1.2 m/s can be achieved in the horizontal test section, which consists of a non-conductive PVC pipe with an inner diameter of 27 mm. A stainless steel sensor casing consisting of a tube, sealed on one side with a conical tip, an outer diameter of 13 mm and a wall thickness of 1 mm, is inserted into the GaInSn loop from the side, such that it is positioned on the axis of the pipe. The ECFM is inserted into the sensor casing and therefore has no direct contact with the liquid metal. The flow velocity of the loop v_{exp} is lower than the flow velocity around the ECFM v_{ECFM} since the sensor casing is effectively reducing the cross section of the test section. Therefore, the volume averaged flow velocity v_{ECFM} is 29.4% higher than v_{exp}. This was confirmed with numerical simulations of the flow inside the GaInSn loop. In the following sections v_{exp} is used as the reference flow velocity, since this is the quantity that is supposed to be measured by the ECFM. The commercially available magnetic-inductive flow rate sensor ABB Copa-XL is used as a reference flow meter for the measurements.

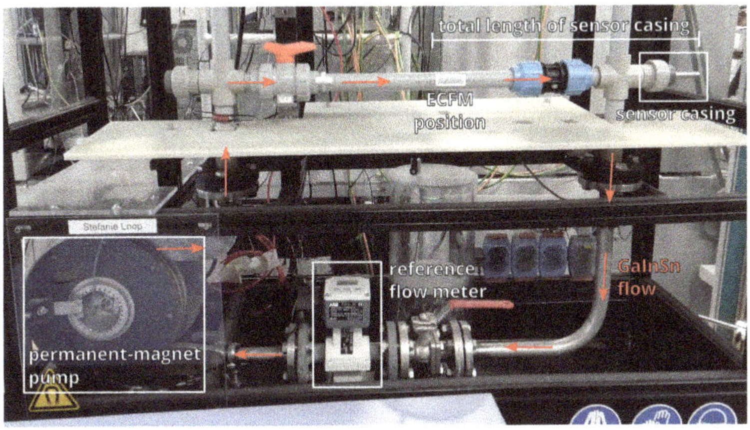

Figure 2. Photograph of the GaInSn loop and measurement setup. The red arrows indicate the flow direction of the liquid metal.

The LIA measures the magnitude r and the phase shift φ of the coil voltage V. The interrelation between these values, using S1 as an example, is as follows:

$$V_{\text{S1}} = r_{\text{S1}} \cdot \mathrm{e}^{j\varphi_{\text{S1}}}. \tag{1}$$

2.2. Calibration of the Numerical Simulation Model

Since a numerical simulation model is used to create the LuT, it is of utmost importance that the representation of the ECFM in the model is as close as possible to the real ECFM. Initially, a very simple 2D axisymmetric model of the sensor (see Figure 1c) is created. It contains the three coils of the ECFM in an infinite volume of air and the stainless steel rod at the centre of the sensor. The dashed red line in Figure 1c indicates the symmetry axis of the model. Using a 2D model has the advantage that the calculations can be performed much faster than with a 3D model, and since there are many thousands of calculations necessary to create the LuT, the model has to be as simple as possible. For comparable setups, the 2D model computes 30 times faster than a 3D model with the same geometry. Obviously, problems or setups that cannot be represented by axisymmetrical models, like bubble detection with the ECFM [27] for example, would require a 3D model. The numerical model is created with the data from the technical drawings and specifications of the ECFM. However, due to manufacturing tolerances, the induced voltages of the ECFM model might be slightly different to those of the real ECFM. Therefore, the model has to be calibrated by adjusting certain geometrical and electrical parameters to match those of the real sensor.

First, the inductance L for each of the three coils is measured and compared to the numerically obtained inductances. At this stage, the ECFM has not been inserted into the test section. Differences between model and real sensor inductances can be corrected by adjusting the number of wire turns n in the numerical model, until the simulated and measured inductances match. In Table 1 the number of turns according to the specifications n_{spec} and the calibrated number of turns n_{cal} are listed.

In the next step of calibration, the excitation current is supplied to the primary coil P1. The induced voltages V_{S1} and V_{S2} in the secondary coils are compared to the induced voltages of the numerical model. Since the inductances of the coils have already been corrected, any remaining voltage differences are most likely caused by a vertical displacement Δd_{P1} of the secondary coils with respect to P1, which can be added to the model. The distances of the secondary coils to P1 are adjusted in the model, until the simulated and measured voltages match.

The last step of the calibration is the compensation of the phase shift between the excitation current and the coil voltages of S1 and S2. Ideally there should be a phase shift of 90° since they are pure induced voltages caused by the primary magnetic field of P1. By using the power amplifier in between the ECFM and the LIA as the signal source, an additional phase shift with respect to the excitation current is introduced, which can be taken into account by setting a reference phase shift φ_{ref} that compensates for the influence of the power amplifier. Furthermore a slight phase shift is introduced by the eddy currents that are induced in the threaded rod at the centre of the sensor. For the sake of simplicity and for an easier comparison of the measurement results, a reference phase shift was chosen which sets the phase shift between induced voltages and excitation current to zero when the ECFM is surrounded by air. This last step is not necessary for the calibration, but it simplifies the handling and evaluation of the results. The adjusted simulation parameters that were obtained through calibration are listed in Table 1. Through this calibration, the mismatch for the induced voltages in S1 and S2 between model and real sensor was reduced below 0.05% for each of the three frequencies (500 Hz, 1 kHz, 2 kHz) that are used for the creation of the LuT. For P1, however, the mismatch lies in the range of several percent, caused by different factors like the internal resistance of the power amplifier as well as the inductance and resistance of the supply wires.

Table 1. Results of the calibration of the three sensor coils.

Coil	L (mH)	n_{spec}	n_{cal}	Δd_{P1} (µm)	φ_{ref} at 0.5, 1, 2 kHz
P1	0.312	290	283	-	-
S1	3.37	1150	1100	69	83.6°, 77.72°, 65.86°
S2	3.41	1150	1108	−4	

3. Results

3.1. Creation of the Look-Up-Table and Numerical Validation of the Method

Now that the ECFM model is calibrated, it can be extended by a numerical model of the GaInSn loop, which adds the test section of the loop, filled with GaInSn and the sensor casing to the numerical model of the ECFM (see Figure 1d). In frame of the simulation model, a constant flow velocity of v_{ECFM} is assumed in the whole test section, but v_{exp} is still used as the reference in the LuT, since this velocity is indicated by the reference flow meter. The electrical conductivity of all stainless steel parts is set to 1.6 MS/m. Due to the simple setup of the test section, it can be represented reasonably well with this 2D axisymmetric model. The calculations are performed with COMSOL Multiphysics 6.1 assuming a constant, homogeneous flow velocity of the liquid metal, parallel to the sensor axis. In any practical application of this method, the velocity v and the electrical conductivity σ of the liquid metal are unknown and influence the induced voltages in the secondary coils of the ECFM, while the voltage of the primary coil is only influenced by σ, albeit by small amounts: for a change in σ of the liquid metal between 3 MS/m and 4 MS/m, the magnitude of V_{P1} changes by 0.2%, whereas the magnitude of each secondary coil changes by nearly 10%. Using the magnitude r_{P1} and phase shift φ_{P1} may still be useful for determining v and σ of the liquid metal, as will be illustrated further below.

In total, the LuT contains the magnitudes and phase shifts of both secondary coils: r_{S1}, r_{S2}, φ_{S1}, φ_{S2} as well as the magnitude and phase shift of P1 and the difference between the secondary coil magnitudes and phase shifts Δr_S and $\Delta\varphi_S$. Traditionally, the voltage difference between the secondary coils is used for determining the flow velocity, since it has the highest sensitivity to changes in v and also cancels out common-mode interference. In this investigation 161 different values of σ between 1 MS/m and 5 MS/m, 81 different values of v between -0.5 m/s and 1.5 m/s at three different frequencies f of the excitation current (500 Hz, 1 kHz, 2 kHz) are used. This results in a total of 39,123 parameter combinations. The calculation of the resulting LuT takes only 45 min on a standard office computer (Intel i5 3 GHz (Intel, Santa Clara, CA, USA), 16 GB RAM). The LuT contains the eight above mentioned phase shifts and magnitudes for all 13,041 parameter combinations of v and σ times the three frequencies that are used. For an unknown parameter combination of v and σ—in this example, one particular combination from the LuT is chosen to be the "unknown" combination—the eight phase shifts and magnitudes are known for each of the three excitation frequencies. In a practical application of this method the voltages V_{P1}, V_{S1} and V_{S2} have to be measured. In order to determine the deviation between the voltages of the unknown parameter combination and all simulated parameter combinations, the mean squared error (MSE) is calculated for each of the eight phase shifts and voltages and three frequencies in the LuT. The MSE is calculated from the respective magnitudes and phase shifts of the unknown parameter combination.

In the following example the MSE for V_{S1} is calculated. The MSE of the voltages with magnitudes and phase shifts for the unknown parameter combination (upc) $V_{S1,upc}$ and all voltages of the LuT $V_{S1,LuT}$ for a given number of different excitation frequencies n_f has to be calculated. Since the induced voltages are strongly dependent on the excitation frequency, appropriate weighting factors w are introduced, which guarantee that each frequency has a similar impact on the total MSE. Otherwise high frequencies would dominate the MSE whereas low f would almost have no impact. The calculation of the MSE is done according to Equation (2) for V_{P1}, V_{S1}, V_{S2} and ΔV_S for all $k = 1, \ldots, 13,041$. Here, V_{S1} is used as an example:

$$MSE^2_{S1,k} = \frac{1}{n_f}\sum_{i=1}^{n_f} \frac{1}{w_{rS1,i}}(r_{S1,upc,i} - r_{S1,LuT,i,k})^2 + \frac{1}{w_{\varphi S1,i}}(\varphi_{S1,upc,i} - \varphi_{S1,LuT,i,k})^2 \quad (2)$$

Weighting the phase shifts is also useful since there is a noticeable difference in the phase shifts for primary and secondary coils, as well as for the voltage difference ΔV_S. The most obvious choice for the weighting factors is the variance (var) of all magnitudes r_{LuT} or phase shifts φ_{LuT} in the LuT with the same coil and frequency. Alternatively, we also consider the squared median absolute deviation (mad) which is known to suppress extreme outliers. Just for comparison, we will also show a few results when using the standard deviation (std) or no weighting at all ($w = 1$). For each of the i frequencies, the weighting factors are calculated according to Equation (3) following the example of the weighting factor w_{rS1} when using the variance for the magnitude of V_{S1}, where $\overline{r}_{S1}$ is the mean value of all r_{S1} in the LuT with the same coil and frequency:

$$w_{rS1,i} = \frac{1}{N-1} \sum_{k=1}^{N=13,041} (r_{S1,LuT,i,k} - \overline{r}_{S1,LuT,i})^2 . \tag{3}$$

The weighting factors for the phase shifts are calculated in the same way by simply replacing the magnitudes by the according phase shifts. For the previous application of this method in [25], weighting by the variance yielded optimal results.

Each of the 13,041 parameter combinations of v and σ in the LuT now has four MSE (for V_{P1}, V_{S1}, V_{S2} and ΔV_S), that include the weighted values for each frequency. The total MSE called MSE_Σ for a given parameter combination is obtained by summing up the four separate MSE according to

$$MSE_{\Sigma,k} = \sqrt{MSE^2_{S1,k} + MSE^2_{S2,k} + MSE^2_{P1,k} + MSE^2_{\Delta S,k}} \quad . \tag{4}$$

Under ideal conditions and assuming that the numerical model has been properly calibrated, the parameter combination with the lowest total MSE_Σ has the maximum likelihood to be the unknown parameter combination. By finding the minimum MSE_Σ in the LuT, the values of v and σ can be inferred. When plotting all MSE_Σ for a given unknown parameter combination (chosen from the LuT for now, not from the measurement) over v or σ it can be observed in Figure 3 that the MSE are converging to a certain minimum. In this case, the minimum of MSE_Σ indicates the correct values of v and σ, since there are no measurement errors or other disturbances that have a negative impact on the results. Although this will certainly look different when using the actual measurements for the calculation of MSE_Σ, it can still be observed that this method is suitable to solve this problem and that the MSE are converging to the sought-after parameters. Measures for how well the MSE are converging are the angle α and the density of MSE_Σ close to the minimum. Both small α and small density facilitate the parameter estimation of v and σ, since there are less parameter combinations with a similar MSE_Σ that can be mistaken for the correct parameter combination in the presence of measurement errors.

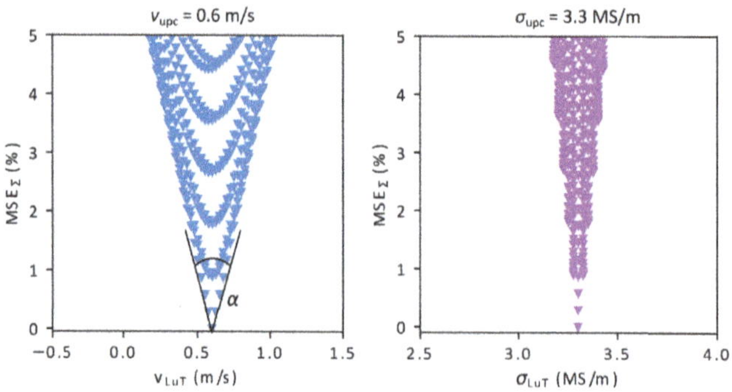

Figure 3. Distribution of the lowest 5% of MSE_Σ for v and σ under ideal conditions, when taking one parameter combination of the LuT as the unknown parameter combination (v = 0.6 m/s, σ = 3.3 MS/m).

In Figure 4 we illustrate how the MSE_Σ for v (blue) and σ (violet) are changing when more and more voltages and frequencies are added, and how the weighting method is affecting the MSE distribution. Which components are used for the calculation is indicated by the text above each diagram. Greyed out components are not used in the calculation of MSE_Σ. The variance is used as the weighting method for the diagrams A–P and T in Figure 4, while diagrams Q, R and S use different weighting methods, which are indicated below each diagram. Starting from the left side in Figure 4, for only a single voltage at a single frequency with using either the magnitude or phase shift, there is no clear minimum of MSE_Σ for v since several combinations of velocity and electrical conductivity can have the same magnitude or phase shift (Figure 4(A,B)). On the other hand, a certain range for σ can already be identified. By combining the (weighted) MSE_Σ for r and φ, it becomes possible to identify v and σ when using a single coil voltage (C). When adding a second voltage (D–F) a clear minimum can be identified, with noticeable differences in MSE_Σ for r and φ. For r, the MSE are converging faster to the minimum (D), which facilitates identification of the correct parameter combination. The density of MSE at the minimum is still comparably high, which means that there are multiple parameter combinations with a similar MSE_Σ, making it harder to identify the correct parameter combination. Adding a second frequency (K) or the voltage of the primary coil P1 (H) to MSE_Σ does not have a significant influence on the results. However, adding the voltage difference of the secondary coils ΔS (G, J, L) improves the quality of the results significantly, since it depends sensitively on v and σ. The density of MSE_Σ is strongly reduced at the minimum and is converging much better than in the previous cases. Adding a third frequency (P) or more voltages (N, O) does not yield additional benefits compared to cases G, J or L.

Regarding the weighting methods, using the variance (O) appears to be the optimal choice, exhibiting the lowest MSE_Σ density and strongest convergence at the minimum. As expected, the worst results are achieved when not weighting at all (S), with the other methods (Q, R) in between where weighting by the squared median absolute deviation has a similar effect as using the variance. The reason why certain coil voltages have a stronger influence on the results than others can be seen in Table 2, where the sensitivity to changes in v and σ are displayed. While the magnitude and phase shift of P1 are barely influenced, ΔS, S1 and S2 are moderately sensitive to changes in σ. Moreover, ΔS is very sensitive to changes in the flow velocity compared to P1, S1 and S2, especially the phase shift. This is why ΔS has such a strong impact on the total MSE.

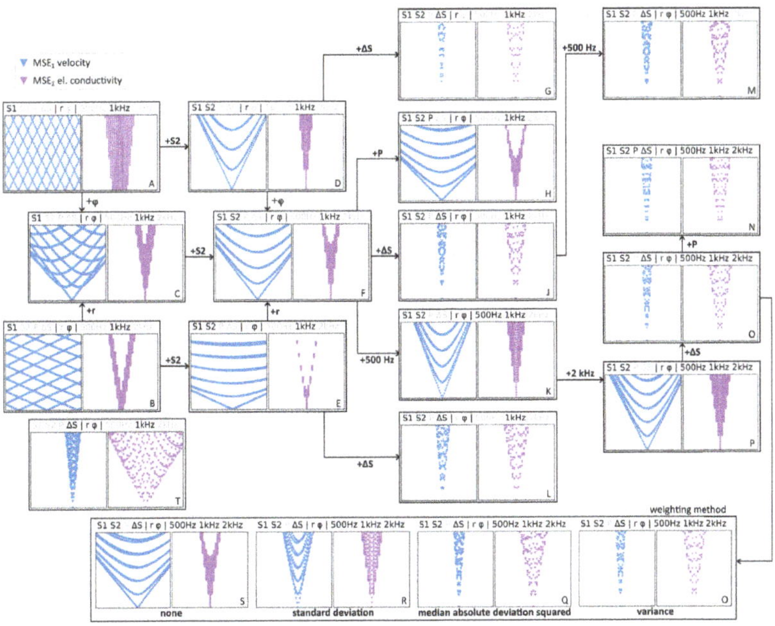

Figure 4. Evolution of the lowest 5% of MSE_Σ for v and σ (similar to Figure 3) when adding more coil voltages, frequencies or changing the weighting method. Blue triangles represent MSE_Σ of the flow velocity, violet triangles represent MSE_Σ of the electrical conductivity.

When using only ΔS for the calculation of MSE_Σ (T), the identification of v and σ is possible, but more accurate results can be achieved when using multiple voltages and frequencies (O). The secondary coils S1 and S2 are more sensitive to changes in σ than ΔS but are considerably less sensitive to changes in v. Therefore, the MSE_Σ distribution has in some cases different shapes, depending on whether v and σ is estimated (A–F). Adding P1 in the calculation of MSE_Σ appears not to provide any improvement for the parameter estimation since its voltage is only slightly influenced by v and σ. This influence is of the same order of magnitude as the measurement accuracy and therefore measurement errors of V_{P1} might significantly distort the results.

Table 2. Relative changes in magnitude and phase shift for the voltages of S1, S2, ΔS and P1 for a constant velocity (0 m/s) and a change in electrical conductivity of 1 MS/m (from 3 to 4 MS/m), as well as for a constant electrical conductivity (3 MS/m) and a change in velocity of 1 m/s (from 0 to 1 m/s) at 1 kHz.

Conditions	**S1, S2**		**ΔS**		**P1**	
	r	φ	r	φ	r	φ
$v = 0$ m/s $\sigma = 3 \rightarrow 4$ MS/m	8.1%	20.6%	5.2%	30.8%	0.2%	1.9%
$v = 0 \rightarrow 1$ m/s $\sigma = 3$ MS/m	0.8%	0.6%	62.1%	111.5%	0.0%	0.0%

Once the LuT has been calculated, it takes less than 0.2 s to run the python script for calculation of the MSE and identification of the values for v and σ. However, in order to determine the unknown v and σ, the simulation model and the real setup have to match as well as possible. In the following section, the practical validation of this method is presented.

3.2. Experimental Validation

In the previous section it was shown that under ideal conditions it is straightforward to determine the unknown v and σ by using multiple voltages and frequencies. In reality, measurement errors, signal noise and the flow profile around the sensor and their influence on the measured voltages have to be considered and can have a negative impact on the results of the parameter estimation. The measurements at the GaInSn loop are conducted for six pre-adjusted flow velocities v_{exp} from 0 m/s to 1 m/s in steps of 0.2 m/s. The electrical conductivity of the liquid metal is assumed to be 3.3 MS/m as is described in [26] for pure GaInSn with a similar composition, although it might be contaminated with oxides or other impurities that have an unknown impact on σ. By using the LuT in combination with the measured voltages, the separate MSE for each magnitude, phase shift of V_{P1}, V_{S1}, V_{S2} and ΔV_{S} and each excitation frequency can be calculated and weighted according to Equation (2). The sum of the separate weighted mean squared errors MSE_Σ does not necessarily have to be calculated from all of the separate MSE. When leaving out, for example, all MSE of a certain frequency or by only using the MSE of the magnitudes or even leaving out a complete coil voltage when calculating MSE_Σ, the quality of the parameter estimation of v and σ can be influenced significantly (see Figure 4). How well a certain combination is performing is evaluated with the deviations Δv between pre-adjusted velocities v_{exp} and the velocity that is obtained by finding the minimum of the MSE_Σ distribution $v_{\mathrm{LuT,min}}$, as well as the deviation $\Delta\sigma$ between the electrical conductivity of the liquid metal σ and the conductivity that is obtained from the LuT $\sigma_{\mathrm{LuT,min}}$:

$$\Delta v = v_{\mathrm{exp}} - v_{\mathrm{LuT,min}} \qquad |\Delta v_{\mathrm{avg}}| = \frac{1}{6}\sum_{m=1}^{6} |v_{\mathrm{exp},m} - v_{\mathrm{LuT,min},m}| \tag{5}$$

$$\Delta\sigma = \sigma_{\mathrm{exp}} - \sigma_{\mathrm{LuT,min}} \qquad |\Delta\sigma_{\mathrm{avg}}| = \frac{1}{6}\sum_{m=1}^{6} |\sigma_{\mathrm{exp}-}\sigma_{\mathrm{LuT,min},m}| \tag{6}$$

Table 3 shows the absolute averages $|\Delta v_{avg}|$ and $|\Delta\sigma_{avg}|$ of all Δv and $\Delta\sigma$ for each of the six v_{exp} for selected cases from Figure 4, when using the measured voltages instead of a set of voltages from the LuT itself and the two best weighting methods. Which of the components were used for the calculation of MSE_{Σ} is indicated by the numbers 1 (=used), 0 (=not used). Here, the column headers indicate the kind of component. S1, S2, P1 and ΔS stand for the different voltages, r and φ stand for the magnitude and phase of the voltages. It is also shown which frequencies were used for the calculation as well as which weighting method (column header w) was used.

Table 3. Results of Δv and $\Delta\sigma$ for the cases G, J, L, M, N, O and T of Figure 4 when weighting by variance (var) and squared median absolute deviation (mad). The color highlights the used components.

Case	$|\Delta v_{avg}|$ (cm/s)	$|\Delta\sigma_{avg}|$ (MS/m)	S1	S2	P1	ΔS	r	φ	500 Hz	1 kHz	2 kHz	w
G	9.3 9.3	0.042 0.041	1	1	0	1	1	0	0	1	0	var mad
J	5.6 4.2	0.064 0.055	1	1	0	1	1	1	0	1	0	var mad
L	11.1 10.9	0.130 0.127	1	1	0	1	0	1	0	1	0	var mad
M	4.3 6.0	0.196 0.185	1	1	0	1	1	1	1	1	0	var mad
N	20.4 26.5	1.173 1.133	1	1	1	1	1	1	1	1	1	var mad
O	7.0 6.6	0.410 0.490	1	1	0	1	1	1	1	1	1	var mad
T	15.8 15.9	0.497 0.487	0	0	0	1	1	1	0	1	0	var mad

Although each of the investigated cases in Table 3 appears similar in Figure 4 when only using numerical data for the investigation of MSE_{Σ}, there are significant differences when using the actual measurement data for calculation of the MSE. By comparing Δv and $\Delta\sigma$ for the different cases, it can be seen that some of them yield more accurate results than others, with cases N and T exhibiting the largest deviations in v_{exp} and σ_{exp}, while cases G, J and M are among the most accurate ones.

Generally, weighting by variance or by squared median absolute deviation give comparable results, with no clear advantage offered by one method over the other. Case T, which represents the bare minimum case where only one voltage and one frequency is used for the estimation of v and σ, gives relatively inaccurate results. Regarding the velocity estimation, case J is more accurate than case T by a factor of 4, and regarding the electrical conductivity accuracy is better by a factor of 7. Notice however, that the electrical conductivity of the liquid metal is not exactly known. From the overall results it can be deduced that at least two voltages, both the magnitude and the phase shift, and one frequency have to be used for the calculation of MSE_{Σ} for obtaining optimal results. Although ΔS has the highest impact for increasing the accuracy of the parameter estimation, it is not sufficient to only use ΔS, rather the individual voltages of S1 and S2 should be used in addition. By contrast, adding the voltage of the primary coil P1 does not give any benefits and would also require more effort since an additional voltage has to be measured. Using one frequency from the range where the ECFM is most sensitive seems to be sufficient, while introducing a second or third frequency appears to offer no advantages for this application and can even be detrimental in some cases (because additional measurement errors are introduced). Using either r or φ is possible but since the two quantities are always measured simultaneously when using a LIA, there are no disadvantages when using both for the calculation of MSE_{Σ}.

In Figure 5 the velocity estimations of case G and case J from Table 3 are compared when using the squared median absolute deviation weighting method. Case G converges faster to the minimum, which is also much lower than for case J. Ultimately, the value of the minimum is unimportant, as long as it can be clearly identified. Due to measurement errors and the differences between the real setup and the simplified simulation model, MSE_{Σ} does not reach zero. When comparing Δv it can be seen that the deviations of case G are distributed more uniformly for different v_{exp} than for case J where the maximum deviation is found at zero velocity and is progressively lower for higher velocities. Regarding the estimation of the electrical conductivity, the MSE_{Σ} in Figure 6 are similar but converge more slowly than for the flow velocity. This is in good agreement with the purely numerical results in Figure 4. While case J yields the lowest Δv overall, the average deviation of σ is smaller for case G (but note again that the "real" value of σ is not well known).

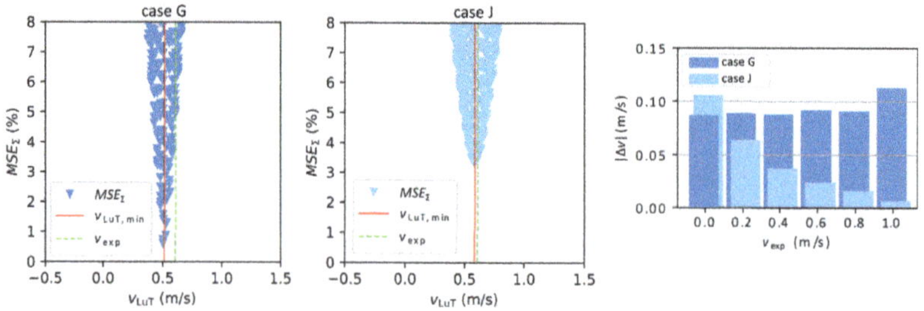

Figure 5. Distribution of MSE_{Σ} of the flow velocity for v_{exp} = 0.605 m/s for cases G and J of Table 3. The red line indicates the minimum of the MSE distribution while the dashed green line shows the pre-adjusted flow velocity. The diagram on the right shows a comparison of Δv for all v_{exp}.

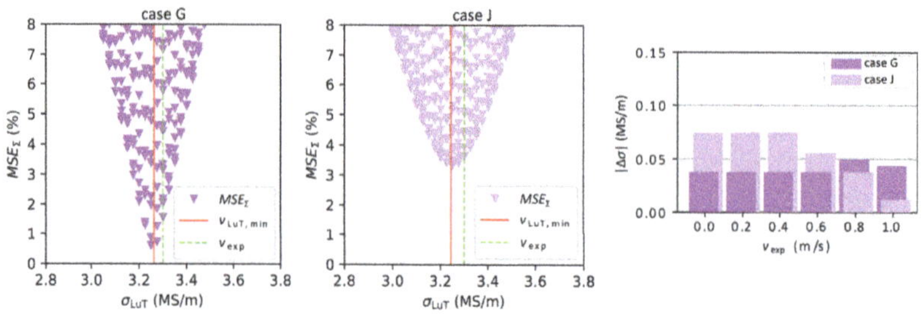

Figure 6. Distribution of MSE_{Σ} of the electrical conductivity for v_{exp} = 0.605 m/s for cases G and J of Table 3. The red line indicates the minimum of the MSE distribution while the dashed green line shows the electrical conductivity of the liquid metal. The diagram on the right shows a comparison of $\Delta\sigma$ for all v_{exp}.

In order to assess the influence of the individual MSE_{Σ} components, the evolution of MSE_{Σ} is traced in Figure 7 by using the measurement data, similar to the more extensive investigation in Figure 4. It can be seen that for case A the flow velocity cannot be identified at all and the estimated conductivity is far too low, even outside of the displayed range of σ. When adding the measurements of the phase shift to MSE_{Σ} a velocity which is far greater than its actual value can be identified, while the σ estimation is now much closer to the actual value. By adding the voltage of the other secondary coil in case F, the velocity estimation is much improved, while the estimated σ is slightly worse. Adding ΔS in case J only has a small influence on the accuracy of the parameter estimation in this case but significantly reduces the MSE_{Σ} density at the minimum and also has the effect that MSE_{Σ} converges faster. When finally adding a second frequency in case M, the velocity estimation as well as the estimation of the electrical conductivity become more accurate.

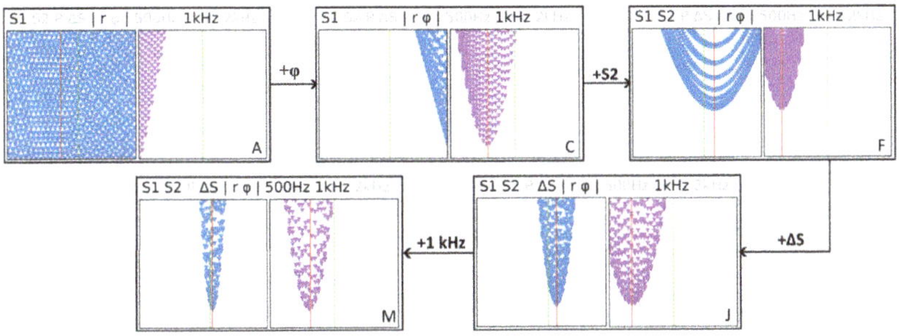

Figure 7. Evolution of MSE_{Σ} from case A to case M when using the measurement data. The red line indicates the minimum of the MSE distribution while the dashed green line shows the pre-adjusted flow velocity or electrical conductivity. Blue triangles represent MSE_{Σ} of the flow velocity, violet triangles represent MSE_{Σ} of the electrical conductivity.

3.3. Measurement Errors

3.3.1. Inductance

The inductances of the sensor coils are measured with a Hameg HM8118 LCR-Bridge. According to the data-sheet, the measurement accuracy depends on the impedance of the coils, and therefore the inductance of P1 has an accuracy of $\pm 0.13\%$, and for S1, S2 an accuracy of $\pm 0.103\%$. For all coils, the resulting measurement error corresponds to a change in less than one turn for the inductance of the coils and is therefore sufficiently accurate for the calibration of the numerical model.

3.3.2. Voltage

For the voltage measurements, a Signal Recovery 7265 DSP Lock-in Amplifier (Signal Recovery, Irvine, CA, USA) is used. According to the data sheet, it has an accuracy of $\pm 0.3\%$ for the measured voltages. Each coil voltage (magnitude and phase shift) is measured for one minute with one data point per second for a constant velocity v_{exp} and frequency. The final voltage, which is used for the calculation of the MSE, is calculated from the mean value of all data points. The measured voltages are almost constant, as can also be seen on the standard deviation of the data series: for every measured voltage it lies below 0.02%, and therefore the random error caused by noise or other electromagnetic disturbances is negligible. In a practical application of this measurement technique, it is sufficient to measure only one data point.

3.3.3. Flow Rate

For measurements of the reference flow rate v_{exp} of the loop, an ABB Copa-XL Electromagnetic Flowmeter (ABB, Zurich Switzerland) is used. According to the data-sheet, the accuracy for the velocity measurements is $\pm 0.5\%$ of the flow velocity when $v_{exp} < 0.7$ m/s, and ± 0.0035 m/s for $v_{exp} > 0.7$ m/s. Like for the coil voltages, v_{exp} is calculated from the mean value of the flow velocities over one minute, with one measurement made every second. For every measured v_{exp}, the standard deviation lies below 0.3%. These values apply for the mean velocity in the pipe. The actual velocity profile of the flow cannot be measured since both the reference flow meter and the ECFM are inductive sensors that are only able to measure the mean flow rate or flow velocity in a certain volume. In addition to the measurement accuracy, the quality of the calibration of the flowmeter has a significant influence on the total measurement error. Since the flow meter was calibrated many years ago when the GaInSn loop was constructed, there might be an additional systematic error of unknown dimensions for all the velocity measurements.

3.3.4. Electrical Conductivity of the Liquid Metal

The electrical conductivity of the eutectic liquid metal alloy GaInSn in the loop is not exactly known, due to the fact that it might contain oxides or other impurities. Depending on the volume concentration of impurities and their electrical conductivities, the total electrical conductivity of the liquid metal might vary according to Equation (1) in [28]. Assuming that the liquid metal has the properties of pure GaInSn as described in [26], it has an electrical conductivity of 3.3 ± 0.3 MS/m at room temperature. It is assumed that there is only a small amount of impurities contained in the liquid metal and their influence on the total electrical conductivity does not exceed the relatively large measurement uncertainty of 0.3 MS/m.

4. Conclusions

By combining numerical simulations and experimental results, we have shown that the flow velocity and the electrical conductivity of a liquid metal can be obtained simultaneously when using an ECFM in combination with a so called Look-Up-Table. This method may not be as accurate as other measurement techniques like a properly calibrated ECFM or an ultrasound doppler velocimetry sensor, but it can be used as an alternative for the time consuming temperature calibration of the ECFM, while also eliminating the need for a temperature measurement of the liquid metal. In order to achieve accurate results, a model for numerical simulations has to be created that matches the real setup as precisely as possible. Additionally, the simulation model of the ECFM has to be calibrated by performing measurements at the real sensor, prior to inserting it into the test section or experimental facility. Once the ECFM is calibrated and the Look-Up-Table has been calculated, the flow velocity and electrical conductivity of the liquid metal can be estimated in less than a second. This time span can be decreased further by reducing the resolution and range of the Look-Up-Table parameters. When the same sensor is put into a different environment, no further calibration of the ECFM is necessary. Only the relevant part of the facility has to be translated into a numerical model, since the calibration of the sensor in air is sufficient. At our liquid metal facility, we were able to achieve an average deviation between real and estimated velocity of 0.04 m/s in a range between 0 m/s and 1 m/s and for the electrical conductivity an average deviation of 4×10^4 S/m. Using multiple excitation frequencies for the sensor is not necessary. Similar results are obtained, regardless of how many excitation frequencies are used. Weighting the mean squared errors is also important, and in this experiment using the variance or squared median absolute deviation provided the best results. Accurate results were achieved when using only one excitation frequency in the sensitive range of the ECFM. Yet, for other applications multiple frequencies might yield better results, depending on the setup and on whether there are components with different electrical conductivities within the range of the sensor. This measurement technique can be applied to similar problems and/or sensors and it is also possible to infer related parameters like the temperature of the liquid metal, for example. As long as the electrical conductivity of the liquid is in the order of magnitude of 0.1 MS/m or more, this method can easily be applied to other liquid metals and alloys.

In summary, the main achievements of this paper are the following:

- First simultaneous measurement of flow velocity and electrical conductivity of a liquid metal using an ECFM.
- Simplified calibration of the sensor compared to conventional methods.
- Parameter estimation of velocity and conductivity takes less than a second, using only one excitation frequency.
- This method can be applied to many different liquid metals and alloys, different parameters and sensors.

Author Contributions: Conceptualization F.S. and N.K. Numerical simulations, experiments and data curation N.K. Writing of the original draft and visualization N.K. Review and editing F.S. All authors have read and agreed to the published version of the manuscript.

Funding: This project was supported by the Euratom research and training programme 2014–2018 in frame of the project ESFR-SMART under the grant agreement No. 754501. It has also received funding from the European Research Council (ERC) under the European Union's Horizon 2020 research and innovation programme (grant agreement No. 787544).

Institutional Review Board Statement: Not applicable.

Informed Consent Statement: Not applicable.

Data Availability Statement: The data that support the findings of this study are available from the corresponding author, F.S., upon reasonable request.

Acknowledgments: We would like to thank William Nash for a critical reading of the revised version.

Conflicts of Interest: The authors declare no conflict of interest.

Abbreviations

The following abbreviations are used in this manuscript:

CIFT	Contactless Inductive Flow Tomography
ECFM	Eddy Current Flow Meter
LIA	Lock-In-Amplifier
LMFBR	Liquid Metal Cooled Fast Breeder Reactors
LuT	Look-Up-Table
mad	Median Absolute Deviation
MSE	Mean Squared Error
std	Standard Deviation
TECFM	Transient Eddy Current Flow Meter
var	Variance
P	Primary coil
S1, S2	Secondary coils
f	Frequency
L	Inductance
n	Number of turns
r	Voltage magnitude
Δr	Difference of voltage magnitudes
v	Flow velocity
V	Voltage
w	Weighting factor
σ	Electrical conductivity
φ	Voltage phase shift
$\Delta\varphi$	Difference of Voltage phase shifts

References

1. Stefani, F.; Gundrum, T.; Gerbeth, G. Contactless Inductive Flow Tomography. *Phys. Rev. E* **2004**, *5*, 056306. [CrossRef] [PubMed]
2. Priede, J.; Buchenau, D.; Gerbeth, G. Contactless electromagnetic phase-shift flowmeter for liquid Metals. *Meas. Sci. Technol.* **2011**, *22*, 055402. [CrossRef]
3. Miralles, S.; Verhille, G.; Plihon, N.; Pinton, J.F. The magnetic-distortion probe: Velocimetry in conducting fluids. *Rev. Sci. Instrum.* **2011**, *82*, 095112. [CrossRef] [PubMed]
4. Thess, A.; Votyakov, E.V.; Kolesnikov, Y. Lorentz force velocimetry. *Phys. Rev. Lett.* **2006**, *96*, 164501. [CrossRef]
5. Pulugundla, G.; Heinicke, C.; Karcher, C.; Thess, A. Lorentz force velocimetry with a small permanent magnet. *Eur. J. Mech. B Fluids* **2013**, *41*, 23–28. [CrossRef]
6. Viré, A.; Knaepen, B.; Thess, A. Lorentz force velocimetry based on time-of-flight measurements. *Phys. Fluids* **2010**, *22*, 125101. [CrossRef]
7. Sokolov, I.; Noskov, V.; Pavlinov, A.; Kolesnikov, Y. Lorentz Force Velocimetry for High Speed Liquid Sodium Flow. *Magnetohydrodynamics* **2016**, *52*, 481–493.
8. Zürner, T.; Vogt, T.; Resagk, C.; Eckert, S.; Schumacher, J. Local Lorentz force and ultrasound Doppler velocimetry in a vertical convection liquid metal flow. *Exp. Fluids* **2017**, *59*, 3. [CrossRef]
9. Lehde, H.; Lang, W.T. Device for Measuring Rate of Fluid. Flow. U.S. Patent 2435043, 27 January 1948.
10. Costello, T.J.; Laubham, R.L.; Miller, W.R.; Smith, C.R. FFTF Probe-Type Eddy-Current Flowmeter: Wet versus dry Performance Evaluation in Sodium. *Nucl. Technol.* **1973**, *19*, 174–180. [CrossRef]

11. Poornapushpakala, S.; Gomathy, C.; Sylvia, J.I.; Krishnakumar, B.; Kalyanasundaram, P. An analysis on eddy current flowmeter—A review. In Proceedings of the Recent Advances in Space Technology Services and Climate Change 2010, Chennai, India, 13–15 November 2010; pp. 185–188.
12. Poornapushpakala, S.; Gomathy, C.; Sylvia, J.I.; Babu, B. Design, development and performance testing of fast response electronics for eddy current flowmeter in monitoring sodium flow. *Flow Meas. Instrum.* **2014**, *38*, 98–107. [CrossRef]
13. Pavlinov, A.; Khalilov, R.; Mamikyn, A.; Kolesnichenko, I. Eddy current flowmeter for sodium flow. *IOP Conf. Ser. Mater. Sci. Eng.* **2017**, *208*, 012031. [CrossRef]
14. Kumar, S.S.; Patri, S.; Sharma, R.K.; Punniamoorthy, R.; Paunikar, V.D.; Cyriac, R.J.; Harishkumaran, S.; Vasudevan, P.; Ramakrishna, R.; Sajish, S.D.; et al. Experimental Studies on Qualification of Structural Integrity of Eddy Current Flow Meter. In Proceedings of the 2nd Quadrennial International Conference on Structural Integrity (ICONS), Chennai, India, 14–17 December 2018; pp. 787–799.
15. Lau, C.; Oleksak, K.; Cetiner, S.M.; Groth, P.; Mauer, C.; Ottinger, D.; Roberts, M.J.; Warmack, B.; Fathy, A.E. Eddy current flow meter model validation with a moving solid rod. *Meas. Sci. Techn.* **2022**, *33*, 075301. [CrossRef]
16. Afflard, A.; Zamansky, R.; Bergez, W.; Tordjeman, P.; Paumel, K. Eddy-Current Flow Meter Response to Spherical Non-Conductive Inclusions Travelling in a Liquid Metal. *Magnetohydrodynamics* **2022**, *58*, 501–507.
17. Shubham; Rajalakshmi, R. Design and Development of Non-Intrusive Eddy Current Flow Meter for High Temperature Liquid Metal Services. In Proceedings of the 1st International Conference on Electrical, Electronics, Information and Communication Technologies (ICEEICT), Trichy, India, 16–18 February 2022; pp. 1–6.
18. Gall, G.; Lau, C.; Varma, V.; Cetiner, S.; Ottinger, D. Measurement of a radial flow profile with eddy current flow meters and deep neural networks. *Meas. Sci. Technol.* **2023**, *34*, 045302. [CrossRef]
19. Buchenau, D.; Eckert, S.; Gerbeth, G.; Stieglitz, R.; Dierckx, M. Measurement technique developments for LBE flows. *J. Nucl. Mat.* **2011**, *415*, 396–403. [CrossRef]
20. Sharma, P.; Suresh Kumar, S.; Nashine, B.K.; Veerasamy, R.; Krishnakumar, B.; Kalyanasundaram, P.; Vaidyanathan, G. Development, computer simulation and performance testing in sodium of an eddy current flowmeter. *Ann. Nucl. Energy* **2010**, *37*, 332–338. [CrossRef]
21. Sureshkumar, S.; Sabih, M.; Narmadha, S.; Ravichandran, N.; Dhanasekharan, R.; Meikandamurthy, C.; Padmakumar, G.; Vijayashree, R.; Prakash, V.; Rajan, K.K. Utilization of eddy current flow meter for sodium flow measurement in FBRs. *Nucl. Eng. Des.* **2013**, *265*, 1223–1231. [CrossRef]
22. Krauter, N.; Stefani, F. Immersed transient eddy current flow metering: A calibration-free velocity measurement technique for liquid metals. *Meas. Sci. Technol.* **2017**, *28*, 105301. [CrossRef]
23. Looney, R.; Priede, J. Alternative transient eddy-current flowmetering methods for liquid metals. *Flow Meas. Instrum.* **2019**, *65*, 150–157. [CrossRef]
24. Krauter, N.; Eckert, S.; Gundrum, T.; Stefani, F.; Wondrak, T.; Frick, P.; Khalilov, R.; Teimurazov, A. Inductive System for Reliable Magnesium Level Detection in a Titanium Reduction Reactor. *Met. Mater. Trans. B* **2018**, *49B*, 2089–2096. [CrossRef]
25. Krauter, N.; Eckert, S.; Gundrum, T.; Stefani, F.; Wondrak, T.; Khalilov, R.; Dimov, I.; Frick, P. Experimental Validation of an Inductive System for Magnesium Level Detection in a Titanium Reduction Reactor. *Sensors* **2020**, *20*, 6798. [CrossRef] [PubMed]
26. Plevachuk, Y.; Sklyarchuk, V.; Eckert, S.; Gerbeth, G.; Novakovic, R. Thermophysical properties of the liquid Ga-In-Sn eutectic alloy. *J. Chem. Eng. Data* **2014**, *59*, 757–763. [CrossRef]
27. Afflard, A.; Zamansky, R.; Paumel, K.; Bergez, W.; Tordjeman, P. Bubble detection in liquid metal by perturbation of eddy currents: Model and experiments. *J. Appl. Phys.* **2023**, *134*, 134502. [CrossRef]
28. Fricke, H. A Mathematical Treatment of the Electric Conductivity and Capacity of Disperse Systems. *Phys. Rev.* **1924**, *24*, 575–587. [CrossRef]

sensors

MDPI

Article

Development of a Simple Setup to Measure Shielding Effectiveness at Microwave Frequencies

Emanuele Cardillo, Fabrizio Lorenzo Carcione, Luigi Ferro *, Elpida Piperopoulos, Emanuela Mastronardo, Graziella Scandurra and Carmine Ciofi

Department of Engineering, University of Messina, 98166 Messina, Italy; ecardillo@unime.it (E.C.); fcarcione@unime.it (F.L.C.); epiperopoulos@unime.it (E.P.); emastronardo@unime.it (E.M.); gscandurra@unime.it (G.S.); cciofi@unime.it (C.C.)

* Correspondence: luferro@unime.it

Abstract: Testing the shielding effectiveness of materials is a key step for many applications, from the industrial to the biomedical field. This task is very relevant for high-sensitivity sensors, whose performance can be greatly affected by electromagnetic fields. However, the available testing procedures often require expensive, bulky, and heavy measurement chambers. In this paper, a cost-effective and reliable measurement procedure for testing the shielding effectiveness of materials is proposed. It exploits a lab-scale anechoic shielded chamber, which is lightweight, compact, and cost-effective if compared to the available commercial solutions. The measurement procedure employs a vector network analyzer to allow an accurate and fast characterization setup. The chamber realization phases and the measurement procedure are described. The shielding capability of the chamber is measured up to 26 GHz, whereas the performance of commercial shielding coatings is tested to demonstrate the measurement's effectiveness.

Keywords: anechoic chamber; microwaves; minehunter vessels; shielding coating; shielding effectiveness

1. Introduction

Until the 20th century, the only electromagnetic emissions on the Earth were due to natural phenomena such as the radiation from the sun, the atmosphere and the Earth itself. However, the continuous and impressive development of new electronic devices and telecommunication systems increased both the quantity and diversity of the electromagnetic emissions in terms of radiated power and frequency [1–3]. Although most of the electromagnetic emissions were intentional and beneficial for society, the coexistence of such a huge number of electromagnetic sources resulted in the onset of electromagnetic interferences (EMIs), which should be carefully considered in order to avoid electromagnetic compatibility (EMC) issues where EMIs interfere with the proper operation of other electronic devices [4,5]. The scientific community has always been prudent concerning the possible effects on human health; several examples have seen the possible effects of EMIs on test equipment, both at low and high frequencies, and for various applications from the biomedical to the industrial field [6–13].

EMI has a relevant impact on the performance of high-sensitivity sensors [14–16]. As an example, electrostatic shielding is a standard procedure for minimizing the capacitive coupling of soil sensors [15]. However, many EMI-sensing studies do not verify the effectiveness of the employed shielding. Moreover, sensors exploited in electronic warfare systems need to be protected from intentional electromagnetic attacks, thus requiring electromagnetic shielding [16].

It is worth noting that the problem, which is under the continuous supervision of the competent authorities, can be addressed and mitigated in a threefold way: (a) by suppressing the emission at the source, (b) by making the coupling path less efficient, (c) by making the receiver less susceptible to the emission [17].

Although the first line of defense is to suppress the emission as much as possible at the source, often this is not a feasible solution due to the intentional nature of the emission, or to the need to deal with EMI sources already present in the market and consequently in the environment.

Therefore, developing lightweight and cost-effective EMI-shielding materials is necessary to mitigate EMI pollution [18–20].

The basic concept behind the shielding effect involves the implementation of EMI-reflection and -absorption processes by means of a suitable material which in turn protects the shielded device from radiation [21,22]. This task can be fulfilled by exploiting conductive materials; to this aim, metal-based enclosure is probably the best-known type of EMI-shielding due to its excellent EMI-shielding effectiveness. However, metals present some limitations related to their weight, rigidity, vulnerability to corrosion and cost, which makes them unsuitable for many applications [23]. On the other hand, conductive polymer composites have shown their effectiveness, and are largely used as shielding materials due to their low cost, strong resistance to corrosion, and light weight [24,25]. Conductive polymer composites usually incorporate conductive fillers in a polymer matrix, thus approaching the electrical conductive properties of the hosting material [26,27].

In view of the availability of shielding materials, measurement procedures and tools are required to properly test their properties. To this aim, different kinds of test methodologies are reported in the scientific literature. In [28], the shielding effectiveness of planar materials has been measured by means of a method based on the ASTM D4935-18 standard [29]. It exploits an absorber box which mitigates the known issues of the ASTM D4935-18 standard. As an example, the 1.5 GHz upper frequency limit has been extended to 18 GHz. Even in the study of Tamburrano et al. [30], a similar technique has been implemented to extend the frequency up to 18 GHz. Moreover, in [31], an alternative method is presented for the SE measurement of planar materials with nonconducting surfaces. In particular, this method overcomes the edge termination problems by absorbing edge-diffracted energy. The results obtained are similar to those obtained using the ASTM D4935 method. Moreover, the method in [31] exploits a simple setup and is cost effective.

Different methods to measure radiated emission exploit reverberation chambers [32]. As an example, in [33], the shielding effectiveness of different materials has been measured in a nested reverberation chamber up to 6 GHz.

In this paper, a measurement setup aimed at testing the shielding effectiveness of material sheets is proposed. The idea has been fueled by the need for a specific application, i.e., the test of shielding panels for minehunter vessels. Indeed, minehunter vessels are required to be highly electromagnetically shielded to prevent the activation of the very sensitive detonator of the mine. The setup proposed in this study employs a lab-made anechoic and shielded chamber, working from 500 MHz to 26 GHz. Although shielded anechoic chambers are commercially available, they are usually very expensive. Through this paper, the realization steps are disclosed in order to make the chamber realization procedure reproducible. The problem of self-realizing anechoic chambers has received a good amount of interest within the scientific literature, with some papers exploiting anechoic chambers for similar purposes, but often with a lower maximum operating frequency [34–37]. Moreover, these works are usually devoted to the characterization of antenna parameters or electromagnetic compatibility tests, whereas the purpose of the proposed chamber differs completely from these applications. The measurement procedure exploits a vector network analyzer (VNA) to allow for an accurate and fast characterization. For the same purpose, microwave signal generators in combination with spectrum analyzers are frequently used in the literature. However, a VNA is able to sweep the incident wave across frequency, thus enabling faster measurement and minimizing the probability of errors. The shielding effectiveness measurement is based on a differential procedure. A face of the chamber is left open to hold the sheet of the material being tested.

Therefore, two measurements of the S21 scattering parameter are performed, with the open face of the chamber covered once by the shielding material being tested and then with the untreated one. Finally, the shielding effectiveness can be obtained by the ratio of these two measurements or by their difference, taking into account that the magnitude of scattering parameters is usually provided in decibels (dB).

Both the shielding effectiveness of the chamber and the performance of commercial shielding coatings have been tested. This work has many cross-cutting applications in addition to the already mentioned ones. As an example, military vessels, vehicles and also satellites benefit from a shielding structure and are able to protect themselves from jamming, and in general from electronic warfare systems [38]. Many shielding accessories are available in the market to protect the human body from external electromagnetic sources, such as shielded clothes to preserve the human body and shielded windows and walls for residential use, or to protect confidential industrial or high-security data.

2. Construction of the Anechoic Shielded Chamber

In this section, the construction of the anechoic chamber is discussed. The structure is composed of a 50 cm × 50 cm × 50 cm aluminum skeleton. These dimensions have been chosen because they fit the size of the shielding panels to be tested, i.e., 50 cm × 50 cm × 0.5 cm. The skeleton has been covered by a proper shielding material, the 3027-217 Flectron Nickel/Copper Polyester Nonwoven by Saint-Gobain S.p.A., which allows one to obtain a high shielding effectiveness with a very low weight. In Figure 1, a picture of the initial construction step of the chamber is shown.

Figure 1. Picture of the anechoic chamber during the first realization step.

Some very critical points in an anechoic chamber, deserving particular care, are the apertures. Although there should not be apertures in the chamber, the antenna inside the chamber must be connected to the instruments outside; thus, a small hole is required to make a path for at least one coaxial cable. In order to preserve the shielding properties, particular attention has been paid to this task. In particular, a steel sheet with a thickness of 5.5 mm was introduced in the rear face of the chamber. A 3.5 mm coaxial panel-connector, suitable to work up to 26 GHz, has been mounted on the steel sheet. By using shielding adhesive tape, it was finally covered with the shielding textile without leaving open slots, thus minimizing leakage. In Figure 2, the rear side of the chamber with the detail of the panel-connector is shown.

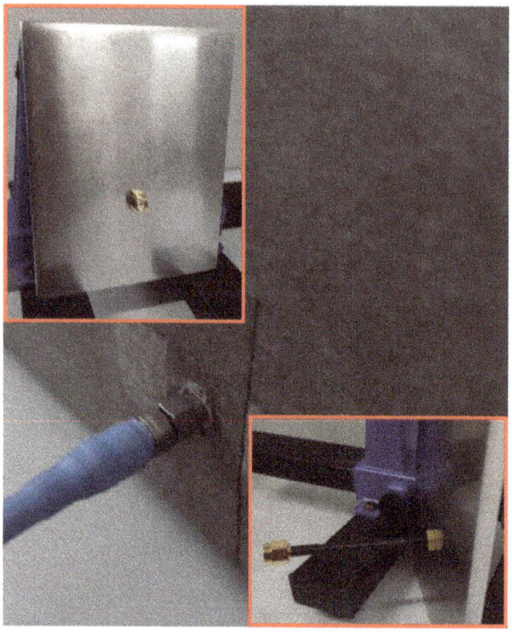

Figure 2. Picture of rear side of the chamber, with the detail of the panel connector in the insets.

The inside of the chamber has been covered by the anechoic material EA-PF3000-XX by Leader Tech. Inc. (Tampla, FL, USA). It is dielectrically loaded polyurethane, shaped in a pyramidal foam to have a smooth transitional impedance via the use of cones, thus providing the required reflection loss level. Figure 3 shows a picture of the realized anechoic chamber.

Figure 3. Picture of the shielding anechoic chamber.

As mentioned in the Introduction, a face of the chamber is left open to hold the sheet of material being tested. However, in order to preliminarily measure the shielding effectiveness of the chamber itself, it needs to be closed; thus, a removable shield and anechoic cover has been designed and realized in the same way as the chamber. In order to ensure a stable closing of the chamber and to minimize leakage, neodymium super magnets have been used to tightly connect the chamber with the cover.

The total raw-material cost of the chamber is about €1500.00, thus about one order of magnitude lower than the cost of a commercial anechoic chamber of similar dimensions, where the design costs have a great impact. In the next section, the details concerning the measurement setup and the related performance of the chamber will be shown.

3. Measurement Setup and Performance of the Chamber

It is known that, considering an N-port network, where V_n^+ is the amplitude of the voltage wave incident on a port n and V_n^- is the amplitude of the voltage wave reflected from a port n, the scattering matrix [S] can be defined in terms of the incident and reflected voltage waves, as shown in (1) [39]

$$S_{ij} = \left.\frac{V_i^-}{V_j^+}\right|_{V_k^+=0\ for\ k\neq j} \tag{1}$$

where i and j are the port numbers.

By considering a two-port network, it is possible to write the scattering parameters in terms of the incident and reflected power wave amplitudes a and b. In particular, the S_{21} parameter is expressed in (2):

$$S_{21} = \left.\frac{b_2}{a_1}\right|_{a_2=0} \tag{2}$$

where the condition $a_2 = 0$ can be straightforwardly obtained by closing port 2 with a matched load.

The shielding effectiveness, SE, of the chamber can be obtained by considering the ratio of the magnitude of the incident electric field, $\vec{E}_I$, to the transmitted electric field, $\vec{E}_T$, or the ratio of the magnitude of the incident magnetic field, $\vec{H}_I$, to the transmitted magnetic field, $\vec{H}_T$, or, alternatively, the ratio of the incident power, $\vec{P}_I$, to the magnitude of the transmitted power, $\vec{P}_T$ [20].

$$SE = \frac{\vec{E}_I}{\vec{E}_T} = \frac{\vec{H}_I}{\vec{H}_T} = \frac{\vec{P}_I}{\vec{P}_T} \tag{3}$$

A block diagram of the setup is shown in Figure 4a. In particular, P_1 and P_2 are the powers generated by the transmitter and received from the receiver, respectively. They represent port 1 and port 2 of a VNA. A_1 and A_2 are the free-space attenuations before and after the shielding interface and A_{SI} is the attenuation caused by the shielding interface, which can be written as in (4).

$$A_{SI} = \frac{1}{SE} \tag{4}$$

By observing that the square module of S_{21} is equal to the ratio of the received power to the transmitted power, it is possible to write S_{21} alongside the shielding interface, i.e., the removable cover, S_{21}^{shield}, as in (5)

$$\left|S_{21}^{shield}\right|^2 = \frac{P_2^{shield}}{P_1} = \frac{A_2 P_T^{shield}}{P_1} = A_2 A_1 A_{SI} \tag{5}$$

where $P_2^{shield} = A_2 P_T^{shield}$, and $P_T^{shield} = A_1 A_{SI} P_1$.

S_{21} can be presented without the shielding interface, S_{21}^{open}, as in (6)

$$\left|S_{21}^{open}\right|^2 = \frac{P_2^{open}}{P_1} = \frac{A_2 P_I}{P_1} = A_2 A_1 \quad (6)$$

where $P_2^{open} = A_2 P_I$, and $P_I = A_1 P_1$. This is because without the shielding interface, A_{SI} can be considered equal to 1.

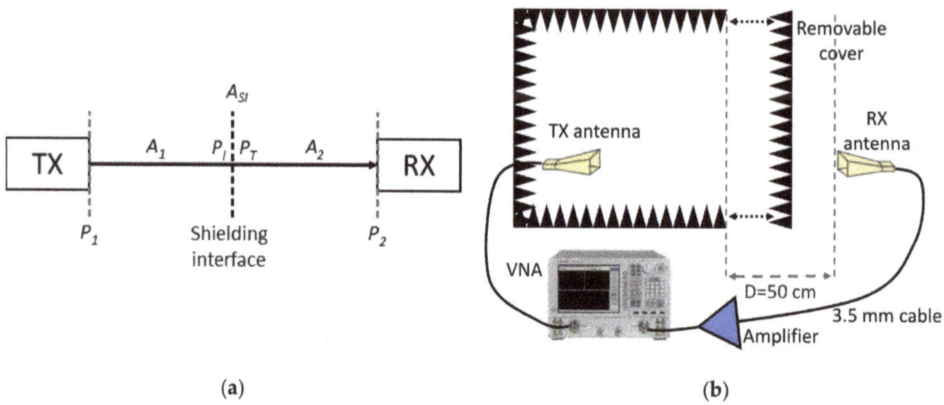

Figure 4. (**a**) Block diagram and (**b**) schematic representation of the measurement setup.

Therefore, SE can be measured by exploiting the scattering parameters, as in (7)

$$SE = \left(\frac{\left|S_{21}^{open}\right|}{\left|S_{21}^{shield}\right|}\right)^2 = S_{21}^{open}\Big|_{dB} - S_{21}^{shield}\Big|_{dB} \quad (7)$$

where S_{21}^{open} and S_{21}^{closed} are S_{21} measured with the chamber open and closed, respectively.

The measurement setup is represented in Figure 4b. It is well known that a source-free environment would be very beneficial to perform shielding or EMC tests, avoiding the detrimental effect of external signals on the measurement's effectiveness. However, the additional signals in the environment usually have very narrow bands compared to the measured bandwidth (500 MHz–26 GHz), e.g., tens of MHz within the 2.4 GHz Wi-Fi band, as much as can be considered a single scattered point which would be easily removed by averaging the measurement.

In particular, a VNA E8364A made by Agilent Technologies, Inc. (Santa Clara, CA, USA), working from 45 MHz to 50 GHz, has been used to straightforwardly measure S_{21}.

Indeed, although the powers of interest here can be also measured by using a microwave spectrum analyzer, it does not provide the option to sweep across the entire frequency range, thus requiring individual measurements, adjusting the frequency center for each tone sent from the signal generator.

Different transmitting (TX) and receiving (RX) antennas have been used to fulfill the whole bandwidth of interest, from 500 MHz to 26 GHz.

The far-field region boundary R_{FF} can be calculated according to (8) [40].

$$R_{FF} = 2\frac{D^2}{\lambda} \quad (8)$$

where λ is the minimum wavelength of the lower frequency range.

The maximum dimension of the transmitting and receiving antennas are 10 cm and 29 cm, respectively. Therefore, in the worst case of the lower-frequency bound of 500 MHz, the far-field regions start at 29 cm and 3.33 cm, respectively. Due to the chamber dimension, and according to the setup of Figure 4, both antennas have always been placed at a distance higher than these limits.

Finally, two microwave amplifiers have been employed to increase the level of the signal; this is particularly important for the case of the measurement with the shielding panel. In particular, the amplifier 310 by Sonoma Instrument, working up to 2 GHz with a gain of around 32 dB and a gain flatness of 0.5 dB, and the 83,018 A by Agilent Technologies, Inc., working up to 26 GHz, with a gain higher than 27 dB at 20 GHz and 23 dB at 26 GHz, have been used.

To measure the SE of the chamber, first, the reference measurement is performed in the absence of the shielding cover, i.e., with the chamber open; afterwards, the measurement is repeated with the shielding cover. The ratio (subtraction in dB) between the S_{21} is extracted and shown in Figure 5, thus representing the total shielding effectiveness of the shielding and anechoic chamber, as illustrated in Figure 4. Due to the noisy nature of the measurement, both the measured raw and the averaged data are shown.

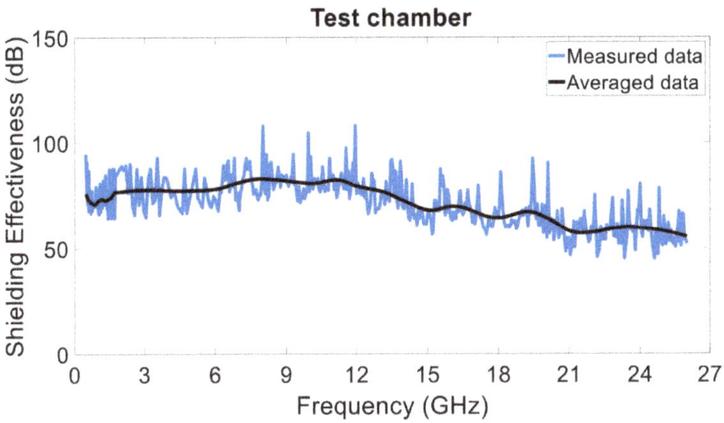

Figure 5. Shielding effectiveness of the anechoic chamber. Raw measured (blue line) and averaged (black line) data.

It is worth noting that, as a preliminary check, the sensitivity of the setup, i.e., the minimum measurable power, has been tested by performing different measurements while decreasing the VNA output power. As long as the effect of the output power decrease has been observed in the received power. The setup worked linearly, thus setting the dynamic range and validating the observed shielding effectiveness.

4. Test of Commercial Coating

The shielding effectiveness of panels, treated with commercial coating, has been measured and the results are reported in this section.

As stated in the Introduction, the main application of this kind of panels consists in the shielding of minehunter vessels.

Two different composite panels have been considered and filled with basalt or glass fibers. The single-skin laminates have been prepared by infusion at INTERMARINE SpA (Sarzana, La Spezia, Italy). Both materials have been painted by using the commercial coating MAX54 by YSHIELD GmbH & Co. (Ruhstorf an der Rott, Germany), and CuPro-Cote by Less EMF Inc. (Latham, NY, USA). The first is an acrylic-based paint, enriched with micrometric graphite and carbon black (>40 wt%). The second is a latex-based paint, filled with micrometric copper particles (47 wt%). Both paints have been dissolved in a solvent (acetone) to morphologically characterize the present fillers, using a scanning electron microscope (SEM) (FEI, QUANTA FEG 450, ThermoFisher Scientific, Waltham,

MA, USA) operating under high vacuum and 20 kV, equipped with Energy Dispersive X-ray Spectrometry EDS (EDAX, Ametek, Tokyo, Japan).

As shown in Figure 6, both types of fillers exhibit a flake morphology (Figure 6a,b), facilitating easy electrical continuity within the matrix. As confirmation of this, for MAX54 paint, only C and O (attributed to partial oxidation of the graphite) peaks are identified through EDS analysis (Figure 6c), while for the CuPro-Cote coating, the spectrum reveals the presence of Ag in addition to the Cu peaks (Figure 6d).

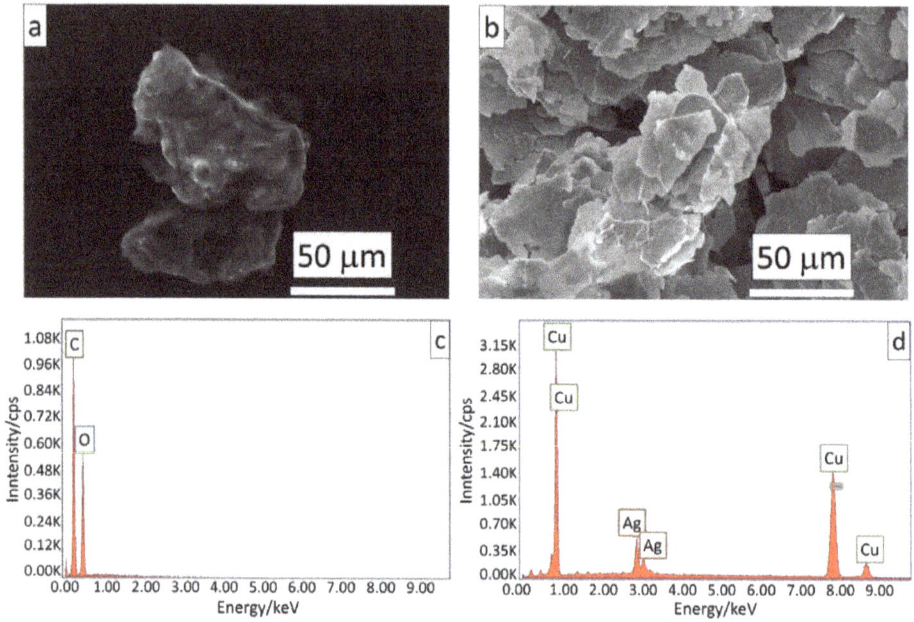

Figure 6. SEM images and EDS spectrum of MAX54 (**a**,**c**) and CuPro-Cote (**b**,**d**) fillers.

All investigated laminates have been coated with a layer of 2 mils, 50 µm of paint. Clearly, different behaviors are expected, not only depending on the type of coating but also on the treated material. Indeed, due to their better absorption capacity, the panels based on basalt fiber are expected to perform better than the ones based on glass fibers. Figure 7 shows the non-treated and treated basalt and glass fiber panels.

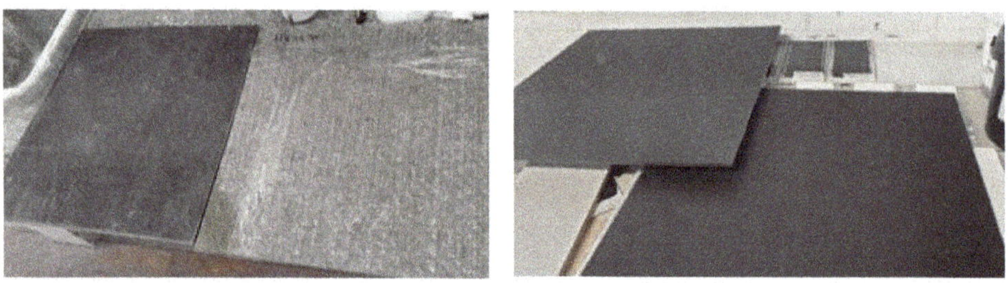

(**a**) (**b**)

Figure 7. (**a**) Non-treated and (**b**) treated basalt and glass fiber panels on the left and right, respectively.

The measurement of the SE of the panel being tested can be obtained in a similar way, i.e., by measuring the S_{21} parameter once with the non-shielding panel and then with the shielding panel, to finally calculate the ratio between the two measurements, or the difference in dB.

It is worth noting that the SE of the chamber itself must be higher than the SE of the panel being tested. Indeed, if the panels have a higher SE, the receiving antenna placed outside the chamber will receive the largest signal contribution from the chamber faces rather than from the panel being tested. As a matter of fact, this is not the case for the proposed setup; this might be easily verified based on the obtained results. Certainly, if the SE of the panel being tested shows values comparable to the SE of the chamber itself, the measured results might be considered dubious. In fact, if the SE of the panels being tested were higher, it could not be measured due to the higher leakages of the chamber itself.

Figures 8 and 9 show the shielding effectiveness of the glass and basalt fiber panels, respectively. The measured SE is lower than the SE of the chamber; thus, the measurement can be considered valid. As expected, the basalt fiber panels show a higher SE, particularly at higher frequencies. Moreover, the CuPro-Cote coating's performance is slightly better than the MAX43's performance.

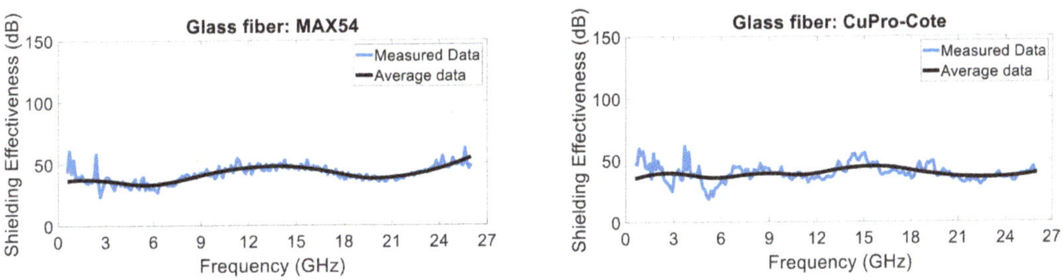

(a) (b)

Figure 8. Glass fiber panels treated with (**a**) MAX54 and (**b**) CuPro Cote coatings.

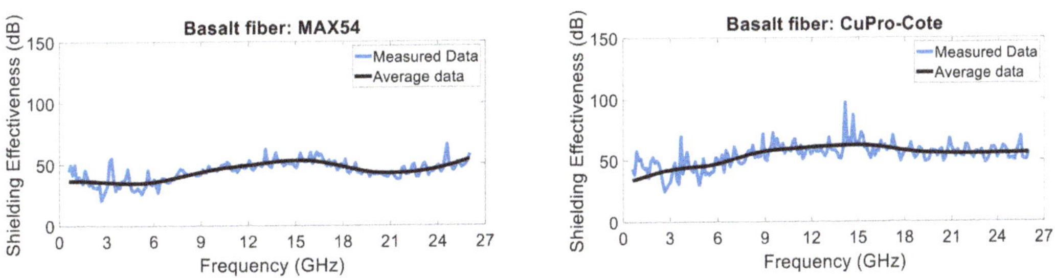

(a) (b)

Figure 9. Basalt fiber panels treated with (**a**) MAX54 and (**b**) CuPro-Cote coatings.

Figure 10 highlights the difference in shielding effectiveness between the tested materials. The performance of the glass fiber materials can be considered quite comparable, except for in the higher frequency range, i.e., from 21 GHz to 26 GHz, wherein the MAX54 fiber shows a higher attenuation. On the other hand, the basalt fiber CuPro-Cote exhibits a considerably higher shielding effectiveness throughout all of the frequency range.

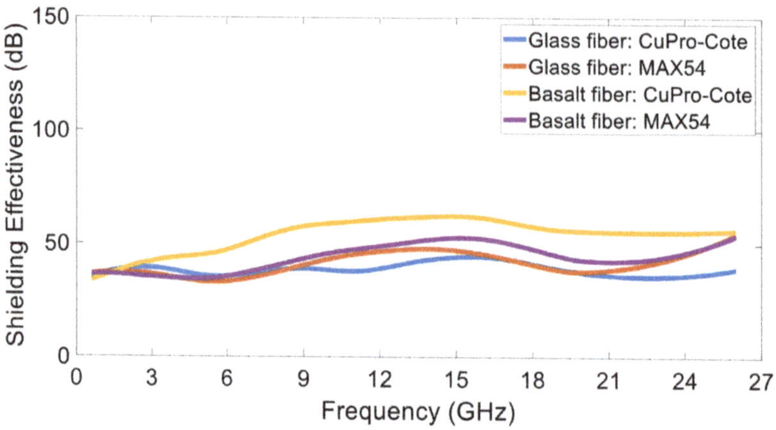

Figure 10. Shielding effectiveness of the tested coatings: glass fiber Cu-Pro-Cote (blue line), MAX54 (orange line), basalt fiber CuPro-Cote (yellow line) and MAX54 (purple line).

5. Conclusions

This contribution dealt with a measurement procedure aimed at testing the shielding effectiveness of coated panels. The proposed setup exploited a lab-made anechoic shielded chamber whose realization steps have been disclosed. The cost-effectiveness and the ease of the measurement procedure are among the main advantages here. The shielding capability of the chamber was measured up to 26 GHz, whereas the performance of commercial shielding coatings was measured by testing different materials and paintings.

Author Contributions: Conceptualization, E.C., F.L.C. and L.F.; methodology, E.C., F.L.C. and L.F.; software, E.C., F.L.C. and L.F.; validation, E.C., F.L.C. and L.F.; formal analysis, E.C., F.L.C. and L.F.; investigation, E.C., F.L.C. and L.F.; resources, E.C., F.L.C., L.F., E.P., E.M., G.S. and C.C.; data curation, E.C., F.L.C. and L.F.; writing—original draft, E.C.; writing—review and editing, E.C., F.L.C., L.F., E.P., E.M., G.S. and C.C.; supervision, E.C., F.L.C., L.F., E.P., E.M., G.S. and C.C. All authors have read and agreed to the published version of the manuscript.

Funding: This research was supported by MISE PON "Imprese e competitività" 2014–2020 FESR, Asse I—Innovazione, project F/190001/01-03/X44 DAS PHANTOMSHIFFE—Development of Ahead Systems and Processes for Highly AdvaNced TechnOlogies for low Magnetic Signature and Highly eFFicient Electromagnetic shielded eco-friendly vessel, CUP B46G19000600008.

Institutional Review Board Statement: Not applicable.

Informed Consent Statement: Not applicable.

Data Availability Statement: Data are contained within the article.

Conflicts of Interest: The authors declare no conflicts of interest.

References

1. Ardila-Rey, J.A.; De Castro, B.A.; Rozas-Valderrama, R.; Orellana, L.; Boya, C.; Muhammad-Sukki, F.; Mas'ud, A.A. Variation in the spectral content of UHF PD signals due to the presence of obstacles in the measurement environment. *IEEE Sens. J.* **2023**, *23*, 22620–22629. [CrossRef]
2. Cardillo, E.; Ferro, L.; Sapienza, G.; Li, C. Reliable eye-blinking detection with millimeter-wave radar glasses. *IEEE Trans. Microw. Theory Tech.* **2024**, *72*, 771–779. [CrossRef]
3. Ahn, J.; Woo, S.; Kim, H.; Song, K.; Huh, S.; Hong, S.E.; Kim, J.; Choi, H.D.; Ahn, S. An out-of-phase wireless power transfer system for implantable medical devices to reduce human exposure to electromagnetic field and increase power transfer efficiency. *IEEE Trans. Biomed. Circuits Syst.* **2022**, *16*, 1166–1180. [CrossRef]
4. Yao, C.-Y.; Liao, W.-J. An estimation method for EMI radiated emissions using measured source voltages. *IEEE Trans. Electromagn. Compat.* **2023**, *65*, 770–779. [CrossRef]

5. Matsumoto, Y.; Gotoh, K.; Yamanaka, Y. Effects of increasing number of disturbance sources on the setting of emission limits. *IEEE Trans. Electromagn. Compat.* **2023**, *65*, 982–996. [CrossRef]
6. Li, R.; Yue, Y.; Song, X.; Ji, S. Measurement and prediction of indoor wideband electric field radiation. *IEEE Trans. Instrum. Meas.* **2021**, *70*, 6504612. [CrossRef]
7. Scandurra, G.; Arena, A.; Cardillo, E.; Giusi, G.; Ciofi, C. Portable and highly versatile impedance meter for very low frequency measurements. *Appl. Sci.* **2021**, *11*, 8234. [CrossRef]
8. Cardillo, E.; Sapienza, G.; Ferro, L.; Li, C.; Caddemi, A. Radar assistive system for people with neurodegenerative disorders through head motion and eyes blinking detection. In Proceedings of the 2023 IEEE/MTT-S International Microwave Symposium—IMS 2023, San Diego, CA, USA, 11–16 June 2023; pp. 979–982. [CrossRef]
9. Scandurra, G.; Cardillo, E.; Giusi, G.; Ciofi, C.; Alonso, E.; Giannetti, R. Portable knee health monitoring system by impedance spectroscopy based on audio-board. *Electronics* **2021**, *10*, 460. [CrossRef]
10. Cardillo, E.; Ferro, L.; Li, C. Microwave and millimeter-wave radar circuits for the next generation contact-less in-cabin detection. In Proceedings of the APMC 2022, Yokohama, Japan, 29 November–2 December 2022; pp. 231–233. [CrossRef]
11. Cardillo, E.; Li, C.; Caddemi, A. Heating, ventilation, and air conditioning control by range-Doppler and micro-Doppler radar sensor. In Proceedings of the European Radar Conference, London, UK, 5–7 April 2022; pp. 21–24. [CrossRef]
12. Campi, T.; Cruciani, S.; Maradei, F.; Feliziani, M. Electromagnetic interference in cardiac implantable electronic devices due to dynamic wireless power systems for electric vehicles. *Energies* **2023**, *16*, 3822. [CrossRef]
13. Anzaldi, G.; Silva, F.; Fernandez, M.; Quilez, M.; Riu, P.J. Initial analysis of SAR from a cell phone inside a vehicle by numerical computation. *IEEE Trans. Biomed. Eng.* **2007**, *54*, 921–930. [CrossRef]
14. Narayanan, S.; Zhang, Y.; Aslani, F. Prediction Models of Shielding Effectiveness of Carbon Fibre Reinforced Cement-Based Composites against Electromagnetic Interference. *Sensors* **2023**, *23*, 2084. [CrossRef]
15. Špikić, D.; Švraka, M.; Vasić, D. Effectiveness of Electrostatic Shielding in High-Frequency Electromagnetic Induction Soil Sensing. *Sensors* **2022**, *22*, 3000. [CrossRef]
16. Youn, S.; Lim, T.H.; Kang, E.; Lee, D.H.; Kim, K.B.; Choo, H. Design of a Miniaturized Rectangular Multiturn Loop Antenna for Shielding Effectiveness Measurement. *Sensors* **2020**, *20*, 3178. [CrossRef]
17. Clayton, R.P. *Introduction to Electromagnetic Compatibility*, 2nd ed.; Wiley-Interscience: Hoboken, NJ, USA, 2006.
18. Sankhla, S.; Warhekar, P.; Bhattacharya, A.; Neogi, S. Effective electromagnetic interference shields based on cellulose nanofibers-carbon nanotubes layered composites. *Mater. Chem. Phys.* **2023**, *310*, 128483. [CrossRef]
19. Qian, Y.; Kuang, Y.; Zhang, Y.; Wei, Y.; Liu, Y.; Wang, C.; Chen, G. High-value utilization of pineapple leaf fibers towards high-performance electromagnetic shielding materials. *Mater. Today Nano* **2023**, *24*, 100393. [CrossRef]
20. Gouda, A.; Ghosh, S. An analytical study of curved frequency selective surfaces for shielding applications. *Microw. Opt. Technol. Lett.* **2023**, *65*, 3139–3146. [CrossRef]
21. Jung, M.; Lee, Y.-S.; Hong, S.-G. Effect of incident area size on estimation of EMI shielding effectiveness for ultra-high-performance concrete with carbon nanotubes. *IEEE Access* **2019**, *7*, 183105–183117. [CrossRef]
22. Bellucci, S.; Bistarelli, S.; Cataldo, A.; Micciulla, F.; Kranauskaite, I.; Macutkevic, J.; Banys, J.; Volynets, N.; Paddubskaya, A.; Bychanok, D.; et al. Broadband dielectric spectroscopy of composites filled with various carbon materials. *IEEE Trans. Microw. Theory Tech.* **2015**, *63*, 2024–2031. [CrossRef]
23. Sankaran, S.; Deshmukh, K.; Basheer Ahamed, M.; Khadheer Pasha, S.K. Recent advances in electromagnetic interference shielding properties of metal and carbon filler reinforced flexible polymer composites: A review. *Compos. Part A Appl. Sci. Manuf.* **2018**, *114*, 49–71. [CrossRef]
24. Modak, P.; Nandanwar, D.V. A review on graphene and its derivatives based polymer nanocomposites for electromagnetic interference shielding. *Int. J. Adv. Sci. Eng. Technol.* **2015**, *1*, 212–214.
25. Zhang, C.S.; Ni, Q.Q.; Fu, S.Y.; Kurashiki, K. Electromagnetic interference shielding effect of nanocomposites with carbon nanotube and shape memory polymer. *Compos. Sci. Technol.* **2007**, *67*, 2973–2980. [CrossRef]
26. Thomassin, J.M.; Jerome, C.; Pardoen, T.; Bailly, C.; Huynen, I.; Detrembleur, C. Polymer/carbon based composites as electromagnetic interference (EMI) shielding materials. *Mater. Sci. Eng. R Rep.* **2023**, *64*, 211–232. [CrossRef]
27. Hsiao, S.T.; Ma, C.C.M.; Tien, H.W.; Liao, W.H.; Wang, Y.S.; Li, S.M.; Huang, Y.C. Using a non-covalent modification to prepare a high electromagnetic interference shielding performance graphene nanosheet. *Carbon* **2013**, *60*, 57–66. [CrossRef]
28. Amaro, A.; Suarez, A.; Torres, J.; Martinez, P.A.; Herraiz, R.; Alcarria, A.; Benedito, A.; Ruiz, R.; Galvez, P.; Penades, A. Shielding effectiveness measurement method for planar nanomaterial samples based on CNT materials up to 18 GHz. *Magnetochemistry* **2023**, *9*, 114. [CrossRef]
29. *IEEE Std 299-2006 (Revision of IEEE Std 299-1997)*; IEEE Standard Method for Measuring the Effectiveness of Electromagnetic Shielding Enclosures. IEEE: Piscataway, NJ, USA, **2007**; pp. 1–52. [CrossRef]
30. Tamburrano, A.; Desideri, D.; Maschio, A.; Sarto, M.S. Coaxial waveguide methods for shielding effectiveness measurement of planar materials up to 18 GHz. *IEEE Trans. Electromagn. Compat.* **2014**, *56*, 1386–1395. [CrossRef]
31. Marvin, A.C.; Dawson, L.; Flintoft, I.D.; Dawson, J.F. A method for the measurement of shielding effectiveness of planar samples requiring no sample edge preparation or contact. *IEEE Trans. Electromagn. Compat.* **2009**, *51*, 255–262. [CrossRef]
32. De Leo, A.; Russo, P.; Primiani, V.M. Emission test method for source stirred reverberation chambers with scalar measurements. *IEEE Access* **2023**, *11*, 16749–16756. [CrossRef]

33. Qi, W.; Chen, K.; Shen, X.; Zhao, Y.; Xu, Q.; Loh, T.H.; Huang, Y. Statistical analysis for shielding effectiveness measurement of materials using reverberation chambers. *IEEE Trans. Electromagn. Compat.* **2023**, *65*, 17–27. [CrossRef]
34. Valente, R.; De Ruijter, C.; Vlasveld, D.; Van Der Zwaag, S.; Groen, P. Setup for EMI shielding effectiveness tests of electrically conductive polymer composites at frequencies up to 3.0 GHz. *IEEE Access* **2017**, *5*, 16665–16675. [CrossRef]
35. Expósito, I.; García Sánchez, M.; Cuiñas, I. Uncertainty assessment of a small rectangular anechoic chamber: From design to operation. *IEEE Trans. Antennas Propag.* **2020**, *68*, 4871–4880. [CrossRef]
36. Khalid, T.; Albasha, L.; Qaddoumi, N.; Yehia, S. Feasibility study of using electrically conductive concrete for electromagnetic shielding applications as a substitute for carbon-laced polyurethane absorbers in anechoic chambers. *IEEE Trans. Antennas Propag.* **2017**, *65*, 2428–2435. [CrossRef]
37. Munteanu, I.; Kakerow, R. Simulation methodology for the assessment of field uniformity in a large anechoic chamber. *IEEE Trans. Magn.* **2014**, *50*, 213–216. [CrossRef]
38. Cardillo, E.; Cananzi, R.; Vita, P.; Caddemi, A. Dual-conversion microwave down converter for nanosatellite electronic warfare systems. *Appl. Sci.* **2022**, *12*, 1524. [CrossRef]
39. Pozar, D.M. *Microwave Engineering*, 4th ed.; Wiley: New York, NY, USA, 2011.
40. Balanis, C.A. *Antenna Theory: Analysis and Design*, 4th ed.; Wiley: New York, NY, USA, 2016.

sensors

MDPI

Article

Thickness Measurements with EMAT Based on Fuzzy Logic

Yingjie Shi [1,2], Shihui Tian [1], Jiahong Jiang [1], Tairan Lei [1], Shun Wang [1], Xiaobo Lin [2] and Ke Xu [1,*]

1 Collaborative Innovation Center of Steel Technology, University of Science and Technology Beijing, Beijing 100083, China; b20200538@xs.ustb.edu.cn (Y.S.); b20200539@xs.ustb.edu.cn (S.T.); m202111260@xs.ustb.edu.cn (J.J.); m202221331@xs.ustb.edu.cn (T.L.); d202110589@xs.ustb.edu.cn (S.W.)
2 Institute of Acoustics, Chinese Academy of Sciences, Beijing 100190, China; linxiaobo@mail.ioa.ac.cn
* Correspondence: xuke@ustb.edu.cn; Tel.: +86-10-62332159; Fax: +86-10-62332947

Abstract: Metal thickness measurements are essential in various industrial applications, yet current non-contact ultrasonic methods face limitations in range and accuracy, hindering the widespread adoption of electromagnetic ultrasonics. This study introduces a novel combined thickness measurement method employing fuzzy logic, with the aim of broadening the applicational scope of the EMAT. Leveraging minimal hardware, this method utilizes the short pulse time-of-flight (TOF) technique for initial thickness estimation, followed by secondary measurements guided by fuzzy logic principles. The integration of measurements from the resonance, short pulse echo, and linear frequency modulation echo extends the measurement range while enhancing accuracy. Rigorous experimental validation validates the method's effectiveness, demonstrating a measurement range of 0.3–1000.0 mm with a median error within ±0.5 mm. Outperforming traditional methods like short pulse echoes, this approach holds significant industrial potential.

Keywords: electromagnetic acoustic transducer; thickness measurement; fuzzy logic; pulse compression

1. Introduction

The Electromagnetic Acoustic Transducer (EMAT) is extensively employed in metal thickness measurements due to its non-contact operation, absence of couplant requirement, and high-temperature resistance [1]. However, the notably low conversion efficiency of the EMAT [2] necessitates high-performance excitation equipment and the amplification of weak signals. The low signal-to-noise ratio (SNR) primarily contributes to the complexity or even failure of signal processing. Despite these challenges, compared to piezoelectric ultrasonics, exploring the EMAT for thickness measurements remains a worthwhile research topic.

The accurate measurement of thickness presents substantial challenges when dealing with extremely thin or thick specimens, severe corrosion, high temperatures [3], or specimens with thick insulation layers [4]. As a result, considerable research has been undertaken to tackle these issues, with a focus on equipment [5], transducers [6], and methods for measuring thickness [7]. Companies like Ritec and Sonemat have notably contributed by creating linear power amplifiers, Class D power amplifiers, and related devices for the amplification of weak signals, offering invaluable assistance to research institutions. Some established commercial teams have developed narrowband handheld EMAT devices for thickness measurements, which exhibit stable performance, albeit within a conventional thickness range. Table 1 outlines the primary parameters of current EMAT thickness measurement equipment. In comparison, the fuzzy logic method introduced in this paper exhibits a broader measurement range and superior accuracy.

Regarding transducers [8], circular transducers are associated with spiral coils [9], while rectangular magnets correspond to racetrack or butterfly coils [10]. While these probes are indeed relatively advanced, significant potential remains for further exploration in the areas of transducer structure, coil layout, and magnetic field enhancement. For instance, Cai Z et al. [11] employed a longitudinal wave transducer and amplified the echo amplitude by optimizing parameters with the arrangement of permanent magnets in a Halbach array. Pei C et al. [12] introduced a modified meander-line-coil EMAT design with an innovative magnetic configuration aimed at increasing the peak flux density to enhance the generation and detection of surface waves. Kang L et al. [13] examined the effect of varying different EMAT parameters on surface waves using the orthogonal test method, resulting in a 25.2% increase in the signal amplitude for the optimized EMAT. Xu X et al. [14] assessed the impact of different coil shapes and excitation frequencies on the SNR of thickness measurement results and explored the operational range of the coil.

Table 1. Comparison of EMAT thickness measurement equipment.

Manufacturer	Frequency (MHz)	Range (mm)	Thickness Measurement Error (mm)
Oktanta EM1401	3.5–5.5	2–200	±0.1 (thickness 2–200)
ACS A1270	3.0	1–100	±0.1
Orisonic ETGmini-X2	4.0	1.5–300	±0.05 (thickness 1.5–10) ±0.01 + H/200 (thickness 10–500)
Ours	0.5–5.0	0.3–1000	±0.03 (thickness 0.3–10) ±0.01 + H/300 (thickness 10–1000)

In terms of thickness measurement strategies, T. Takagi et al. [7] explored the impact of the Electromagnetic Acoustic Resonance (EMAR) method on thickness measurements in the context of steel corrosion thinning. S. Dixon et al. [15] utilized the conventional short pulse echo method to measure the thickness of low carbon steel and stainless steel under high-temperature conditions. Sheng-Bin Wang et al. [16] employed a pulse compression method to enhance the SNR. Zhao [17] used a laser source to generate shear waves and implemented electromagnetic ultrasonic receptions to measure the thickness of thin steel plates, approximately 0.4 mm thick, achieving impressive results. Qiu J et al. [18] managed to instantly excite a pulsed magnetic field by controlling the excitation current, which enabled the measurement of the thickness of 50 mm-thick steel plates with a small adsorption force.

EMAT equipment, transducers [19], and measurement methods have reached relative maturity. Current thickness measurement schemes mainly include the following: (1) building specific experimental systems for specimens' thickness, and (2) using commercial thickness gauges. However, these methods require corresponding instruments and have their own limitations. Generally, the short pulse (Toneburst) echo method has a measurement range of 2–300 mm with moderate accuracy and is unable to measure thin or thick specimens beyond this range. Resonance methods can measure thin specimens with high precision but require large equipment and have high performance requirements, with few cost-effective commercial systems available. A linear frequency modulation echo is suitable for thick object measurement but requires wideband transmission and reception, with higher equipment demands and limited applicational scenarios. Hence, it is essential to explore a method to extend both the measurement range and accuracy under specific hardware conditions. Specifically, this study employs a strategy that utilizes the short pulse time-of-flight (TOF) method for initial thickness estimation, followed by secondary measurements guided by fuzzy logic principles. The integration of measurements from the resonance, short pulse echo, and linear frequency modulation echo extends the measurement range while enhancing accuracy. Therefore, the proposed method demonstrates potential for industrial applications.

This paper is structured as follows: Section 2 outlines various thickness measurement methods and analyzes limitations in constant hardware systems [20]. Based on the above analysis, Section 3 designs fuzzy logic and describes the thickness measurement process, which constitutes the main innovation of this study. Section 4 constructs an experimental system and validates it with 30 thickness samples, analyzing the measurement range and errors through fuzzy logic fusion, along with an analysis of measurement uncertainty.

Section 5 concludes the paper, summarizing the findings and discussing potential avenues for future research.

2. Theory

2.1. Transduction Mechanism

The transduction mechanisms of the EMAT include the Lorentz force, magnetostrictive force, and magnetization force. Among them, the Lorentz force is primarily applicable to non-magnetic materials, while both the Lorentz force and magnetostrictive force are applicable to magnetic materials. A permanent magnet generates a static magnetic field B_s in the direction perpendicular to the specimen, and an alternating current J_c is passed through the coil, creating a time-varying magnetic field B_d around the coil. This induces an oppositely directed eddy current J_e in the sample. The dynamic magnetic field B_d flows horizontally on the surface of the specimen in the eddy current region. As a result, the eddy currents J_e produce a horizontal Lorentz force F_s and a vertical Lorentz force F_d under the influence of the static and dynamic magnetic fields, respectively. Additionally, in magnetic materials, the magnetostrictive forces F_M are horizontal vibrations. Figure 1 shows a schematic diagram of these transduction mechanisms.

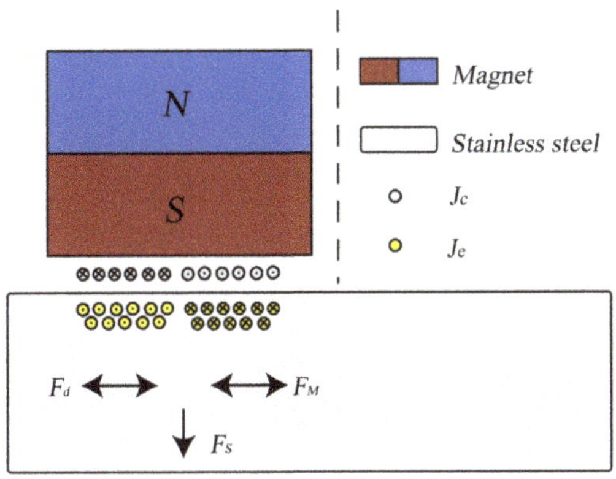

Figure 1. A schematic of the EMAT transduction mechanisms.

In this thickness measurement application, since the amplitude of the static magnetic field is much greater than that of the dynamic magnetic field, the designed EMAT primarily relies on the static Lorentz force. The specimen is subjected to a force in the horizontal direction within the range of skin depth, while the propagation direction of ultrasonic shear waves is vertical. The transduction processes described above can be represented by the following equations. The volume force of the Lorentz F_s force arises according to Equation (1). Equation (2) describes the strain generated by the magnetostrictive effect, where ε represents the strain tensor, σ represents the stress tensor, and H represents the magnetic field intensity.

$$F_s = J_e \times B_s \tag{1}$$

$$\varepsilon = f(\sigma, H) \tag{2}$$

Equation (3) illustrates that the reception of ultrasonic waves is the reverse process of the generation of static Lorentz forces. The particles cut the magnetic inductance lines, creating an induced electromotive force. For a continuous length circular coil, it can be expressed by the following equation, where J_r represents the total induced current received in the coil, μ represents the particle displacement, η represents the electromagnetic conversion efficiency, σ represents the electrical conductivity of stainless steel, and l represents the integral length of the coil. Equation (4) represents the inverse magnetostrictive effect, which is the process of stress generating magnetic induction, where B represents the magnetic flux density.

$$Jr = \int_0^L \eta\sigma\left(\frac{\partial\mu}{\partial t}\right) \times B_d dl \tag{3}$$

$$B = y(\sigma, H) \tag{4}$$

2.2. TOF Measurement

When ultrasonic waves encounter an interface, they are reflected. Given the speed of sound, the distance that the ultrasonic waves travel within the specimen can be calculated by measuring the time from emission to reception. This process is known as time of flight (TOF). The specimen thickness can be calculated using the following formula, where c_s represents the shear wave velocity, d represents the thickness, Δt_i represents the time difference between the i-th and (i-1)-th echoes, and N represents the average number of valid echoes:

$$d = \frac{\sum c_s \Delta t_i}{2N} \quad (i = 1, 2, \ldots N) \tag{5}$$

Continuous wave (CW) and linear frequency modulation (LFM) are commonly used as excitation waveforms, as depicted in Equations (6) and (7), respectively. The frequency of the CW signal is f_c, while A, f_0, B, and τ represent the amplitude, starting frequency, bandwidth, and pulse width of the LFM signal, respectively. Depending on the number of waves (M) in the pulse train, CW signals can be divided into short pulse and long pulse, which are used for thickness measurements by the TOF and resonance method, respectively. Typically, Toneburst signals consist of fewer than 5 waveform cycles, and windowing is applied. In contrast, resonant sweep signals contain a large number of cycles.

$$u_{\mathrm{cw}}(t) = A * \sin(2\pi f_c t)\ 0 \le t \le M/f_c \tag{6}$$

$$u_{\mathrm{in}}(t) = A * \sin\left(2\pi\left(f_0 t + \frac{Bt^2}{2\tau}\right)\right)\ 0 \le t \le \tau \tag{7}$$

The basic processing flow of the echo method algorithm includes the following: noise reduction [21], pulse compression (which is a specific method for LFM) [22], envelope extraction, and the peak detection of echoes [3], as well as the calculation of average time intervals between multiple peaks.

2.3. Resonance Measurement

The EMAR [23] utilizes the principle of echo superposition and employs a sweep frequency method for excitation. When the resonance enhancement frequency corresponds to the thickness, the spectral response sharply protrudes. The thickness of the specimen can be derived by calculating the frequency difference between adjacent spectral peaks, where f_N represents the frequency of the N-th harmonic peak.

$$d = \frac{\sum \frac{c_s}{2(f_{N+1} - f_N)}}{N - 1} \quad (i = 1, 2, \ldots (N-1)) \tag{8}$$

Equation (8) is employed to determine the thickness of the specimen. The thickness is computed by measuring the harmonic frequencies of various orders, obtaining frequency differences, and calculating the thickness from these values. Contrary to the analog heterodyne spectrum scanning method used in RITEC-5000 equipment, this study employs ADC to directly acquire the original signal. Through a frequency domain analysis, the amplitude of the corresponding frequency at each sweep point is obtained. Full digital processing is also a part of this work.

It should be noted that the excitation length is generally the same, and the signal is subjected to an FFT analysis for an equal duration after the blind zone. It shows resonance achieved at 3.36 MHz with a 30 μs excitation for a 1.4 mm-thick aluminum plate. Figure 2a,b depict the time–domain and frequency–domain results, respectively. Since the maximum blind zone in the measured sweep occurs around 30 μs after the signal ends, this study uses 163.84 μs (8192 points) of data starting from 70 μs for the FFT analysis. The amplitude of the energy at the frequency axis of 3.36 MHz is 1.56, which represents a single resonance amplitude value. By performing a sweep from 0.5 MHz to 5 MHz with a 39.6 kHz interval, the sweep results shown in Figure 2c can be obtained.

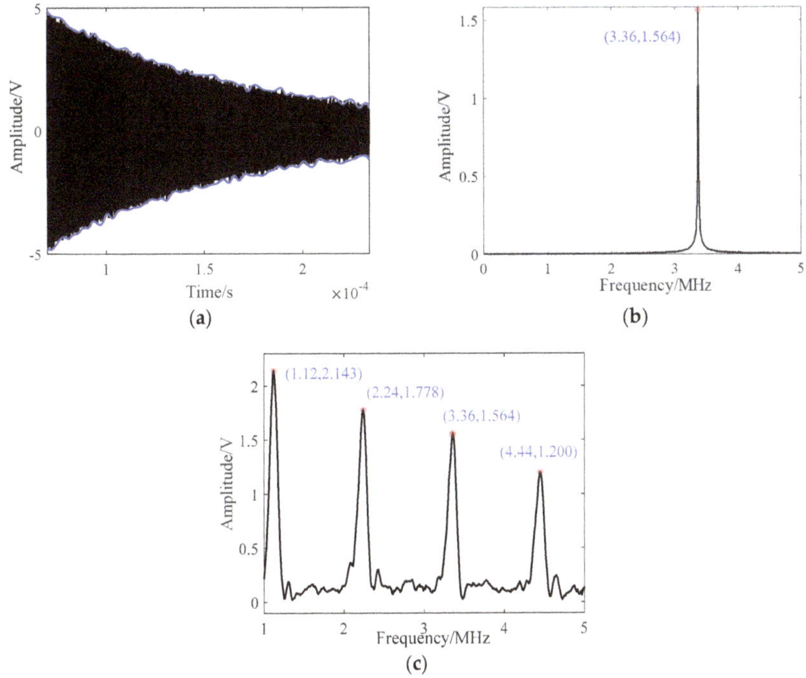

Figure 2. The digital resonance analysis: (**a**) time domain, (**b**) frequency domain, and (**c**) the sweep results.

2.4. Limitations and Solutions

The main limitation of current Toneburst-based EMAT thickness measurement systems is their limited range, leading to reduced accuracy and stability beyond this scope. As the thickness of the specimen decreases, it becomes difficult to accurately distinguish the time intervals between echoes in the time domain using TOF [24], which is mainly due to the following two aspects:

(1) The electronic system: The duplex circuit between the EMAT high-voltage pulse transmission and weak signal amplification results in a system blind spot. When the time interval between echoes is short, the echo with a higher signal intensity mainly exists in this time period, which decreases the SNR. Due to the existence of the system blind spot, it may not be possible to accurately detect the echo signal.

(2) The ultrasonic propagation distance and wavelength: As the thickness of the specimen decreases, the ultrasonic propagation distance approaches or becomes smaller than the transmitted wavelength, resulting in the superposition of multiple echoes in the time domain. When multiple echoes overlap, the temporal characteristics of the echo signal change, making it difficult to resolve and measure the echoes.

In Figure 3a, the blue dashed box indicates the saturation state. The filter and oscillation attenuation circuit mitigate the superposition of oscillations at specific frequencies, thereby reducing the blind zone. In Figure 3b, three echoes are present. The short intervals between echoes lead to an overlap within the blue box. The uncertain superposition effect, due to the correlation between the echo time and wavelength, hinders the differentiation of echoes through signal processing.

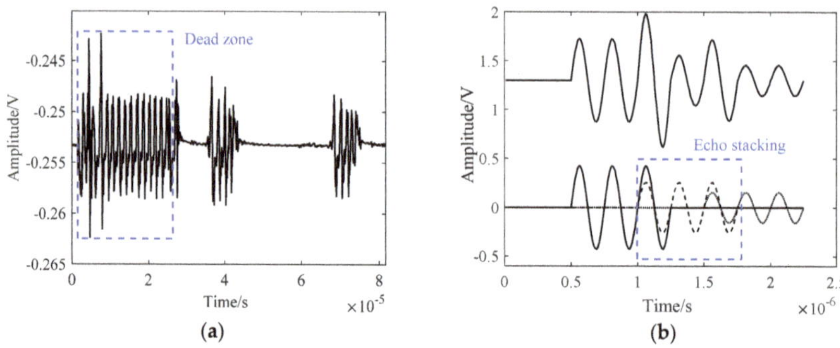

Figure 3. Reasons for failure of the TOF method: (**a**) blind zone and (**b**) echo overlap.

In any EMAT system, these issues are unavoidable, and special attention needs to be paid to these limitations in the case of thin thickness measurements. To overcome these issues, other methods or techniques may need to be adopted, such as using higher frequencies, improving circuits, or combining other measurement methods.

In this study, the authors propose the use of the resonant method [25] on existing hardware to solve the problem of thin specimen thickness measurements and reduce the blind zone. The resonant method requires at least one spectral peak to fall within a flat transmission–reception frequency range to ensure valid thickness measurement results. The ZYNQ board is used for DAC waveform generation and ADC sampling. The power amplifier and weak signal amplifier are limited in frequency bands due to electronic equipment, with the main frequency range shown in Table 2. The flat frequency range is designed to be 0.5 MHz to 5 MHz due to the influence of parameters of the electronic system and coil. By employing Equation (8), the theoretical minimum thickness is ascertained to be 0.323 mm. It should be noted that, under identical conditions of the frequency scanning range and resolution, an increase in thickness results in a decline in measurement precision; thus, the resonance method is ideally suited for gauging the thickness of thin specimens.

Table 2. Hardware limitations.

Hardware	Limitation	Frequency (MHz)	Problems
Power amplifier	Frequency	0.05–20	High frequency energy attenuation is serious
Weak signal amplifier	Filter	0.5–5	Bandpass filter
DAC	Refresh rate	50	The actual sine wave frequency is lower than 5 MHz
ADC	Sample rate	50	Restricted Nyquist sampling theorem
EMAT's coil	Impedance matching frequency	3	Narrowband matches are flat around the center frequency

When the specimen thickness is large, the increased propagation distance of ultrasonic waves results in significant signal attenuation, which poses a challenge, especially for EMAT systems with a low SNR. Traditional methods, such as increasing the number of CW signals, cannot simultaneously enhance the signal intensity and distance resolution capability. In this case, using LFM signals is a better choice. By selecting an appropriate bandwidth and pulse width for linear frequency modulation signals, the signal intensity and distance resolution capability can be increased. The broadband nature of LFM allows them to propagate over a larger thickness range and provides better signal intensity. Additionally, pulse compression processing can further improve the temporal resolution and peak amplitude of the signals. Therefore, using LFM signals can achieve better performance in measuring the thickness of thick specimens, enhancing the signal intensity and distance resolution and achieving accurate thickness measurements.

Based on the above analysis, the three thickness measurement methods differ in range: the EMAR method is better suited for thin pieces [26], Toneburst for medium thicknesses, and LFM for thick measurements. Under limited hardware conditions, this paper employs fuzzy logic to integrate these three methods, autonomously selecting the appropriate measurement method, thereby broadening the range.

3. Fuzzy Logic Measurement

3.1. Fuzzy Logic Design

Fuzzy logic [27], as a mathematical tool, is used to address the uncertainty and ambiguity present in problems. Data fusion, based on fuzzy logic, can resolve differences in attributes from various data sources [28]. Given that different sensors or measurement methods have varying resolutions, ranges, and accuracies, a more comprehensive and accurate dataset can be obtained through fuzzy logic fusion [29]. Fuzzy logic is applicable to the data fusion of various EMAT thickness measurement methods; hence, it is introduced in this paper.

Based on the theoretical analysis in Section 2, the extensive experimental data in Section 4, previously published papers, and the range intervals provided by commercial thickness measurement devices, this paper concludes that the resonance method has an accuracy range of 0.3–4.0 mm, the Toneburst method ranges from 2.0 to 150.0 mm, and LFM exceeds 20.0 mm. These estimates are based on experience. Considering the uneven distribution of the measurement range and the requirement for values to start from 0, the normalized value D_{i} is obtained as shown in the following formula.

$$D_i = log2(d_i + 1) \tag{9}$$

In the above formula, d_i represents the thickness (in mm) measured by different methods. The relationship between D_{i} and $(d_i + 1)$ is logarithms, ensuring a uniform distribution of each thickness measurement method along the X-axis. This design rationale is demonstrated in Figures 4, 10 and 11. Introduce the average value D of the effective normalization results of three methods as per the following formula.

$$D = \frac{\sum_{i=1}^{N} D_i}{N}, \ N = 1,2,3. \tag{10}$$

In the above formula, N represents the total number of effective measurement methods at this thickness. For example, both the EMAR and Toneburst method can effectively measure an aluminum plate of 1.201 mm in thickness. In this scenario, $d_1 = 1.216$, $d_2 = 1.175$, N = 2, $D_1 = 1.148$, $D_2 = 1.121$, and D = 1.1345.

The membership function used in the fuzzy rules is a Gaussian function, and the equation is as follows.

$$k_i = e^{-\alpha_i |D_i - X_i|^2}, i = 1,2,3. \tag{11}$$

where $\alpha = 4/\gamma^2$, γ is a constant determining the distribution of the rule. The γ values selection principle ensures good superposition between different logical determinations and is adjusted based on the thickness measurement ranges of different methods. In this section, the α values used are all 1.2. X_i represents the median value in the design, which is, respectively, taken as 0, 4, and 6 in the three methods. The value of X_i is determined by the measurement range of different methods, which is typically chosen at the center of a well-defined range. Based on experimental data from existing test blocks, this paper selects the thickness center as the value of X_i, ensuring maximum weighting at the center. Specifically, for the EMAR method, the center is set at 0 mm, so $X_1 = D_1(0+1) = 0$. For the Toneburst method, d_2 is set to 15, corresponding to the center of effective specimens ranging from 2 to 150 mm, and $X_2 = D_2(15+1) = 4$. Take $X_3 = 6$ as an integer. The choice of X_i can be based on either theoretical or experimental measurement ranges. This paper determines the fuzzy logic parameters through limited experimental data to achieve better results.

The Gaussian function is normalized to obtain the normalized weight values, and this is to ensure that the result is true.

$$c_i = \frac{k_i}{\sum_1^3 k_i}, i = 1,2,3. \tag{12}$$

Figure 4 displays the membership function, with the horizontal axis representing D on a logarithmic scale. The vertical axis represents the amplitude of the normalized weight c_i. The red, blue, and green curves in this figure represent the measurement result weights of the resonance method, the Toneburst method, and the LFM method, respectively.

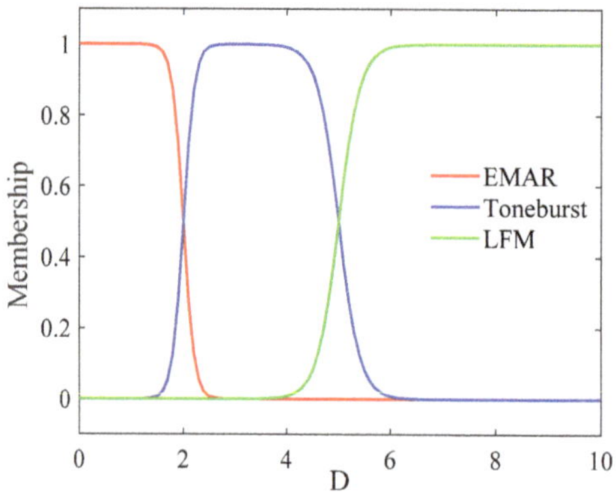

Figure 4. Membership function diagram.

The fuzzy logic [30] rules are presented in Table 3, which is categorized into small (S), small medium (SM), medium (M), medium large (ML), and large (L).

Table 3. Fuzzy logic decision rules.

D	Fuzzy Logic
0.3~1.5	S
1.5~2.5	SM
2.5~4.8	M
4.8~6	ML
≥6	L

The calculation method for the thickness d varies when the normalized thickness D belongs to different rules, where R_1 represents the first rule, and so on. For instance, when $D = 1.1345$, the fuzzy logic complies with the first rule and is determined as S. In this case, $d = d_1 = 1.216$ mm. When the nominal thickness value is 6.0 mm, $D = 2.6411$, and it is determined as SM. At this time, according to Formulas (9)–(12), $c_1 = 0.0021$, $c_2 = 0.9979$, and $d = 0.0021 * 5.437 + 0.9979 * 6.033 = 6.032$ mm.

$$\begin{cases} R_1 : IF(D \in S) & THEN(d = d_1) \\ R_2 : IF(D \in SM) & THEN(d = c_1 d_1 + c_2 d_2) \\ R_3 : IF(D \in M) & THEN(d = d_2) \\ R_4 : IF(D \in ML) & THEN(c_2 d_2 + c_3 d_3) \\ R_5 : IF(D \in L) & THEN(d = d_3) \end{cases} \tag{13}$$

The above design of fuzzy logic rules is abstract. To help readers grasp this concept more vividly, this section provides illustrative examples by substituting partial data from Table 4 in Section 4 into selected formulas.

3.2. Principles of Fuzzy Logic and Thickness Measurement Process

The thickness measurement procedure, as illustrated in Figure 5, is described as follows:

(1) The measurement process begins with the excitation of a 3-cycle 4 MHz pulse. The echo signal processing algorithm is detailed in Section 2.

(2) Based on the initial thickness measurement results and fuzzy logic, a secondary measurement method is selected: (a) For thicknesses less than or equal to 10 mm, or those that are undetectable, the resonant method is applied. A frequency sweep is conducted at 39.6 kHz intervals within the 0.5 MHz to 5 MHz range. The frequency difference is calculated from the envelope of the spectral response curve to determine the measured thickness. (b) For initial thickness measurements greater than or equal to 20 mm, or those that are undetectable, the LFM method is used.

(3) Fuzzy logic is applied to merge the secondary measurement results, yielding the final thickness measurement.

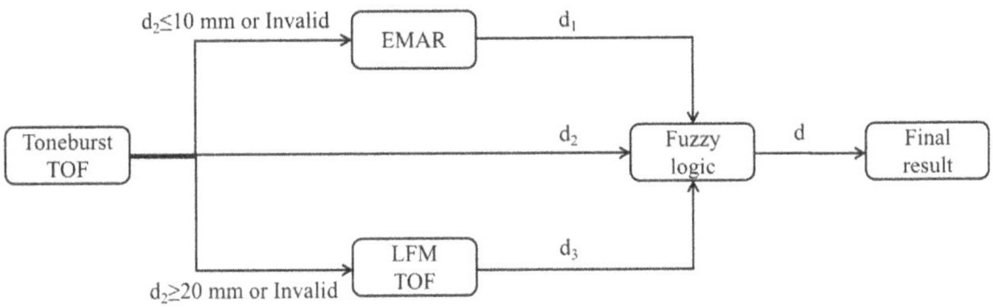

Figure 5. Flow chart of the thickness measurements.

4. Experiments

4.1. Experimental Setup

The experimental setup, as depicted in Figure 6, encompasses a variety of components including a ZYNQ board, a power amplifier, a weak signal amplifier, a duplexer, and an EMAT. At the core of the experimental setup is the ZYNQ7020 (Xilinx, San Jose, CA, USA) board, a specialized circuit designed around a heterogeneous FPGA. This board generates TTL Gate and Sin signals through DAC902 (TI, Dallas, TX, USA) and utilizes AD9226 (ADI, Wilmington, MA, USA) for sampling, with a refresh rate of 50 MHz. The entire system operates on a Linux platform running Jupyter and employs Python for programming. It controls the transmission of signals to the transducer through a power amplifier while concurrently capturing the incoming analog echo signals. The Ga2500 (RITEC, Warwick, RI, USA) power amplifier can deliver up to 5 KW of power output. Additionally, the signal amplifier, specifically RITEC's BR640A (RITEC, Warwick, RI, USA), comes equipped with a hardware filter ranging from 100 kHz to 5 MHz and offers a gain of 68 dB. The duplexer was meticulously designed to achieve broadband impedance matching with the probe.

The specimens for the experiment included a variety of samples such as thin aluminum plates, carbon steel step blocks, long carbon steel bars, and other types. These samples, with thicknesses ranging from 0.2 to 1000.0 mm, were measured in accordance with the study's requirements. This diverse selection of specimens allowed for a comprehensive and robust evaluation of the system's performance.

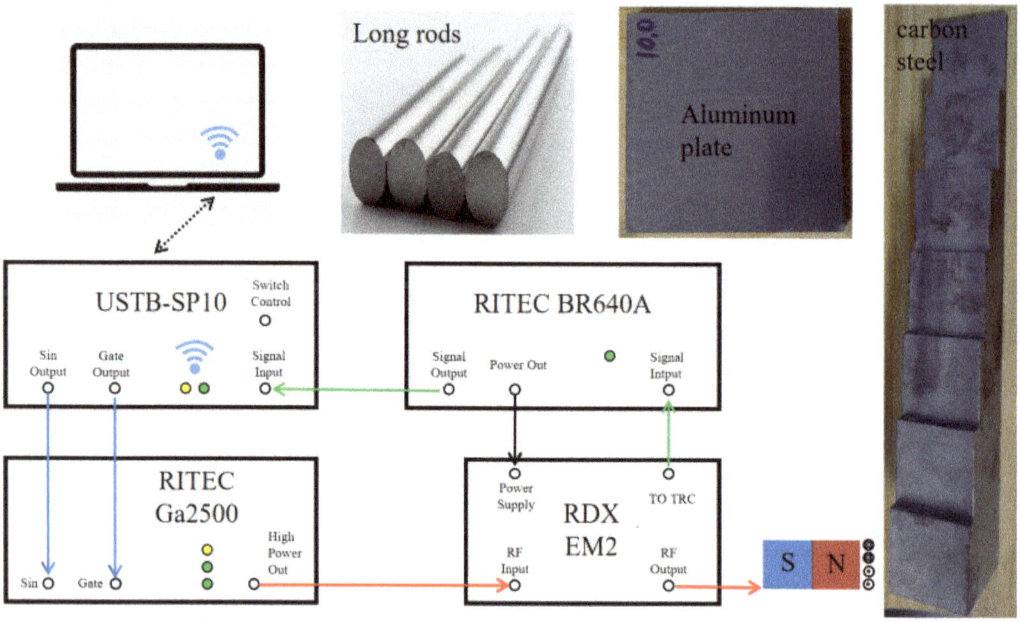

Figure 6. Diagram of experimental system.

4.2. Experimental Verification

Drawing upon the theoretical analysis in Section 2.4, this paper conducted thickness measurements on thin aluminum plates ranging from 0.2 mm to 10 mm using the resonance method. The echo waveform is shown in Figure 5. The thickness values were calculated according to the method outlined in Section 2.3 and recorded in Table 4.

The partial results of the resonance scan are illustrated in Figure 7, where 0.3, 0.5, 0.8, 6.0, 8.0, and 10.0 mm are plotted, representing the minimum and maximum cases, respectively. At the nominal value of 0.3 mm (at the system's measurable bandwidth boundary), the resonance fundamental frequency is 5 MHz, resulting in a calculated thickness of 0.323 mm. It can be observed that the thickness measurement accuracy is poor at this limit. At the nominal value of 0.5 mm, the resonance fundamental frequency is 3.210 MHz, yielding a calculated thickness of 0.503 mm, where the error is small. For a large thickness value of 6 mm, the average frequency difference of adjacent resonance harmonics is $(2.140 - 1.546)/2 = 0.297$ MHz, giving a thickness value of 5.437 mm. With a frequency difference of 0.278 MHz (obtained from 1.824 MHz minus 1.546 MHz), the corresponding thickness is calculated to be 5.809 mm. It can be seen that the thickness measurement error is larger when the thickness is larger, which aligns with the theoretical analysis. This is primarily due to the increase in the ratio of frequency resolution to frequency difference. If the resonance method is still used for thickness measurements at this time, the accuracy is significantly compromised.

Measurements of varying thicknesses (12.0, 24.0, 36.0, . . ., 155.0, and 200.0 mm) were conducted using the Toneburst method, as depicted in Figure 8. As the propagation distance increased, the amplitude of the echo signal diminished. For example, at 155 mm, it became challenging to discern the secondary echo. In the case of larger thicknesses, the short pulse echo method's limitation becomes apparent, as errors cannot be mitigated by averaging the intervals between multiple echoes.

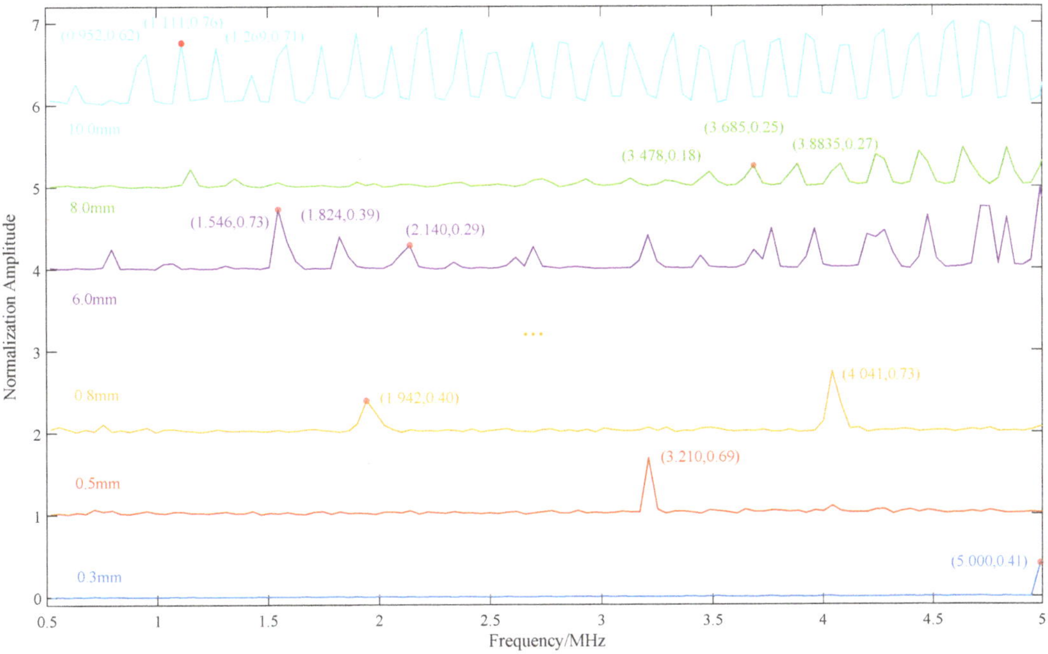

Figure 7. Resonance scanning.

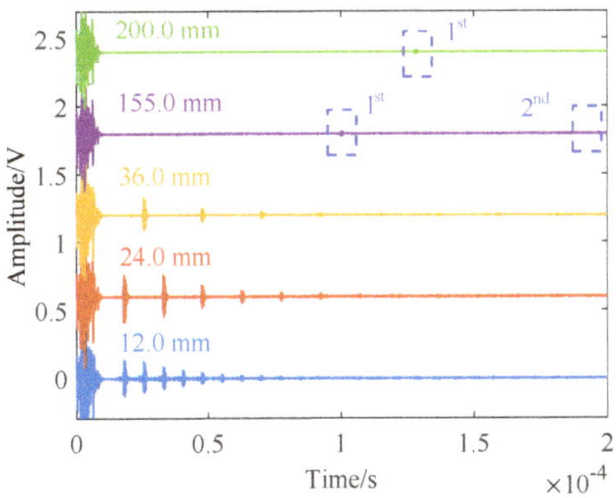

Figure 8. Thickness measurements using Toneburst pulses.

Taking a 100.0 mm test block as an example, thickness is measured using the LFM method. Unlike the Toneburst method, which involves multiple acquisitions and time–domain averaging, the pulse compression of a single acquisition datum from Figure 9a yields Figure 9b. The thickness value of 100.13 mm is calculated based on the time difference of 62.0 μs between two adjacent reflected pulse peaks (186.0 − 124.0 = 62.0 μs). Compared to the single acquisition in the time domain of Figure 8, the SNR is significantly improved. To observe effective signals, the time axis is set to 100–250 μs.

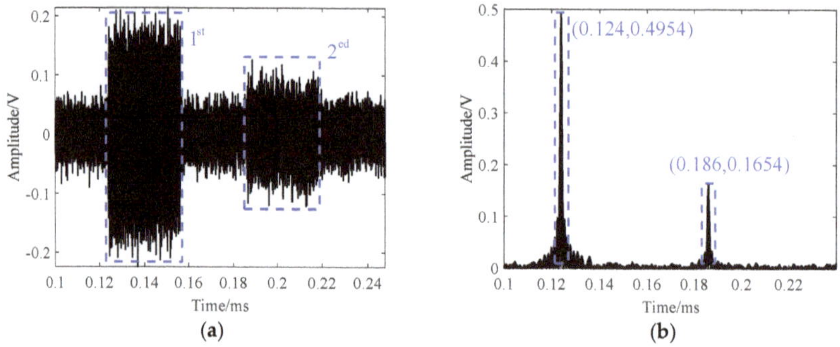

Figure 9. Thickness measurements using the LFM method. (**a**) Time–domain echo. (**b**) Results of pulse compression.

4.3. Fuzzy Logic Fusion

The thickness of 30 test samples was individually measured. For instance, if the EMAR and Toneburst method are applicable to the 2.0 mm test sample, these methods are used to measure the effective thickness value ten times, respectively. The averages of these measurements, 2.038 mm and 2.024 mm, respectively, are recorded in Table 4.

The second column of the table displays the thickness indicated at the time of purchase for each specimen. However, some blocks are not standard. To validate the "actual thickness" against the EMAT measurement results, a high-precision micrometer is employed for calibrating test blocks of 10 mm and below, while vernier calipers are used for blocks larger than 10.0 mm. The thickness tolerance of the workpiece is minimized by averaging five measurements, as recorded in column 3. Columns 4, 5, and 6 present thickness measurements obtained from three distinct methods. A dash "-" signifies that the method was not utilized. The seventh column represents the fuzzy logic determination based on the principles outlined in Table 3, while the eighth column records the final thickness measurement result, derived from the fusion of different methods via fuzzy logic. The ninth column represents the error between the thickness measurements obtained using fuzzy logic and those obtained with a micrometer.

The experimental results show that the resonance method is only effective for measuring relatively small thicknesses. The traditional echo method has a blind spot for small thicknesses and cannot measure large thicknesses. The LFM method can measure thick samples and demonstrates better accuracy and range.

Table 4. Thickness measurement results of three methods and fuzzy logic fusion.

Material	Thickness (mm)	Micrometer d (mm)	EMAR d_1 (mm)	Toneburst d_2 (mm)	LFM d_3 (mm)	Fuzzy Judgment	Fuzzy Thickness Result (mm)	Fuzzy Error (mm)
Aluminum	0.2	0.198	-	-	-	-	0.000	-
	0.3	0.312	0.323	-	-	S	0.323	0.011
	0.5	0.504	0.503	-	-	S	0.503	−0.001
	0.8	0.803	0.799	0.632	-	S	0.799	−0.004
	1.0	0.996	0.997	0.984	-	S	0.997	0.001
	1.2	1.201	1.216	1.175	-	S	1.216	0.015
	1.4	1.394	1.402	1.416	-	S	1.402	0.008
	1.5	1.483	1.478	1.365	-	S	1.478	−0.005
	2.0	2.035	2.038	2.024	-	SM	2.038	0.003
	3.0	3.003	3.007	2.985	-	SM	2.996	−0.007
	4.0	4.029	4.049	4.001	-	SM	4.003	−0.026
	5.0	5.030	4.924	5.048	-	M	5.048	0.018
	6.0	6.045	5.437	6.033	-	M	6.033	−0.012
	8.0	8.004	7.966	8.001	-	M	8.001	−0.003
	10.0	10.037	10.188	10.063	-	M	10.063	0.026

Table 4. *Cont.*

Material	Thickness (mm)	Micrometer *d* (mm)	EMAR d_1 (mm)	Toneburst d_2 (mm)	LFM d_3 (mm)	Fuzzy Judgment	Fuzzy Thickness Result (mm)	Fuzzy Error (mm)
Carbon steel	12.0	12.00	-	11.92	-	M	11.920	−0.08
	15.0	14.98	-	14.92	-	M	14.920	−0.06
	19.0	19.04	-	19.12	-	M	19.120	0.08
	24.0	23.98	-	23.95	23.97	M	23.950	−0.03
	30.0	30.02	-	29.89	29.96	ML	29.921	−0.099
	36.0	36.00	-	35.96	36.00	ML	35.989	−0.011
	42.0	42.00	-	42.02	42.01	ML	42.011	0.011
	48.0	48.00	-	48.03	48.02	ML	48.020	0.02
	80.0	80.04	-	80.09	80.03	L	80.030	−0.01
	100.0	100.06	-	100.04	100.08	L	100.080	0.02
	155.0	155.69	-	156.30	155.99	L	155.990	0.3
	200.0	200.32	-	201.40	200.21	L	200.210	−0.11
	400.0	401.90	-	-	401.56	L	401.560	−0.34
	600.0	598.30	-	-	599.04	L	599.040	0.74
	1000.0	1000.40	-	-	1000.17	L	1000.170	−0.23

In the above table, 0.2 mm falls outside the measurement range, rendering the actual result incorrect. The measurement range of different methods is depicted in Figure 10, with the X-axis representing the logarithm of the actual thickness. The Y-axis heights represent the four distinct methods, with each method corresponding to a specific height. The presence of marks signifies the effective measurement of thickness using these methods. The figure illustrates that the three measurement methods have overlapping and complementary ranges. The fuzzy logic thickness measurement results encompass the entire measurement area, thereby enhancing the thickness measurement range.

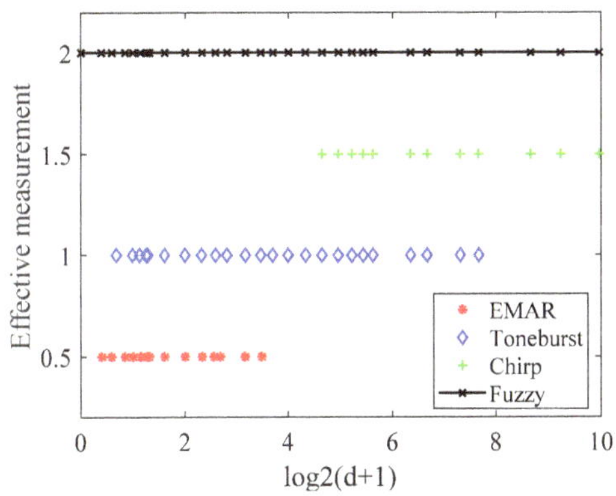

Figure 10. The range of different methods.

The comparison of various thickness measurement methods is presented in Table 4, with corresponding errors illustrated in Figure 11. This figure displays the errors of the EMAR, Toneburst, LFM, and fuzzy methods, represented by red, blue, green, and magenta lines, respectively. Generally, single-method measurements exhibit substantial errors at the range extremes. Specifically, (1) the Toneburst method demonstrates poor accuracy and stability for measurements below 2.0 mm, with errors reaching 21.3% at 0.8 mm. Conversely, the EMAR method performs well within this range, and fuzzy logic accurately categorizes it as "small". (2) The EMAR method exhibits significant errors beyond 5.0 mm, attributable to a single sweep interval. However, in the overlapping region with Toneburst, fuzzy logic designates it as "M", yielding accurate measurements. (3) The Toneburst method incurs substantial errors at or beyond 155 mm and fails to measure thicknesses at 400.0, 600.0, and 1000.0 mm. To extend the measurement range for thicker samples, the LFM method is introduced. Fuzzy logic categorizes this range as "ML" or "L".

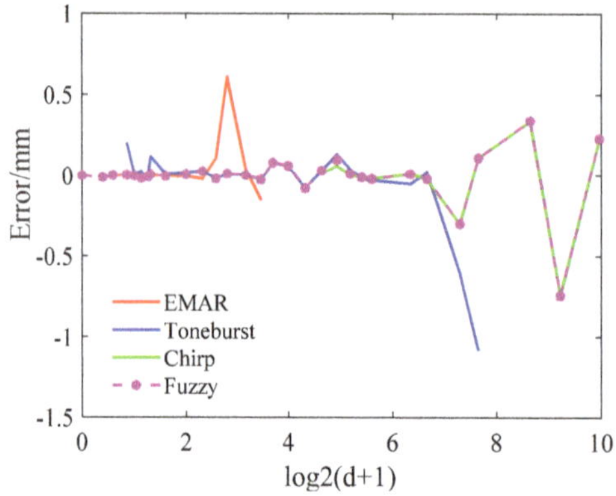

Figure 11. Error analysis of different methods.

In summary, the integration of different methods' measurements through fuzzy logic reduces errors, expands the measurement range, and enhances robustness.

4.4. Measurement Uncertainty

The integration of thickness results via fuzzy logic extends the measurement range. This approach compensates for errors inherent in a single method, thereby reducing measurement inaccuracies. Uncertainties in thickness measurements arise from various factors, including electronic systems and temperature fluctuations. Errors due to the electronic system are quantified and offset by a calibration block, the details of which are not discussed here. To maintain temperature stability, the test blocks were conditioned at room temperature for 24 h prior to experimentation. Each of the 30 types of specimens underwent 10 fuzzy logic thickness measurements, followed by an analysis of measurement uncertainty. Figure 12 illustrates the thickness measurement error, where the X-axis represents the sequence number of different thicknesses (i.e., 1, 2, 3. . ., 30), and the Y-axis denotes the error (in mm) between the fuzzy logic measurement result and the actual thickness (*d*).

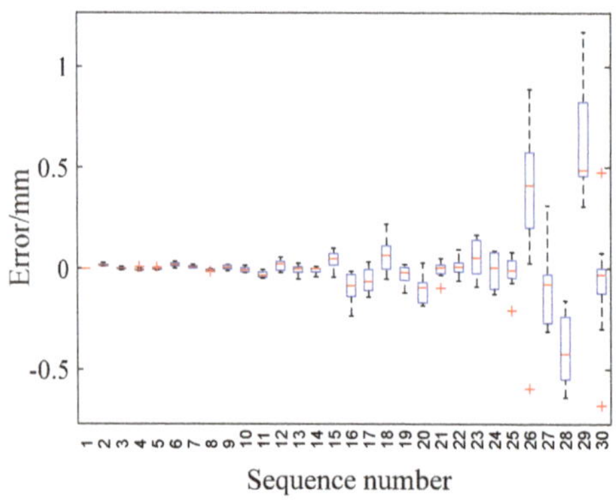

Figure 12. Error analysis of fuzzy logic thickness measurement results.

The figure above demonstrates that the median error for varying thicknesses remains within ±0.5 mm, with all errors falling within ±1.2 mm. This error appears to increase proportionally with thickness, a trend primarily attributed to the differences in thickness measurement methods and the thickness of the specimen itself. For smaller thicknesses, the errors are not only minimal but also tightly clustered, a trend that underscores the stability and high accuracy of the EMAR method. Conversely, larger thicknesses are associated with larger errors and a greater number of outliers, some of which exceed 1 mm. Despite the magnitude of the error increasing with thickness, the error percentage remains relatively small. This discrepancy could potentially be explained by two main factors: (1) the accuracy and range limitations of the vernier caliper and (2) the reduction in the number of echoes within the same sampling time, which weakens the average effect of multiple echo intervals.

5. Conclusions and Future Work

For hardware with a limited frequency bandwidth, different methods exhibit varying thickness measurement ranges and accuracies. By integrating fuzzy logic with the resonance, short pulse echo, and LFM methods, this paper has designed a thickness measurement approach. This approach effectively extends the measurement range, enhances accuracy, and strengthens the robustness of measurements. The proposed approach of a primary conventional evaluation and a secondary fuzzy logic thickness measurement scheme can effectively improve the measurement efficiency. The experimental results demonstrate that by using fuzzy logic, this study achieves a large measurement range of 0.32–1000 mm with an accuracy of 1%, providing a direction for enhancing the performance of existing thickness measurement devices.

Future work will focus on four key areas. (1) Frequency selection: different thickness measurement methods correspond to different receiver frequency bands. By incorporating a switchable frequency selection circuit, the signal-to-noise ratio of the echo signal can be enhanced when using a single method. (2) Data accumulation and fuzzy logic: as thickness measurement data accumulates, prior knowledge can be utilized to refine and enhance the fuzzy logic. (3) Equipment miniaturization: based on existing work, the power amplifier and preamplifier should be miniaturized to achieve overall customization of the equipment. (4) Utilizing accumulated thickness data, this research can explore the performance of lightweight AI algorithms and their potential deployment on small, embedded platforms in the future.

Author Contributions: Conceptualization, Y.S. and X.L.; methodology, Y.S.; software, Y.S.; validation, Y.S., S.T. and J.J.; formal analysis, T.L.; investigation, S.W.; resources, K.X.; data curation, Y.S.; writing—original draft preparation, Y.S.; writing—review and editing, Y.S.; visualization, S.W.; supervision, K.X.; project administration, K.X.; funding acquisition, K.X. All authors have read and agreed to the published version of the manuscript.

Funding: This work was partially funded by the National Natural Science Foundation of China (No. 52174352).

Institutional Review Board Statement: Not applicable.

Informed Consent Statement: Not applicable.

Data Availability Statement: Data are contained within the article.

Conflicts of Interest: The authors declare no conflict of interest.

References

1. Ren, W.; Xu, K.; Dixon, S.; Zhang, C. A Study of Magnetostriction Mechanism of EMAT on Low-Carbon Steel at High Temperature. *NDT E Int.* **2019**, *101*, 34–43. [CrossRef]
2. Jia, X.; Ouyang, Q. Optimal Design of Point-Focusing Shear Vertical Wave Electromagnetic Ultrasonic Transducers Based on Orthogonal Test Method. *IEEE Sens. J.* **2018**, *18*, 8064–8073. [CrossRef]
3. He, M.; Shi, W.; Lu, C.; Chen, G.; Qiu, F.; Zhu, Y.; Liu, Y. Application of Pulse Compression Technique in High-Temperature Carbon Steel Forgings Crack Detection with Angled SV-Wave EMATs. *Sensors* **2023**, *23*, 2685. [CrossRef]
4. Liu, T.; Pei, C.; Cai, R.; Li, Y.; Chen, Z. A Flexible and Noncontact Guided-Wave Transducer Based on Coils-Only EMAT for Pipe Inspection. *Sens. Actuators A Phys.* **2020**, *314*, 112213. [CrossRef]
5. Rieger, K.; Erni, D.; Rueter, D. Examination of the Liquid Volume Inside Metal Tanks Using Noncontact EMATs from Outside. *IEEE Trans. Ultrason. Ferroelectr. Freq. Control* **2021**, *68*, 1314–1327. [CrossRef]

6. Rieger, K.; Erni, D.; Rueter, D. Noncontact Reception of Ultrasound from Soft Magnetic Mild Steel with Zero Applied Bias Field EMATs. *NDT E Int.* **2022**, *125*, 102569. [CrossRef]
7. Miao, H.; Li, F. Shear Horizontal Wave Transducers for Structural Health Monitoring and Nondestructive Testing: A Review. *Ultrasonics* **2021**, *114*, 106355. [CrossRef] [PubMed]
8. Wang, S.; Li, Z.; Kang, L.; Hu, X.; Zhang, X. Modeling and Comparison of Three Bulk Wave EMATs. In Proceedings of the IECON 2011—37th Annual Conference of the IEEE Industrial Electronics Society, Melbourne, VIC, Australia, 7–10 November 2011; pp. 2645–2650. [CrossRef]
9. Diguet, G.; Miyauchi, H.; Takeda, S.; Uchimoto, T.; Mary, N.; Takagi, T.; Abe, H. EMAR Monitoring System Applied to the Thickness Reduction of Carbon Steel in a Corrosive Environment. *Mater. Corros.* **2022**, *73*, 658–668. [CrossRef]
10. Trushkevych, O.; Edwards, R.S. Differential Coil EMAT for Simultaneous Detection of In-Plane and Out-of-Plane Components of Surface Acoustic Waves. *IEEE Sens. J.* **2020**, *20*, 11156–11162. [CrossRef]
11. Cai, Z.; Li, Y. Optimization Design of Electromagnetic Ultrasonic Longitudinal Wave Transducer Based on Halbach Array. *J. Electr. Eng. Technol.* **2021**, *36*, 4408–4417.
12. Pei, C.; Zhao, S.; Xiao, P.; Chen, Z. A Modified Meander-Line-Coil EMAT Design for Signal Amplitude Enhancement. *Sens. Actuators A Phys.* **2016**, *247*, 539–546. [CrossRef]
13. Kang, L.; Dixon, S.; Wang, K.; Dai, J. Enhancement of Signal Amplitude of Surface Wave EMATs Based on 3-D Simulation Analysis and Orthogonal Test Method. *NDT E Int.* **2013**, *59*, 11–17. [CrossRef]
14. Xu, X.; Tu, J.; Zhang, X. Study on the Characteristics of Common Electromagnetic Ultrasonic Coils for Steel Plate Thickness Measurement. *China Test.* **2020**, *46*, 5.
15. Burrows, S.E.; Fan, Y.; Dixon, S. High Temperature Thickness Measurements of Stainless Steel and Low Carbon Steel Using Electromagnetic Acoustic Transducers. *NDT E Int.* **2014**, *68*, 73–77. [CrossRef]
16. Wang, S.; Wang, S.; Liang, B.; Jiang, C.; Zhao, C.; Zhou, W. Research on Linear Frequency Modulation Pulse Compression in Electromagnetic Ultrasonic Thickness Measurement. In Proceedings of the 2020 IEEE Far East NDT New Technology & Application Forum (FENDT), Kunming, China, 20–22 November 2020; pp. 145–149. [CrossRef]
17. Zhao, Y.; Ma, J.; Chen, J.W.; Ouyang, H.; Wang, M.H.; Zhao, P.; Zhang, B. Noncontact Method Applied to Determine Thickness of Thin Layer Based on Laser Ultrasonic Technique. *Mater. Res. Innov.* **2014**, *18*, S2-1081–S2-1085. [CrossRef]
18. Qiu, J.; Ding, H.; Wang, S. Design of Pulse Electromagnet in Electromagnetic Ultrasonic Thickness Measurement Device for Steel Plate. *China Test* **2018**, *44*, 6.
19. Guo, W.; Yu, Z.; Chui, H.-C.; Chen, X. Development of DMPS-EMAT for Long-Distance Monitoring of Broken Rail. *Sensors* **2023**, *23*, 5583. [CrossRef] [PubMed]
20. Suresh, N.; Balasubramaniam, K. Quantifying the Lowest Remnant Thickness Using a Novel Broadband Wavelength and Frequency EMAT Utilizing the Cut-Off Property of Guided Waves. *NDT E Int.* **2020**, *116*, 102313. [CrossRef]
21. Zhao, S.; Zhou, J.; Liu, Y.; Zhang, J.; Cui, J. Application of Adaptive Filtering Based on Variational Mode Decomposition for High-Temperature Electromagnetic Acoustic Transducer Denoising. *Sensors* **2022**, *22*, 7042. [CrossRef]
22. Richards, M.A. *Fundamentals of Radar Signal Processing*, 1st ed.; McGraw-Hill: New York, NY, USA, 2005.
23. Yusa, N.; Song, H.; Iwata, D. Probabilistic Evaluation of EMAR Signals to Evaluate Pipe Wall Thickness and Its Application to Pipe Wall Thinning Management. *NDT E Int.* **2021**, *122*, 102475. [CrossRef]
24. Wang, J.; Wu, X.; Song, Y.; Sun, L. Study of the Influence of the Backplate Position on EMAT Thickness-Measurement Signals. *Sensors* **2022**, *22*, 8741. [CrossRef] [PubMed]
25. Liu, T.; Pei, C.; Cheng, X.; Zhou, H.; Xiao, P.; Chen, Z. Adhesive Debonding Inspection with a Small EMAT in Resonant Mode. *NDT E Int.* **2018**, *98*, 110–116. [CrossRef]
26. Hirao, M.; Ogi, H. *Electromagnetic Acoustic Transducers: Noncontacting Ultrasonic Measurements Using EMATs*; Springer: Tokyo, Japan, 2017. [CrossRef]
27. Ross, T.J. *Fuzzy Logic with Engineering Applications*; John Wiley & Sons: Hoboken, NJ, USA, 2005.
28. Mendel, J.M. Fuzzy Logic Systems for Engineering: A Tutorial. *Proc. IEEE* **1995**, *83*, 345–377. [CrossRef]
29. Kong, L.; Peng, X.; Chen, Y.; Wang, P.; Xu, M. Multi-sensor measurement and data fusion technology for manufacturing process monitoring: A literature review. *Int. J. Extrem. Manuf.* **2020**, *2*, 022001. [CrossRef]
30. Lin, X.; Yu, Y.; Sun, C. A Decoupling Control for Quadrotor UAV Using Dynamic Surface Control and Sliding Mode Disturbance Observer. *Nonlinear Dyn.* **2019**, *97*, 781–795. [CrossRef]

sensors MDPI

Article

Eddy Current Sensor Array for Electromagnetic Sensing and Crack Reconstruction with High Lift-Off in Railway Tracks

Yuchun Shao [1], Zihan Xia [1,*], Yiqing Ding [1], Bob Crocker [2], Scott Saunders [2], Xue Bai [1], Anthony Peyton [1], Daniel Conniffe [1] and Wuliang Yin [1]

[1] School of Electrical and Electronic Engineering, University of Manchester, Manchester M13 9PL, UK; yuchun.shao@manchester.ac.uk (Y.S.); yiqing.ding@postgrad.manchester.ac.uk (Y.D.); xue.bai-10@postgrad.manchester.ac.uk (X.B.); a.peyton@manchester.ac.uk (A.P.); daniel.conniffe@manchester.ac.uk (D.C.); wuliang.yin@manchester.ac.uk (W.Y.)

[2] Sperry Rail International Ltd., Derby DE24 8UP, UK; bobcrocker@btinternet.com (B.C.); ssaunders@sperryrail.com (S.S.)

* Correspondence: zihan.xia-5@postgrad.manchester.ac.uk

Abstract: A reliable and efficient rail track defect detection system is essential for maintaining rail track integrity and avoiding safety hazards and financial losses. Eddy current (EC) testing is a non-destructive technique that can be employed for this purpose. The trade-off between spatial resolution and lift-off should be carefully considered in practical applications to distinguish closely spaced cracks such as those caused by rolling contact fatigue (RCF). A multi-channel eddy current sensor array has been developed to detect defects on rails. Based on the sensor scanning data, defect reconstruction along the rails is achieved using an inverse algorithm that includes both direct and iterative approaches. In experimental evaluations, the EC system with the developed sensor is used to measure defects on a standard test piece of rail with a probe lift-off of 4–6 mm. The reconstruction results clearly reveal cracks at various depths and spacings on the test piece.

Keywords: eddy current testing; electromagnetic induction; planar structure; theoretical calculation; measurement

1. Introduction

Rail safety inspection has gained significant attention across multiple industries due to past incidents, driving the development and exploration of advanced detection methods and maintenance practices. Ratcheting, occurring within a spacing of 0.8 to 20 mm [1–4], may lead to the formation of closely spaced cracks. This process is often initiated by the continuous passage of train wheels over rail tracks, potentially resulting in critical safety hazards.

Eddy current (EC) testing is a widely used non-destructive testing (NDT) technique, distinguished by its capability of identifying surface and subsurface defects within rails without direct contact. This technique can be adapted for delivering instantaneous, real-time feedback and necessitates negligible surface preparation, streamlining the inspection process [1,5]. However, EC testing has limitations that affect the performance of rail inspection, including a reduction in signal and spatial resolution at high liftoff [6,7]. In response to the inherent limitations, many innovative solutions have been studied.

Li et al. developed an EC system with an increased coil gap to compensate for signal attenuation, allowing for accurate measurement of crack depth and inclination at the working frequency of 17 kHz [8]. However, a large coil gap can lead to a reduction in spatial resolution. Xu et al. proposed a differential EC system that can be applied to evaluate surface opening defects up to 8 mm deep with a maximum 15% relative error at a frequency of 60 kHz [9]. However, the spatial resolution is not quantified as the samples have a 100 mm crack spacing.

Kwon et al. proposed a 16-channel eddy current flaw detection system that works at 300 kHz with a lift-off of 1 mm and a high reliability of 1 mm margin of error [10]. However, maintaining a very low lift-off distance during rail inspection can increase the risk of probe damage or rail surface wear. Blitz et al. used differential-coil probes for detecting surface cracks in steel with a high lift-off of up to 12 mm, while there are limitations in accurately resolving closely spaced defects, as evidenced by 100 mm crack spacings [11].

To enhance the spatial resolution of the signal, Sukhanov et al. present a visualization technique for small heterogeneities in flat metallic objects using an eddy current sensor. It employs differential and flat spiral coils. The technique improves image resolution via post-processing. Spatial linear filtration and regularization are key to enhancing signal-to-noise ratios. Experimentally, for metallic objects at a 15 mm range, the method led to an increase in reconstructed signal resolution by up to 2.6 times relative to the raw measured data [12]. The sensor employed is limited to a single channel, necessitating the use of a scanner to comprehensively survey the desired area, which constrains the overall efficiency of the investigation.

Minimizing lift-off effects helps preserve spatial resolution. Meng et al. presented a simplified algorithm. It uses a multifrequency sensor to reduce lift-off impact in eddy current testing. This method tests nonmagnetic coatings on ferromagnetic substrates. It achieves up to 3% accuracy in thickness measurements for lift-offs up to 10 mm [13]. A significant limitation of using multi-frequency sensors for rail defect detection is their challenge in differentiating signals from various defect types due to the distinct interactions of each frequency with the substrate. This issue requires complex signal processing for accurate defect identification, increasing computational demands [14,15]. Additionally, the sensor's performance at different frequencies can be affected by external factors, including temperature and humidity, complicating the defect detection process.

In this study, a multi-channel EC sensor array is produced to examine the cracks on rails. Previous studies explored the development of eddy-current sensors. The adaptive probe was proposed to change measurement resolution through changes in the transducer's dimension by using multi-section excitation coils and the variable geometry of a core [16]. A matrix-type EC probe with a single transmitting coil, multiple receiving coils, and a ferrite core was implemented in the study [17]. These EC probes were designed to detect defects in certain dimensions. To distinguish the cracks on a railway with 5 mm of space, the dimensions of the coils and the space between them need to be much smaller, which was not achieved in previous studies. The EC system operates at 200 kHz with a 4–6 mm probe lift-off. The system can obtain data from any combination of the coil pairs. Therefore, the received data for coil combinations with multiple coil spacings and orientations can be evaluated. From measurements, the defects are reconstructed by solving an inverse problem with direct and iterative image reconstruction algorithms. Its capability to accurately identify and map cracks of various depths and intervals is demonstrated. This design effectively balances spatial resolution and lift-off distance, potentially making it suited for future railway inspection applications, and it can resolve cracks spaced as close as 5 mm.

2. Method

2.1. Measurement Setup for Rail Testing

Images of rail crack testing are shown in Figure 1. The customized multi-channel EC instrument, previously reported [18], is employed to acquire the EC probe signals from the coil array. The probe is moved above the testing sample across the cracks illustrated in Figure 2. This sensor can provide 36 independent measurements from all various coil combinations, featuring multiple coil spacings and sensitive orientations. By scanning once, abundant information about the defects can be collected for further analysis.

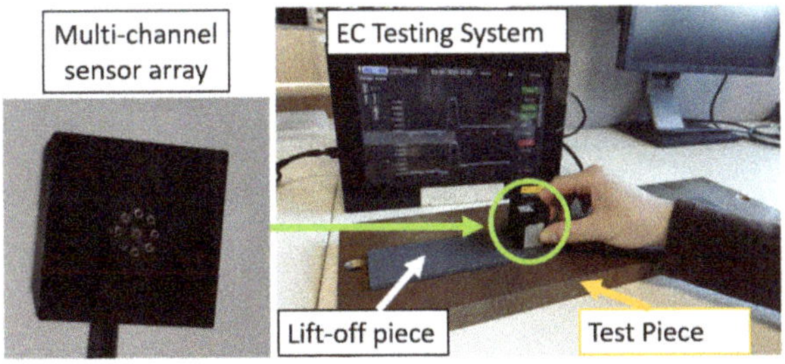

Figure 1. Schematic of rail crack scanning.

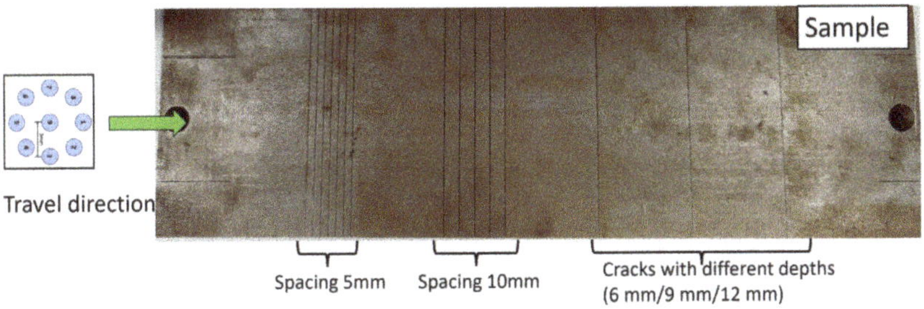

Figure 2. Illustration of line scanning along the sample.

The cracks on the rail could have various orientations, depths, and spaces between them. As illustrated in Figure 2, the observed cracks on the steel sample (grade R220). can be categorized into three distinct groups. The first group exhibits a spacing of 5 mm and a depth of 9 mm. In the second group, there are five cracks with a spacing of 10 mm and a depth of 9 mm. Furthermore, the third group comprises three cracks, each characterized by varying depths (6 mm, 9 mm, and 12 mm). To test the cracks with depth and spacing in the range of $d \in [6, 9]$ mm and $s \in [5, 20]$ mm, respectively, the ferrite-core coil array shown in Figure 3 was designed, which is symmetric and sensitive to cracks in various orientations. For testing a typical rail steel with an electrical conductivity of 4 MS/m and magnetic permeability of 40, an excitation frequency of 200 kHz, relating to the EC skin depth of 0.09 mm, is selected for rail surface crack detection. In principle, maintaining this frequency ensures fast data sampling rates even at high vehicle speeds, and at the same time, the frequency should not go further higher to avoid the reduction in sensor penetration depth caused by the skin effect phenomenon. For testing cracks on rail, the customized EC testing system can provide signals between all independent coil pairs at a speed of 127 frames per second. To ensure space between adjacent frames smaller than 5 mm, the speed should be lower than 0.635 m/s, i.e., 2.28 km/h. Of course, this is just an experimental setup. In a normal operation, 10 pairs of coil measurements are normally adopted, and not all combinations are necessary, which would give a speed of 36 km/h if the distance between adjacent frames of data is 5 mm. Similarly, it can be 72 km/h if the distance between adjacent frames of data is 10 mm.

Each coil consists of 200 turns of copper enameled wire with a diameter of 0.05 mm, as shown in Figure 1. In a rectangular coil cross-section, there are five layers, with 40 turns in each layer. The coil dimensions include an outer radius of 0.7 mm, an inner radius of 0.4 mm, and a height of 2.4 mm. A ferrite core is inserted in each coil to increase flux density. The implemented coil follows the previous study [18]. The amplitude of the excitation current is around 5 mA.

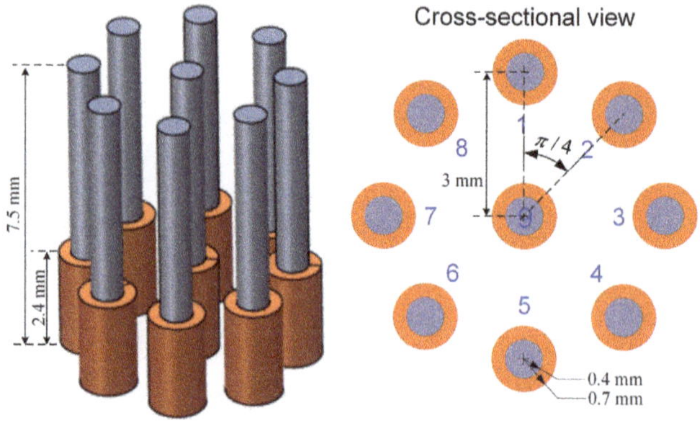

Figure 3. Basic structure of the EC probe.

2.2. *Forward Problem of Rail Testing*

In the EC testing scenario, the governing diffusion equation of magnetic vector potential is as follows:

$$\nabla^2\mathbf{A} - j\omega\mu\sigma\mathbf{A} = \mu_0\mathbf{J}_s \tag{1}$$

where ω is the angular frequency, $\mathbf{J}_s$ is the current density of the source field, μ and σ are magnetic permeability and electrical conductivity of the measurement space, respectively.

The electromagnetic field can be obtained by solving the diffusion equation, e.g., through the finite element method (FEM) and boundary element method (BEM), with the known source field excited by a coil array. For an array consisting of 9 coils, there are 36 independent measurements between coil pairs. The coil pairs have different spacing and orientations. The sensitivity of conductivity variation $\Delta\sigma$ on a plate due to defects, for certain coil pairs with index i, j, has been previously defined [19]:

$$\begin{aligned} \lambda^{ij} &= \int_\Omega S_\sigma^{ij}\Delta\sigma dV \\ S_\sigma^{ij} &= \mathbf{E}^i \cdot \mathbf{E}^j \end{aligned} \tag{2}$$

where λ^{ij} indicates the voltage variation on the j-th coil due to the excitation on the i-th coil, Ω is the region of the plate, and $\mathbf{E}^i$ and $\mathbf{E}^j$ are source fields excited by the i-th and j-th coils, respectively.

The integral in Equation (2) can be approximated by discretizing the plate domain into a variety of homogeneous voxels:

$$\lambda^{ij} \approx \mathbf{S}_\sigma^{ij}\Delta\boldsymbol{\sigma}$$

where $\mathbf{S}_\sigma^{ij} \in C^{1\times n}$, $\Delta\boldsymbol{\sigma} \in R^{n\times 1}$ is the conductivity distribution, n indicates the number of voxels.

Considering all independent measurements between k coils, with the number of measurements being $m = k(k-1)/2$, the matrix form holds for the following:

$$\boldsymbol{\lambda} \approx \mathbf{S}_\sigma\Delta\boldsymbol{\sigma} \tag{3}$$

where $\boldsymbol{\lambda} = \left(\lambda^{ij}\right) \in C^{m\times 1}$ is the measurement vector, $\mathbf{S}_\sigma = \left(\mathbf{S}_\sigma^{ij}\right) \in C^{m\times n}$ is the sensitivity matrix. For simplicity, the sensitivity $\mathbf{S}_\sigma$ and conductivity variation $\Delta\boldsymbol{\sigma}$ are represented by $\mathbf{S}$ and $\boldsymbol{\sigma}$ hereinafter.

On the surface of the tested plane shown in Figure 4, the sensitivity of conductivity $\mathbf{S}_\sigma^{ij}$ between representative coil pairs is shown in Figure 5. The electric field intensity in Equation (1) is calculated by the COMSOL 6.0 software, in which the electrical conductivity and relative magnetic permeability of the tested plate are 2 MS/m and 100, and the relative magnetic permeability of ferrite cores is 2000. There exists a high-value region of sensitivity between the transmitting (Tx) and receiving (Rx) coils. The sensitivity maps show that the measurements of different coil pairs contain complementary areas on the plane. By combining these measurements, the details of the cracks could be reconstructed.

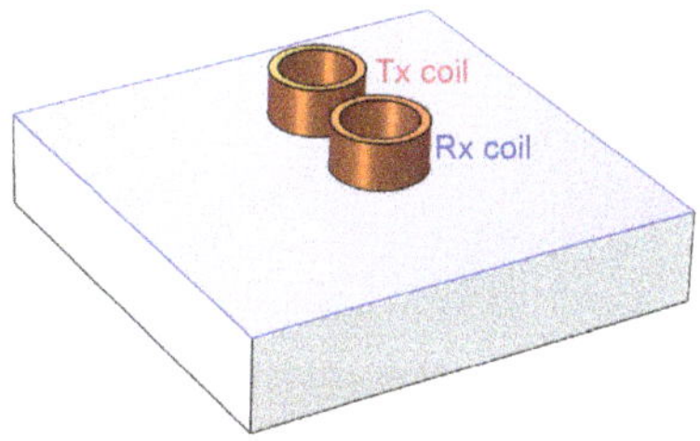

Figure 4. Surface of a conductive plane to evaluate the sensitivity between Tx and Rx coils.

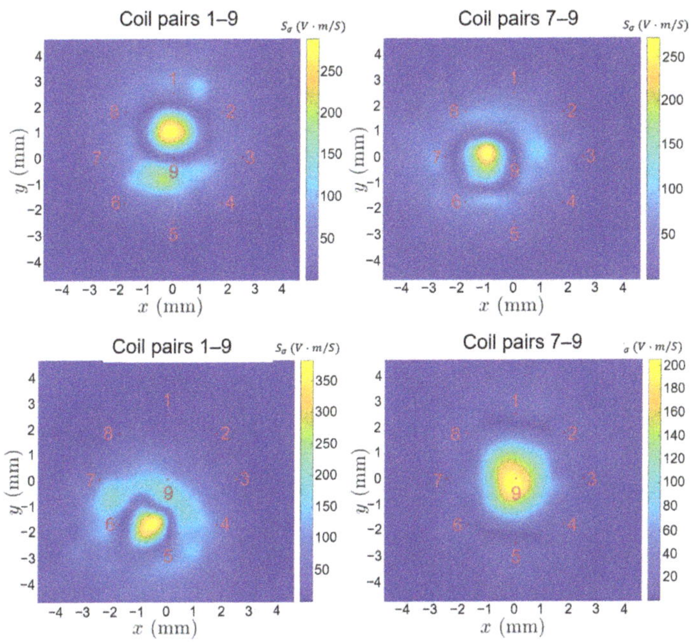

Figure 5. Sensitivity of conductivity for $\mathbf{S}_\sigma^{19}$, $\mathbf{S}_\sigma^{79}$, $\mathbf{S}_\sigma^{56}$, and $\mathbf{S}_\sigma^{15}$ (colour map indicates the value of the sensitivity with unit volt·m/S, number is the index of coils). Red points indicate the positions of ferrite-core coils.

In general, at lower lift-offs, the closely placed pairs work well, but as we increase lift-off, the far-placed pairs work well due to their larger sensitivity range.

This study implemented all independent measurements between various coil pairs (36 measurements from the 9-coil probe) to reconstruct the cracks. Each coil pair has a distinct sensitive region, as shown in Figure 5, which contains the unique prior information for image reconstruction.

Without prior knowledge of the lift-off and crack, which is the case in real applications, it is best to select all combination coil pairs for crack reconstruction, as each of them may contribute to reconstruction.

2.3. Inverse Problem of Crack Reconstruction

Direct and iterative approaches based on sensitivity can be employed to reconstruct the conductivity variation, and reflect the distribution of defects on rail. Several representative direct methods, i.e., linear back projection (LBP), truncated singular value decomposition (TSVD), Tikhonov regularization, iterative algorithms, simultaneous iterative reconstruction technique (SIRT), Newton–Raphson (NR), and conjugate gradient (CG), are investigated for reconstruction of the relative change of conductivity.

The LBP algorithm reconstructs the conductivity distribution through a linear mapping from the measurement vector space to the conductivity distribution space:

$$\mathbf{g} = \mathbf{S}^{\mathrm{T}}\boldsymbol{\lambda}$$

where $\mathbf{g}$ indicates the reconstructed conductivity distribution.

The singular value decomposition provides a means to obtain the solution of conductivity distribution directly. If the rank of sensitivity matrix $\mathbf{S}$ is p, $p = \min\{m, n\}$, it can be decomposed into $\mathbf{S} = \mathbf{U}\Lambda\mathbf{V}^{\mathrm{T}}$, $\mathbf{U} = [\mathbf{u}_1, \mathbf{u}_2, \ldots, \mathbf{u}_m] \in C^{m\times m}$, $\mathbf{V} = [\mathbf{v}_1, \mathbf{v}_2, \ldots, \mathbf{v}_n] \in C^{n\times n}$ and $\Lambda = \mathrm{diag}(\delta_1, \delta_2, \ldots, \delta_p) \in R^{m\times n}$ with all components zero except for those diagonal components, $\mathrm{diag}(\cdot)$ indicates the vector-to-matrix diag operator.

The TSVD algorithm employs

$$\mathbf{g} = \mathbf{S}^{+}\boldsymbol{\lambda}, \mathbf{S}^{+} = \mathbf{V}\overset{\sim}{\Lambda}^{-1}\mathbf{U}^{\mathrm{T}}$$

where $\overset{\sim}{\Lambda}^{-1} = \mathrm{diag}(w_1/\delta_1, w_2/\delta_2, \ldots, w_p/\delta_p)$, w_1, w_2 are filtering parameters to reduce the influence of small singular values.

The Tikhonov method is one of the most commonly used universal regularization tools for solving ill-posed inverse problems. Based on the standard Tikhonov regularization procedure, the solution of equation $\boldsymbol{\lambda} + \mathbf{e} = \mathbf{S}\mathbf{g}$, $\mathbf{e}$ indicates the measurement error vector, is expressed as follows:

$$\mathbf{g} = \left(\mathbf{S}^{\mathrm{T}}\mathbf{S} + \beta\mathbf{I}\right)^{-1}\mathbf{S}^{\mathrm{T}}\boldsymbol{\lambda}$$

where $\mathbf{I}$ is the identity matrix and β is the regularization parameter. More details on direct methods can be found in [20,21].

The iterative approaches optimize the reconstructed conductivity in an iterative manner:

$$\mathbf{g}_{k+1} = \mathbf{g}_k + \Delta\mathbf{g}_k \tag{4}$$

where k is the iterative step.

The NR algorithm can find the root of the non-linear function for the least squares problem $\|\mathbf{S}\mathbf{g} - \boldsymbol{\lambda}\|_2^2$, by approaching the local minimum iteratively. An appropriate initial guess is required to reach the global minimum by the NR algorithm. SIRT is commonly used for image reconstruction in X-ray computerized tomography and was introduced to electromagnetic tomography. The algorithm usually requires a large number of iterations, and its convergence rate is relatively low. The updating directions of the NR and SIRT algorithms are shown in Table 1.

Table 1. Updating directions of the NR and SIRT algorithms for conductivity reconstruction.

Algorithm	Updating direction, $\Delta\mathbf{g}_k = \mathbf{d}_k(\mathbf{S}\mathbf{g}_k - \boldsymbol{\lambda})$, $\mathbf{d}_k \in C^{n\times m}$
NR	$\mathbf{d}_k = -\left(\mathbf{S}^{\mathrm{T}}\mathbf{S}\right)^{-1}\mathbf{S}^{\mathrm{T}}$
SIRT	$d_k = -\gamma\mathbf{S}^{\mathrm{T}}\mathrm{diag}(\mathrm{diag}*\left(\mathbf{S}\mathbf{S}^{\mathrm{T}}\right)^{-1})(\mathbf{S}\mathbf{g}_k - \boldsymbol{\lambda}), \gamma = 10^{-3}/\max\{\mathrm{diag}*\left(\mathbf{S}\mathbf{S}^{\mathrm{T}}\right)\}$, $\mathrm{diag}*(\cdot)$ indicates the matrix-to-vector diag operator

The CG algorithm updates the variable following the orthogonality of residuals and the conjugacy of the search directions. The algorithm is characterized by its fast convergence rate while suffering from measurement noise. Setting the initial conditions $\mathbf{g}_0 = \mathbf{S}^{\mathrm{T}}\boldsymbol{\lambda}$, $\mathbf{r}_0 = \mathbf{S}^{\mathrm{T}}\boldsymbol{\lambda} - \mathbf{A}\mathbf{g}_0$ and $\mathbf{p}_0 = \mathbf{r}_0$, the implementation of the CG algorithm for conductivity reconstruction is expressed by the following:

$$\Delta\mathbf{g}_k = \alpha_k\mathbf{p}_k$$

$$\alpha_k = \frac{\mathbf{r}_k{}^{\mathrm{T}}\mathbf{r}_k}{\mathbf{p}_k{}^{\mathrm{T}}\mathbf{A}\mathbf{p}_k}$$

$$\mathbf{r}_{k+1} = \mathbf{r}_k - \alpha_k\mathbf{A}\mathbf{p}_k$$

$$\beta_k = \frac{\mathbf{r}_{k+1}{}^{\mathrm{T}}\mathbf{r}_{k+1}}{\mathbf{r}_k{}^{\mathrm{T}}\mathbf{r}_k}$$

$$\mathbf{p}_{k+1} = \mathbf{r}_{k+1} + \beta_k\mathbf{p}_k$$

The optimization repeats until the iterative step reaches its maximum value or the residues fall below the threshold. Comprehensive analysis of the CG, NR, and SIRT reconstruction algorithms can be found in the previous studies [22,23], which is briefly described here to avoid redundancy.

To obtain the 1-D scanning results of defect distribution, the arithmetic mean of 2-D reconstruction results is aligned into a vector with elements corresponding to each testing point of the probe, $\mathbf{v} = v_i \in R^{l\times 1}$, l represents the number of testing points.

The flowchart of data processing is shown in Figure 6 below.

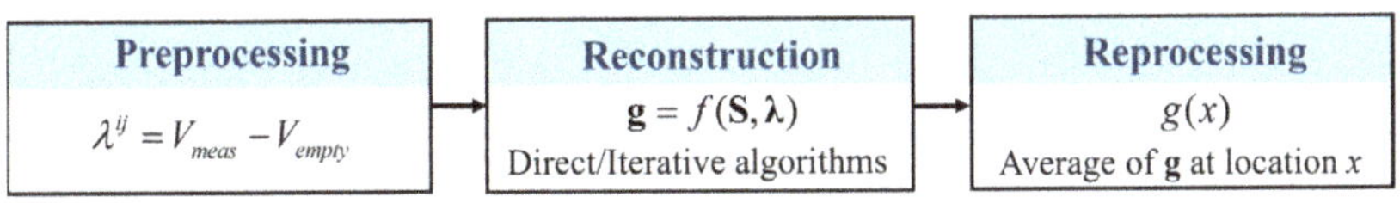

Figure 6. The flowchart of data processing.

The employed field-quantity method for the forward problem and direct and iterative algorithms for the inverse problem are presented in a concise manner.

The methods we proposed are effective. Furthermore, the realization of crack reconstruction in combination with the miniature probe and customized EC system is one of the main highlights of the paper. We make it possible to obtain the signals that clearly distinguish the closely spaced cracks and variations in depth for the first time.

3. Results and Discussion

All potential transmitter and receiver coil pairings (shown in Appendix A) were thoroughly evaluated, with data collected throughout the experiment. The influences of probe lift-off, spatial resolution of defects, and the effectiveness of the implemented inverse algorithm for crack reconstruction are discussed in detail.

3.1. Received Data

Theoretically, if adjacent cracks on a rail are closely spaced, the response signals from these cracks can merge together, resulting in a single signal that displays as a continuous high value range due to low spatial resolution. Figure 7 presents the normalized voltage signals obtained by scanning across seven cracks spaced 5 mm apart, with a 5 mm lift-off, at a frequency of 200 kHz. The seven peaks of the response signal are represented as the small peaks on the bump. This figure presents data from 15 of the possible 36 transmitter (T) and receiver (R) coil combinations contained in the multichannel EC array. Each plot within the figure shows the response signal for seven cracks, appearing as distinct peaks atop the main bump. The combinations from T9–R1 to T9–R8 have a consistent coil spacing of 3 mm. The pairings T1–R2 and T1–R8 feature the narrowest coil spacing at 2.296 mm, while T1–R5 has the widest spacing at 6 mm, as referenced in Figure 3. At a 5 mm lift-off distance, all coil spacings can detect a group of cracks. However, identifying the exact number of cracks requires further analysis through algorithms due to the fewer distinct peaks on the bump.

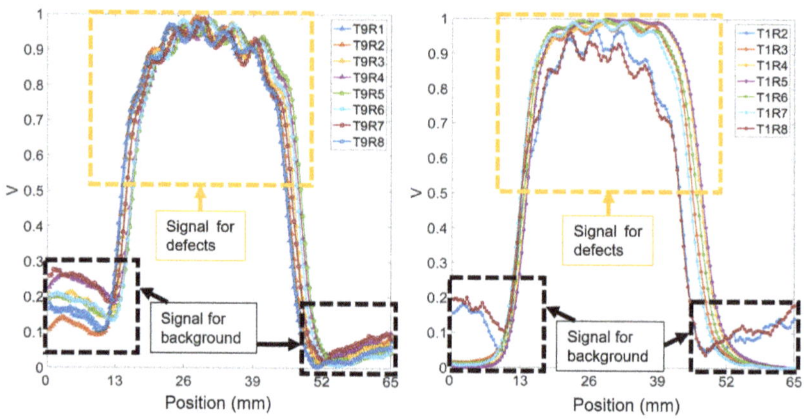

Figure 7. Normalized received signals of 15 T–R pairings for seven cracks at 5 mm spacing, at 5 mm lift-off, and 200 kHz.

Reducing the probe lift-off improves its spatial resolution. This study examines the balance between spatial resolution and lift-off effects by scanning a sample featuring cracks spaced 5 mm apart, with lift-offs varying between 4 mm and 6 mm. The raw data for all transmitter–receiver (T–R) pairings at 4 mm, 5 mm, and 6 mm lift-offs are illustrated in Figures A1–A3 in the Appendix A. At a 4 mm lift-off, the peaks corresponding to individual cracks are more obvious and can be observed visually. Signal amplitude is greater than at lift-offs of 5 mm and 6 mm. At a 6 mm lift-off, the signal for T–R pairings with smaller spacings becomes noisy, obscuring the peaks on the bump and reducing the amount of inspectable information.

The sensitive region varies for coil pairs in different spacing and locations, which mainly concentrates near the linking region between T–R pairs as shown in Figure 5. From the measurement signals in Figure A3, it is clear that the coil pairs with relatively small space are more sensitive to the closely spaced cracks, while those pairs with large spacing are less sensitive to distinguishing the closely spaced cracks but have a far wider range for deeper and individual cracks.

The data obtained from scanning five adjacent cracks, each separated by a 10 mm spacing, and three cracks with varying depths (6 mm, 9 mm, and 12 mm), with a 5 mm lift-off, are presented in Figures A4 and A5 in the Appendix A. These results are encouraging for direct crack detection, as the cracks can be distinctly identified through five distinct peaks in the raw data. Furthermore, the capability to discern cracks of different depths is demonstrated by the amplitude variation of the response signal, where a shallower crack corresponds to a lower signal amplitude. This correlation between crack depth and signal amplitude indicates the effectiveness of the detection method in distinguishing cracks of varying severities directly from the raw data.

Distinguishing the response signals corresponding to specific cracks from the peak of the bump becomes challenging when the lift-off is set to 6 mm. In addition, there is a noticeable decrease in the amplitude of the crack signal peaks compared with 4 mm and 5 mm lift-off. The measurement signals at the 6 mm lift-off can detect the region of cracks, while the individual cracks can hardly be distinguished at this lift-off.

The signal-to-noise ratio (SNR) of different coil pairs is mainly influenced by factors including coil spacing (determining the amplitude of signals), symmetry of ferrite cores, fabrication of the handmade miniature probe, and consistency among various measurement channels in the customized EC system.

Although a smaller liftoff provides superior spatial resolution and signal strength, it poses a higher risk of damage to the delicate sensor probe, potentially increasing maintenance costs. Conversely, a larger liftoff introduces noise into the signal and captures less information. Consequently, a 5 mm lift-off is selected to balance spatial resolution and signal strength, mitigating the risk of probe damage and maintaining data quality.

3.2. *Evaluation of Results Using Inverse Problems*

There are differences in the results from different inverse algorithms, although they all try to find the solution method of the least-squares problem. Image reconstruction in electrical tomography, for example, has a similar pursuit where different algorithms have been developed over the years [20,21]. The inverse problem is non-linear, underdetermined, ill-conditioned, and ill-posed, which makes it difficult to obtain "the solution". It is not a case of finding a simple solution in the least-squares sense.

The performance of various algorithms is different regarding the reconstructed crack depth, space, and lift-off. Adjusting parameters in the algorithm is another way to improve the reconstruction accuracy. The main point of this study is to reconstruct the distribution of cracks on rail samples by utilizing multiple combinations of measurements of the coil array, and the reconstruction method is implemented to improve the spatial resolution of crack reconstruction.

In this section, results using different inverse problem methods are evaluated and compared.

3.2.1. Seven Cracks with 5 mm Spacing

The processed data utilizing all the algorithms can be observed in Figure 8. All 36 pairings of received signals are employed when applying inverse method algorithms. It can be noted that the signals for seven cracks, each spaced 5 mm apart, are distinguished from a significant bump into seven peaks when the lift-off is set to 4 mm with inverse algorithms applied. At a 5 mm lift-off, the Newton–Raphson, CG, and LBP methods continue to exhibit good results. However, when the lift-off increases to 6 mm, the crack signals are obscured by noise, rendering the plots devoid of useful information. The criterion for cracking is the visibility of clear individual peaks with clear intervals above the noise signals. Therefore, the preferred algorithmic methods are Newton–Raphson, CG, and LBP, with the processed signals at a lift-off of 4 mm, 5 mm, and 6 mm.

Distinguishing the response signals corresponding to specific cracks from the peak of the bump becomes challenging when the lift-off is set to 6 mm. This difficulty is evident in Figure A3 in the Appendix A, where the received signals from coil pairs with shorter spacings are processed. The underlying issue is that the reduced distance between coils results in a diminished penetration depth, adversely affecting signal clarity and the ability to identify cracks accurately. Accordingly, pairings 10 to 14, characterized by their wider spacings, were selected for analysis across all algorithms. The outcomes of this evaluation are depicted in Figure 9, presented below. The analysis indicates a reduction in noise interference, allowing for the detection of cracks in the plots. Yet, determining the exact number of cracks remains challenging. Additionally, there is a noticeable decrease in the amplitude of the crack signal peaks at 4 mm and 5 mm lift-off, likely attributable to the diminished quantity of input signals available for data processing.

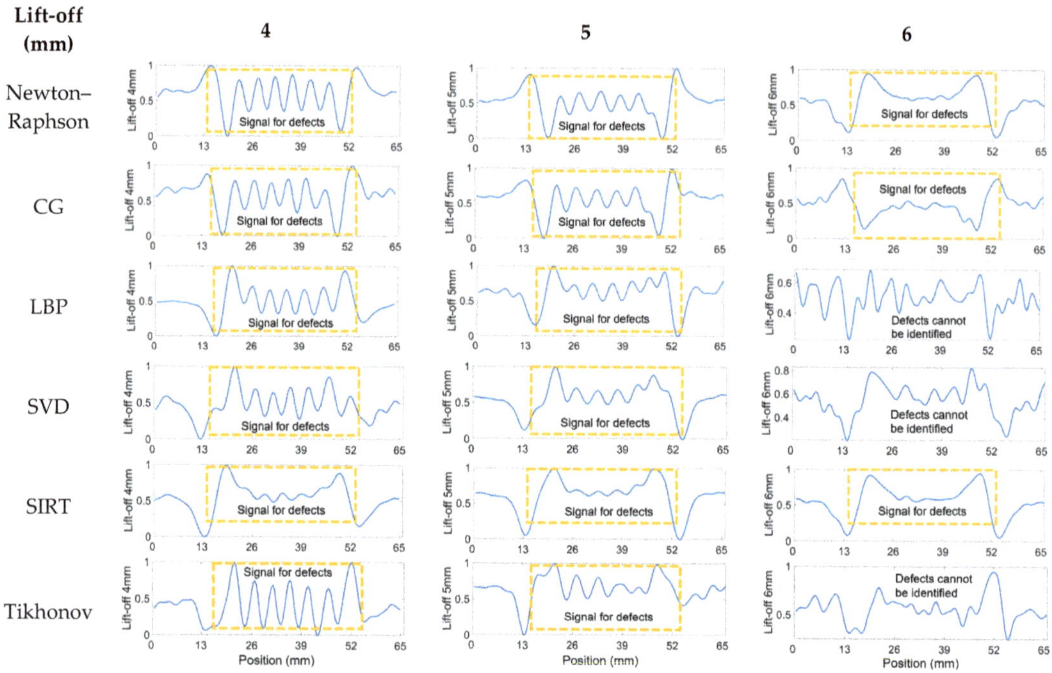

Figure 8. Processed signals of 36 T–R pairings for seven cracks at 4, 5, and 6 mm spacing and 5 mm lift-off.

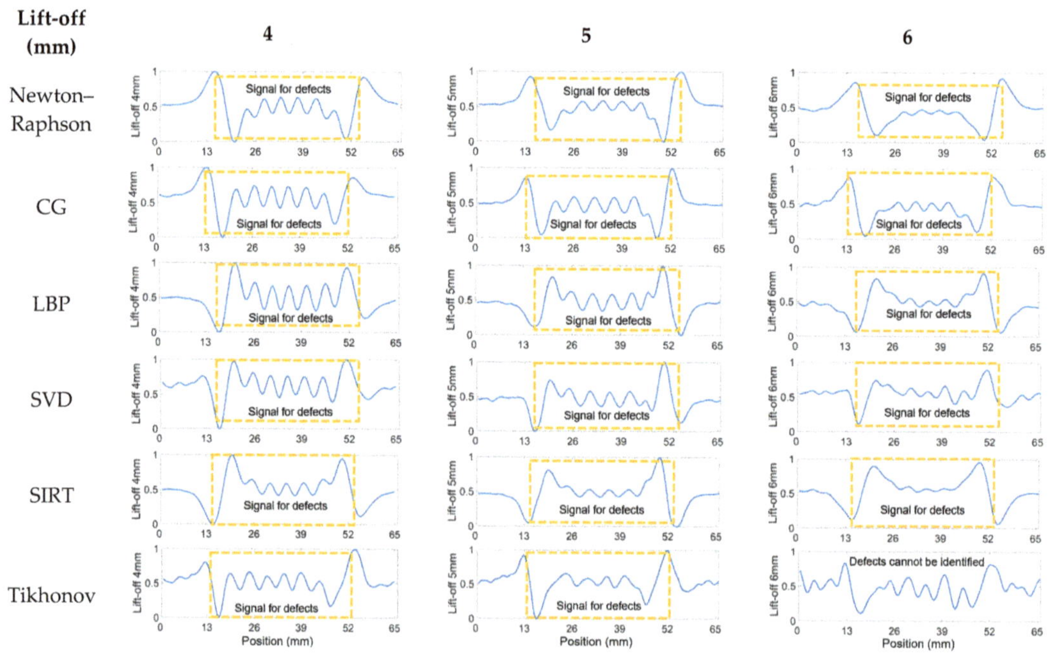

Figure 9. Processed signals of 10–14 T–R pairings for seven cracks at 4, 5, and 6 mm spacing and 5 mm lift-off.

3.2.2. Five Cracks with 10 mm Spacing

Given that the response signal can clearly identify cracks with a 10 mm spacing, it is crucial to assess whether the processing algorithms might alter the raw data. Figure 10 illustrates this point effectively, showing that all five cracks can be distinguished by their respective peaks in the processed data. This indicates the algorithms preserve the integrity of the raw signal, allowing for accurate identification of cracks, a vital aspect of the reliability of the detection method.

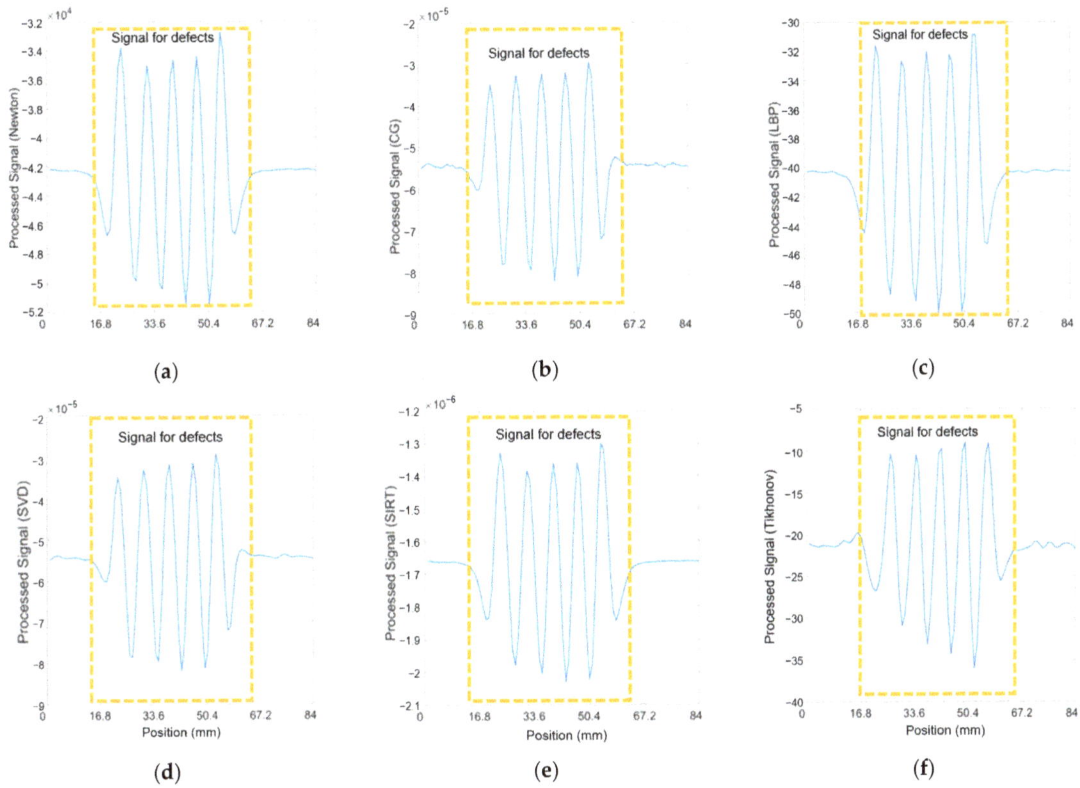

Figure 10. Processed signals of 36 T–R pairings for 5 cracks at 10 mm spacing, 5 mm lift-off, and 200 kHz using (**a**) Newton–Raphson (**b**) CG (**c**) LBP (**d**) SVD (**e**) SIRT (**f**) Tikhonov methods.

3.2.3. Three Cracks with 6 mm, 9 mm, and 12 mm Depths

The algorithms were also applied to signals received from three cracks of varying depths, aiming to determine if data processing impacts the signal's sensitivity to crack depth. Figure 11 demonstrates that the peak amplitude of the signal for each crack correlates nearly directly with the crack's depth. This correlation suggests that the algorithms effectively enable the assessment of cracks of different depths on railway tracks, providing a reliable means to gauge the severity of such defects based on signal amplitude analysis.

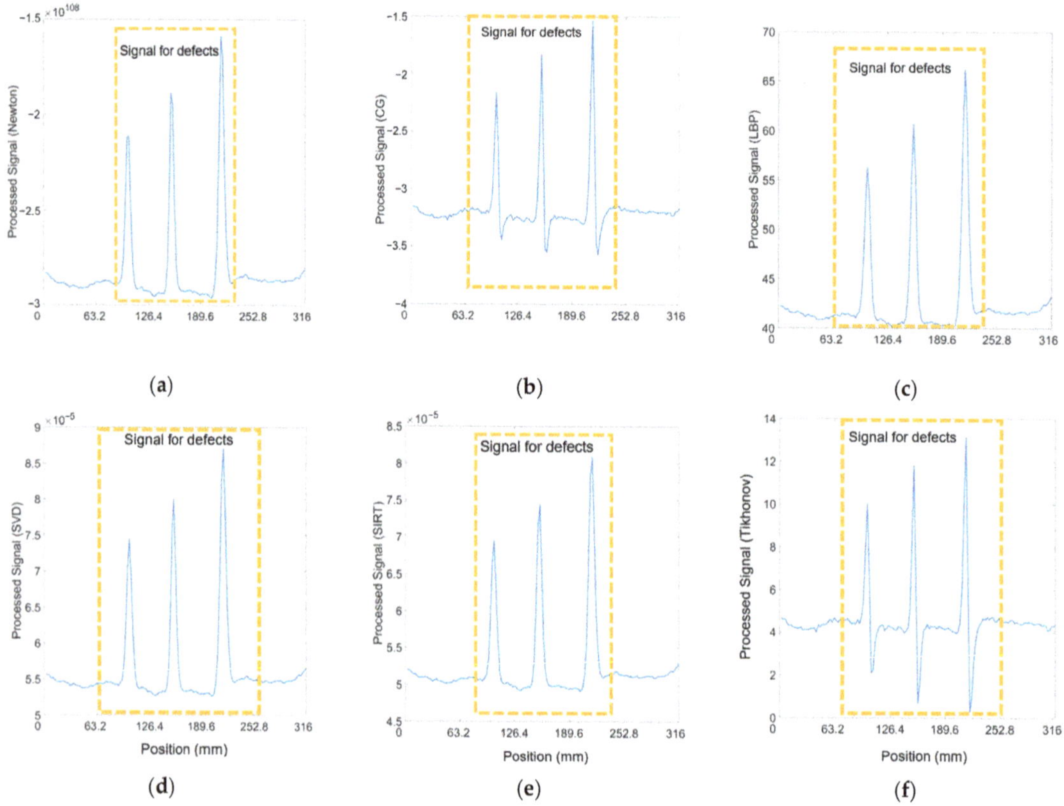

Figure 11. Processed signals of 36 T–R pairings for three cracks with 6, 9, and 12 mm depths at 5 mm lift-off and 200 kHz using (**a**) Newton–Raphson (**b**) CG (**c**) LBP (**d**) SVD (**e**) SIRT (**f**) Tikhonov methods.

4. Conclusions

In conclusion, this research introduces an innovative approach to railway track crack detection, incorporating a nine-coil sensor array, a multi-channel instrument, and sophisticated inverse algorithms, specifically the Newton–Raphson, CG, and LBP algorithms. Operating at a frequency of 200 kHz and with a lift-off of 5 mm, this system achieves an optimal balance between lift-off and spatial resolution. It significantly enhances the detection of closely spaced cracks (as narrow as 5 mm apart) through refined data processing techniques, facilitating more accurate crack identification. The carefully chosen probe lift-off not only minimizes the risk of damage to the sensitive sensor probe but also ensures data quality. This approach does not compromise the system's ability to detail intricate information about the defects. The processed data demonstrates the system's success in distinguishing between cracks of various depths and intervals, highlighting the algorithms' capacity to extract useful information from the raw signal. These findings show promise for realizing non-contact railway defect detection capabilities.

Future research will expand upon this work by exploring the integration of this system into railway test vehicles, assessing its performance in practical rail conditions, and further refining the sensing system and algorithms for increased performance.

Author Contributions: Conceptualization, Y.S., Z.X., B.C., S.S., A.P. and W.Y.; methodology, Y.S., Z.X., B.C., S.S., A.P. and W.Y.; software, Y.S. and Z.X.; validation, D.C., Y.S., Z.X., A.P. and W.Y.; formal analysis, Y.S., Z.X., B.C., S.S., A.P. and W.Y.; investigation, Y.S., Z.X., B.C., S.S., A.P. and W.Y.; resources, B.C., S.S., A.P. and W.Y.; data curation, Y.S. and Z.X.; writing—original draft preparation, Y.S. and Z.X.; writing—review and editing, Y.S., Z.X., B.C., S.S., X.B., A.P. and W.Y.; visualization, Y.S. and Y.D.; supervision, A.P. and W.Y.; project administration, A.P. All authors have read and agreed to the published version of the manuscript.

Funding: This research was funded by the University of Manchester and Sperry Rail under the EPSRC Impact Accelerator Grant Reference IAA-298.

Institutional Review Board Statement: Not applicable.

Informed Consent Statement: Not applicable.

Data Availability Statement: The datasets presented in this article are not readily available because the data are part of an ongoing study. Requests to access the datasets should be directed to the corresponding author, Zihan Xia (zihan.xia@manchester.ac.uk).

Conflicts of Interest: Authors Bob Crocker and Scott Saunders were employed by the Sperry Rail International Ltd. The remaining authors declare that the research was conducted in the absence of any commercial or financial relationships that could be construed as a potential conflict of interest.

Appendix A

This appendix presents a comprehensive illustration of the raw data collected by the sensing system, which is employed for data processing.

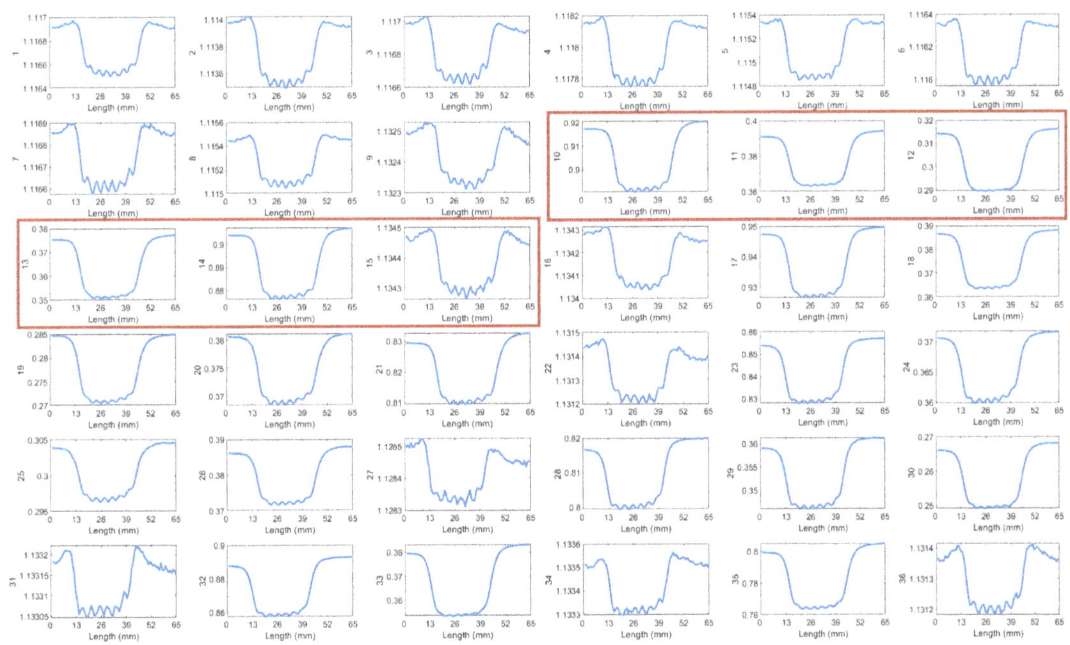

Figure A1. Received signals of 36 T–R pairings for seven cracks at 5 mm spacing, 4 mm lift-off, and 200 kHz. The plots in the red box are utilized for further data processing, with the results displayed in Figure 9.

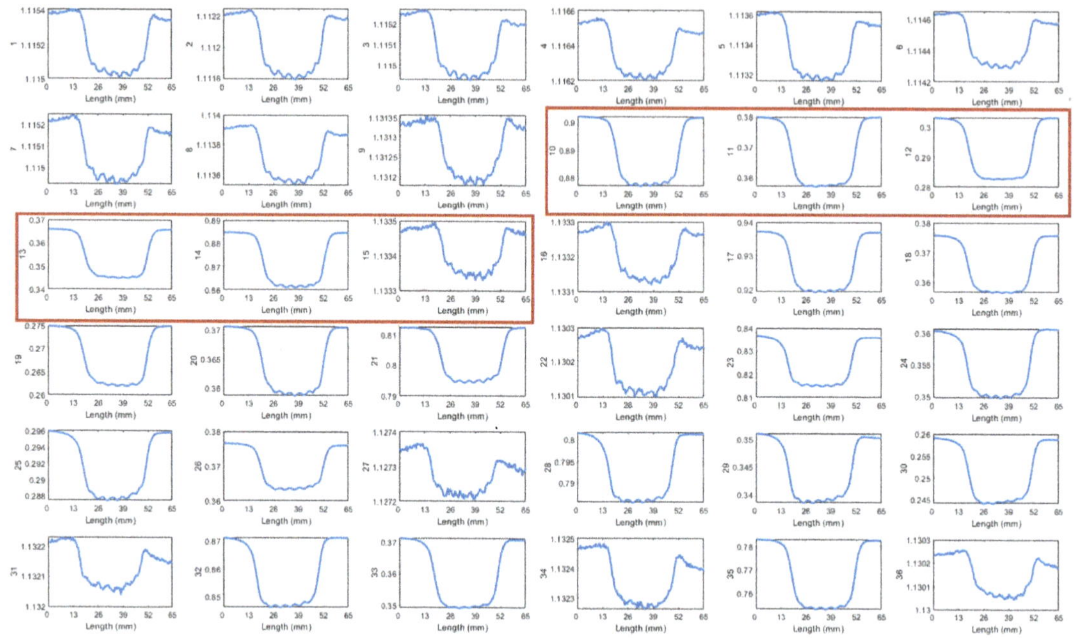

Figure A2. Received signals of 36 T–R pairings for 5 cracks at 5 mm spacing, 5 mm lift-off. and 200kHz. The plots in the red box are utilized for further data processing, with the results displayed in Figure 9.

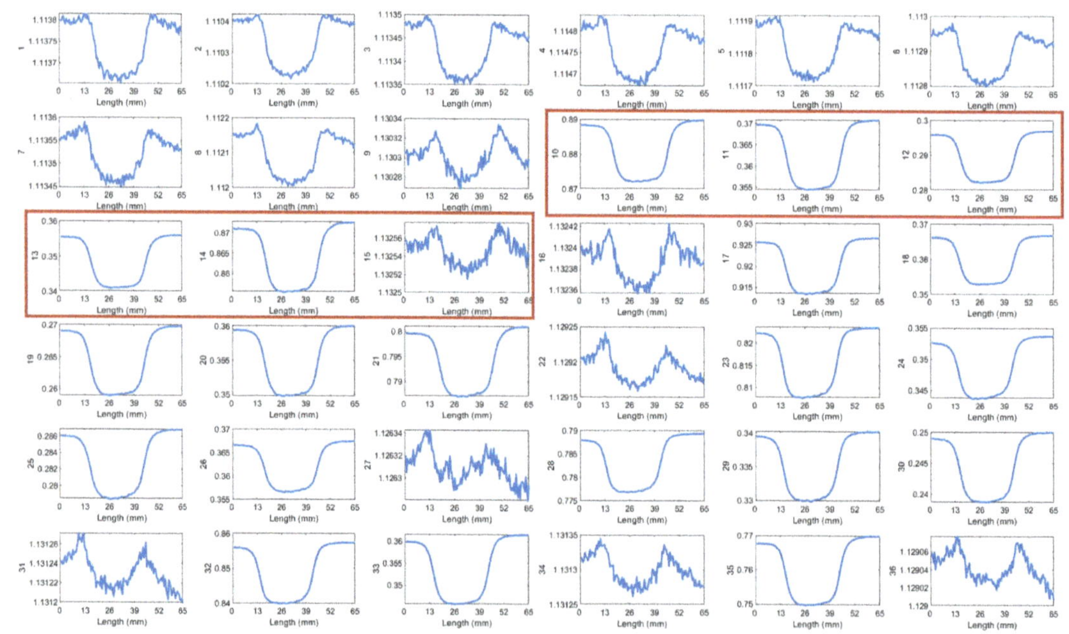

Figure A3. Received signals of 36 T–R pairings for seven cracks at 5 mm spacing, 6 mm lift-off. and 200 kHz. The plots in the red box are utilized for further data processing, with the results displayed in Figure 9.

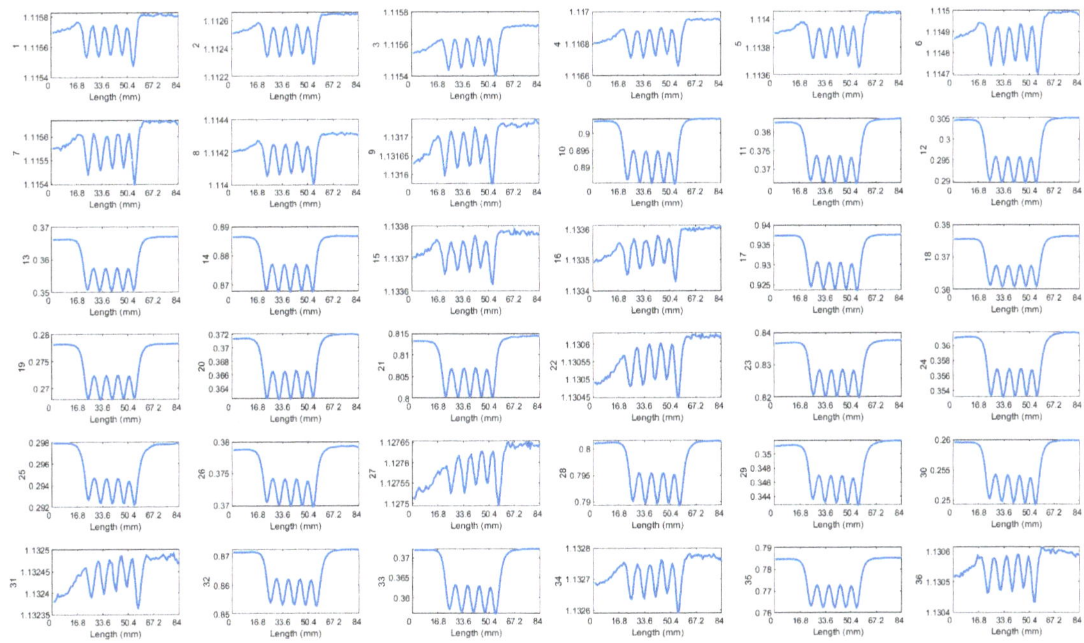

Figure A4. Received signals of 36 T–R pairings for 5 cracks at 10 mm spacing, 5 mm lift-off. and 200kHz.

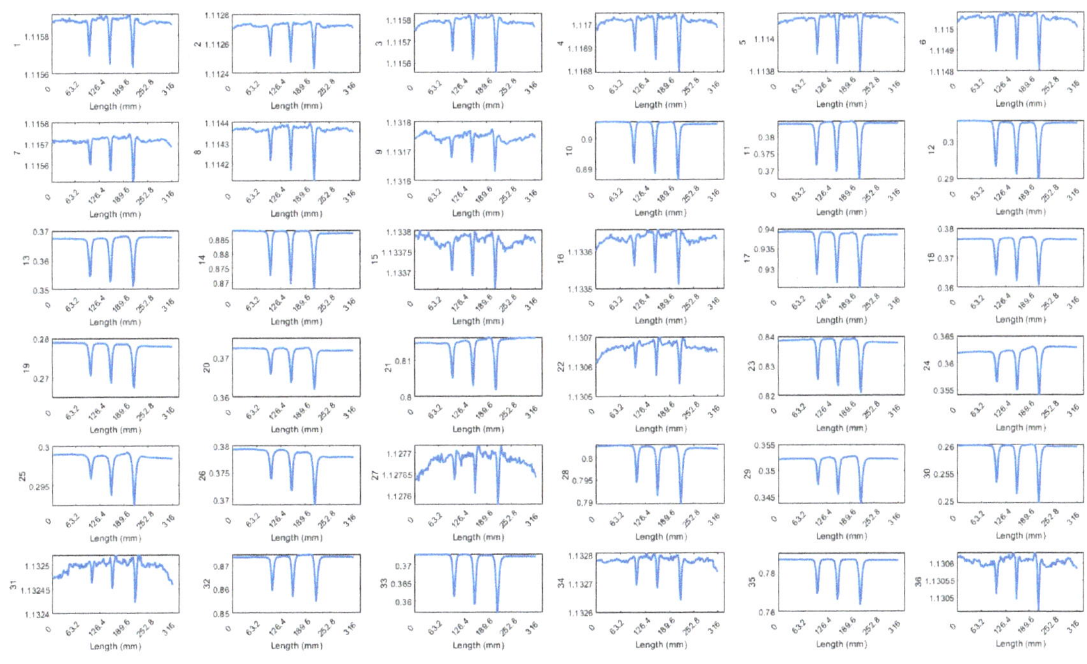

Figure A5. Received signals of 36 T–R pairings for three cracks with depths of 6, 9, 12 mm, 5 mm lift-off, and 200 kHz.

References

1. Saunders, S.; Crocker, R.A. Pedestrian and Vehicle-Mounted System for Detecting RCF in Rail using Eddy Currents. In Proceedings of the 19th World Conference on Non-Destructive Testing, Munich, Germany, 13–17 June 2016.
2. Haidemenopoulos, G.N.; Sarafoglou, P.I.; Christopoulos, P.; Zervaki, A.D. Rolling contact fatigue cracking in rails subjected to in-service loading. *Fatigue Fract. Eng. Mater. Struct.* **2016**, *39*, 1161–1172. [CrossRef]
3. Shen, J.; Zhou, L.; Warnett, J.; Williams, M.; Rowshandel, H.; Nicholson, G.; Davis, C. The influence of RCF crack propagation angle and crack shape on the ACFM signal. In Proceedings of the 19th World Conference on Non-Destructive Testing, Munich, Germany, 13–17 June 2016; pp. 1–9.
4. Tillberg, J.; Larsson, F.; Runesson, K. A study of multiple crack interaction at rolling contact fatigue loading of rails. *Proc. Inst. Mech. Eng. Part F J. Rail Rapid Transit* **2009**, *223*, 319–330. [CrossRef]
5. Rajamäki, J.; Vippola, M.; Nurmikolu, A.; Viitala, T. Limitations of eddy current inspection in railway rail evaluation. *Proc. Inst. Mech. Eng. Part F J. Rail Rapid Transit* **2018**, *232*, 121–129. [CrossRef]
6. Mandache, C.; Brothers, M.; Lefebvre, V. Time domain lift-off compensation method for eddy current testing. *NDT. Net* **2005**, *10*, 1–7.
7. Van Drunen, G.; Cecco, V.S. Recognizing limitations in eddy-current testing. *NDT Int.* **1984**, *17*, 9–17. [CrossRef]
8. Li, X.; Tian, G.; Li, K.; Zhang, Q.; Lu, X. Investigation of rolling contact fatigue cracks using the transmitter-receiver eddy current testing under moving conditions. *Nondestruct. Test. Eval.* **2024**, *39*, 614–633. [CrossRef]
9. Xu, P.; Zhu, C.L.; Zeng, H.M.; Wang, P. Rail crack detection and evaluation at high speed based on differential ECT system. *Measurement* **2020**, *166*, 108152. [CrossRef]
10. Kwon, S.G.; Lee, T.G.; Park, S.J.; Park, J.W.; Seo, J.M. Natural Rail Surface Defect Inspection and Analysis Using 16-Channel Eddy Current System. *Appl. Sci.* **2021**, *11*, 8107. [CrossRef]
11. Blitz, J.; Willsttäter, V.J.; Oaten, S.R.; Hajian, N.T. Eddy-current surface-crack sizing in steel with high lift-off. *NDT Int.* **1987**, *20*, 105–110. [CrossRef]
12. Sukhanov, D.; Zavyalova, K.; Kadurina, A. Method for enhancement of spatial resolution of eddy current imaging. *Meas. Sci. Technol.* **2019**, *30*, 065402. [CrossRef]
13. Meng, X.; Lu, M.; Yin, W.; Bennecer, A.; Kirk, K.J. Evaluation of coating thickness using lift-off insensitivity of eddy current sensor. *Sensors* **2021**, *21*, 419. [CrossRef] [PubMed]
14. Xu, P.; Zeng, H.; Qian, T.; Liu, L. Research on defect detection of high-speed rail based on multi-frequency excitation composite electromagnetic method. *Measurement* **2022**, *187*, 110351. [CrossRef]
15. Xu, P.; Qian, T. High-speed rail defect detection using multi-frequency exciting 3D ACFM. *Measurement* **2024**, *227*, 114160. [CrossRef]
16. Chady, T.; Lopato, P. Multi-Resolution Eddy Current Measurements, AIP Conference Proceedings. *Am. Inst. Phys.* **2007**, *894*, 370–377. [CrossRef]
17. Chady, T.; Enokizono, M.; Todaka, T.; Tsuchida, Y.; Sikora, R. A Family of Matrix Type Sensors for Detection of Slight Flaws in Conducting Plates. *IEEE Trans. Magn.* **1999**, *35*, 3655–3657. [CrossRef]
18. Shao, Y.; Meng, T.; Yu, K.; Xia, Z.; Huang, R.; Tao, Y.; Chen, Z.; Avila, J.R.S.; Yin, W. Automatic detection and imaging of rivet hole defects for aircraft structures with optimized sensor array using eddy current method and image analysis. *IEEE Sens. J.* **2022**, *23*, 4597–4606. [CrossRef]
19. Yin, W.; Peyton, A.J. Sensitivity formulation including velocity effects for electromagnetic induction systems. *IEEE Trans. Magn.* **2009**, *46*, 1172–1176. [CrossRef]
20. Cui, Z.; Wang, Q.; Xue, Q.; Fan, W.; Zhang, L.; Cao, Z.; Sun, B.; Wang, H.; Yang, W. A review on image reconstruction algorithms for electrical capacitance/resistance tomography. *Sens. Rev.* **2016**, *36*, 429–445. [CrossRef]
21. Yang, W.; Peng, L. Image reconstruction algorithms for electrical capacitance tomography. *Meas. Sci. Technol.* **2002**, *14*, R1. [CrossRef]
22. Su, B.; Zhang, Y.; Peng, L.; Yao, D.; Zhang, B. The use of simultaneous iterative reconstruction technique for electrical capacitance tomography. *Chem. Eng. J.* **2000**, *77*, 37–41.
23. Liu, X.; Wang, Y. An improved conjugate gradient image reconstruction algorithm for electromagnetic tomography. *Sens. and Imag.* **2022**, *23*, 5. [CrossRef]

sensors

MDPI

Article

Pulsed-Mode Magnetic Field Measurements with a Single Stretched Wire System

Joseph Vella Wallbank [1,*], Marco Buzio [2], Alessandro Parrella [2], Carlo Petrone [2] and Nicholas Sammut [1]

1 Department of Microelectronics and Nanoelectronics, Universita ta' Malta, MSD 2080 Msida, Malta
2 Technology Department, CERN—European Organization for Nuclear Research, 1217 Geneva, Switzerland; marco.buzio@cern.ch (M.B.); carlo.petrone@cern.ch (C.P.)
* Correspondence: joevellawalb@gmail.com

Abstract: In synchrotrons, accurate knowledge of the magnetic field generated by bending dipole magnets is essential to ensure beam stability. Measurement campaigns are necessary to characterize the field. The choice of the measurement method for such campaigns is determined by the combination of magnet dimensions and operating conditions and typically require a trade-off between accuracy and versatility. The single stretched wire (SSW) is a well-known, polyvalent method to measure the integral field of magnets having a wide range of geometries. It, however, requires steady-state excitation. This work presents a novel implementation of this method called pulsed SSW, which allows the system to measure rapidly time-varying magnetic fields, as is often needed, to save power or gain beam time. We first introduce the measurement principle of the pulsed SSW, followed by a combined strategy to calculate the absolute magnetic field by incorporating the classic DC SSW method. Using a bending magnet from the Proton Synchrotron Booster located at the European Organization for Nuclear Research as a case study, we validate the pulsed SSW method and compare its dynamic measurement capabilities to a fixed induction coil, showing thereby how the coil calibration must be adjusted according to the field level. Finally, we assess the method's measurement accuracy using the standard SSW as a reference and present an analysis of the primary noise contributors.

Keywords: single stretched wire; magnetic field measurements; particle accelerators; bending magnet; induction coil calibration

1. Introduction

In synchrotrons, large sets of dipole electromagnets are used to bend the particle beam in the horizontal plane along a ring-shaped trajectory, according to the Lorentz force generated by a vertical magnetic field [1]. Over a magnetic cycle, the dipole field changes proportionately to the beam energy to ensure that the beam remains centered within the vacuum chamber. We consider a single bending magnet, as represented schematically in Figure 1, where the beam circulates along the longitudinal axis of the magnet z, and x and y represent the transverse and vertical directions, respectively. The vertical field component:

$$B_\mathrm{y}(x, y, z, I(t)) \tag{1}$$

is a function of the position and magnet excitation current $I(t)$. The magnetic field is approximately proportional to the current, which is typically very uniform in the gap between the magnet poles and rolls off to zero rapidly in the surrounding fringe region. The main property of interest for synchrotron operation is the longitudinal integral of the field or, equivalently, its average $\bar{B}$:

$$\bar{B}(I(t)) = \frac{1}{l_\mathrm{m}} \int_{-\infty}^{+\infty} B_\mathrm{y}(0, 0, z, I(t))\, dz \tag{2}$$

where l_m is a reference length, defined as the length of the circular arc traced by the beam inside the magnet, under the assumption of constant field (hard-edge model, [1]):

$$l_m = \frac{2\pi R}{N_B} \quad (3)$$

where R is the bending radius of the synchrotron and N_B is the number of bending magnets powered in series. The reference length is usually close, but not necessarily identical, to the length of the magnet's iron yoke. We also note that the total length of the accelerator ring is always greater than $2\pi R$, because the beam trajectory between dipoles is generally composed of additional straight sections corresponding to different kinds of components, such as focusing quadrupoles or RF cavities. In practice, integration in (2) can be truncated to a short distance from both ends of the iron yoke, with 2.5 times the height of the magnet aperture being more than sufficient [2].

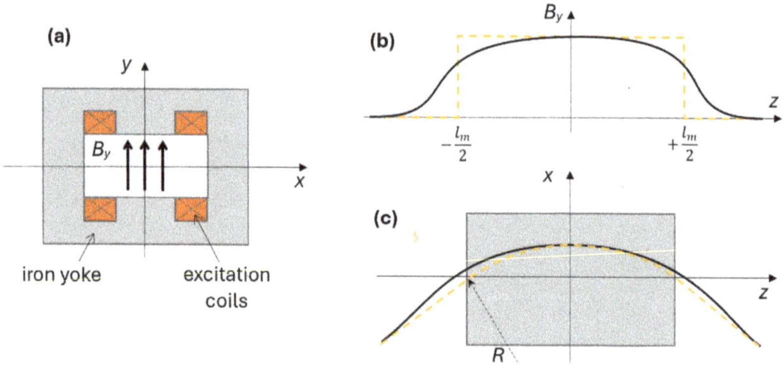

Figure 1. Schematic representation of the beam path in a dipole magnet, assuming small bending angle (i.e., $l_m \ll R$) so that the arc length of the path can be replaced by its projection onto the longitudinal axis z. Dashed curves refer to the hard-edge model. (**a**) Vertical cross-section of the magnet; (**b**) longitudinal field profile; (**c**) beam path (top view).

To program the current excitation cycles, machine operators need knowledge of the current-to-field relationship (2) with a relative precision of 10^{-4}. However, the magnet response is usually affected (up to several percent) by nonlinear phenomena such as magnetic saturation, hysteresis, and eddy currents, which cannot easily be predicted by means of FE models or measured with beam-based techniques during operation. The most accurate and cost-effective approach is to measure the integral of the magnetic field directly using a reference magnet. Here, the two most widely used measurement methods are based on induction coils and single stretched wire (SSW) systems, which have the added benefit of providing information on the quality and uniformity of the main field component [3].

Induction coils represent one of the most basic types of magnetic field sensors. A coil typically consists of multiple loops of conducting wire, wound around a rectangular core, and positioned along the longitudinal axis of a magnet's aperture [4]. The output voltage is proportional to the number of turns and the rate of change of the linked magnetic flux, according to Faraday's law. A fixed coil is therefore suitable for measuring time-varying magnetic fields, in which case, however, it can register only their variation. Conversely, a coil rotating or translating in a stationary field coil can be used for absolute measurements but at the cost of a more complex mechanical setup. In both cases, very-high-quality results are possible, provided that the coil is adapted to match a magnet's specific geometry. Unfortunately, manufacturing uncertainties increase with the length and number of turns, making accurate calibration of integral coils more difficult, as is discussed in Section 2.4.

With respect to induction coils, the single stretched wire (SSW) method presents complementary advantages and drawbacks. In a dipole magnet, the classic static SSW (s-SSW) method is used to measure the absolute integral of the magnetic field by displacing a single taut conducting wire perpendicularly to the field direction. A key advantage of this method lies in the flexibility to cover a wide range of magnet lengths and aperture widths with the same hardware. Moreover, the width of the flux linkage area swept by the stretched wire is defined with exquisite precision by the translation stages, which is often better than the micrometer level and in stark contrast with typical induction coils. As a result, the static SSW method is often used as a reference to calibrate the width of integral induction coils, as discussed in the next section. The limitations of this method are twofold; First, the setup can only measure steady-state magnetic fields. Secondly, as the method is equivalent to using a single-turn variable-geometry coil, the output voltage depends mainly on the magnetic field strength [5]. Other SSW methods exist where the magnet or wire is excited using an AC waveform. However, we shall not cover them here as they are mainly used to determine the magnetic axis of quadrupole magnets. Overall, both measurement methods are induction-based and, therefore, inherently linear to the field level, which is a key advantage over other types of magnetic sensors, such as Hall probes [6].

In the following, we shall compare the performance of these measurement methods by taking a bending magnet currently in operation at the European Organization for Nuclear Research (CERN) as a case study. At CERN, the Large Hadron Collider (LHC) proton injector chain includes four synchrotrons that progressively increase the beam energy from 160 MeV to 6.5 TeV, thus limiting nonlinear effects that grow exponentially with the dynamic range of each ring. We focus on the first synchrotron in the chain, the Proton Synchrotron Booster (PSB), which accelerates protons up to 2 GeV. This synchrotron consists of $N_\mathrm{b} = 32$ dipoles that bend the beam with a radius $R = 8.3$ m, corresponding to a reference magnetic length of $l_\mathrm{m} = 1.630$ m, as per Equation (3). The PSB bending dipoles are excited by the 1.2 s long current cycle depicted in Figure 2, which reaches a current level $I_\mathrm{max} = 5400$ A at the extraction flat-top [7]. Here, the magnets produce an integrated field of $l_\mathrm{m}\bar{B}_\mathrm{max} = 1.835$ Tm, which we will use throughout the following analysis as a normalization reference. The magnetic cycle is characterized by a high ramp rate, with a ramp-up time of around 0.5 s that peaks at about $5\,\mathrm{T\,s^{-1}}$. This rapid cycling allows for large numbers of protons to be accelerated while limiting thermal dissipation in the magnet's excitation coils, which were originally designed to operate at the much lower level of $\sim$2300 A [8]. Unfortunately, high ramp rates induce substantial eddy currents in the iron yoke of an accelerator magnet, degrading the field uniformity and causing the magnetic field to lag behind the current waveform [9]. In the PSB, the eddy currents delay the magnetic field's response by as much as 0.7 ms at beam extraction, corresponding to a field error of about 0.2 %. This highly dynamic nature of the magnetic field compounds the need for accurate measurements while restricting the usage of the static SSW, thereby making the precise calibration of induction coils more difficult.

For this reason, we have developed the so-called pulsed SSW (p-SSW) method; a novel implementation of the SSW setup that enables the system to measure fast time-varying magnetic fields by combining dynamic measurements taken at different wire positions. In the context of our case study, the pulsed SSW setup serves mainly as a method to complement and cross-calibrate fixed coil measurements. The availability of suitable induction coils, especially with lengths on the order of meters, is limited by the lack of commercial offers and the level of resources needed to design, manufacture, and calibrate them. Due to its inherent flexibility in adapting to a wide range of geometries, the pulsed SSW method represents a very effective alternative.

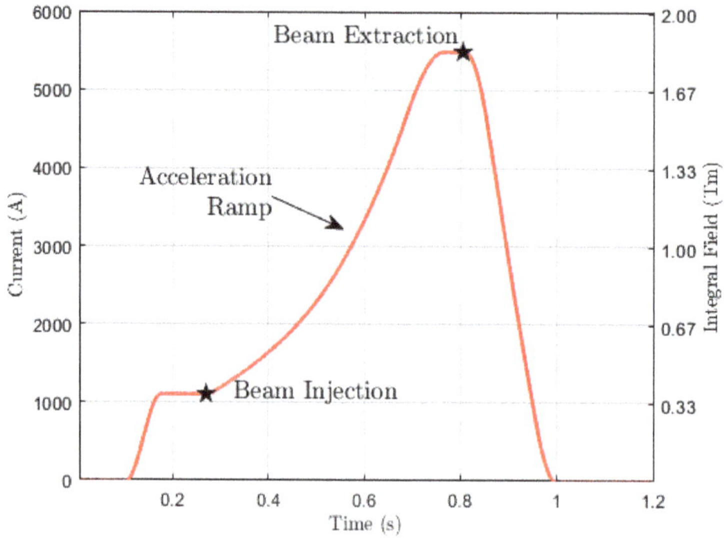

Figure 2. Current waveform powering the PSB bending magnets during a 2 GeV magnetic cycle.

This paper is structured as follows: In Section 2, we first introduce the operating principles of both the induction coil and the static and pulsed SSW implementations; then, in Section 3, we detail the test setup and measurement procedure of the PSB bending dipole. In Section 4, we present, analyze, and compare the results obtained with all three methods, highlighting how the pulsed SSW method has been used to calibrate a fixed induction coil. In Section 5, we make concluding remarks and outline planned developments.

2. Measurement Principle

In this section, we describe the measurement principle of the three methods used in this work. The symbol Φ shall denote absolute flux linked through an induction loop, whether an induction coil or a wire. The symbol $\Delta\Phi$ shall be used for space or time flux differences (or, equivalently, excitation current).

2.1. Static SSW

The general operating principle of the static SSW method is well documented [10–12]. A conducting wire is stretched along the axis of the magnet and then displaced along a known trajectory while the magnet current I remains constant. The path of displacement of the wire depends on the magnet type and the harmonics of the field that are measured, ranging from a simple segment to a complex elliptical rotation [13]. For the present case study, we shall consider a straight dipole magnet, as shown in Figure 3. The wire is stretched between two symmetric linear stage supports over a length ℓ parallel to the longitudinal z axis. The wire lies in the magnet's midplane $y = 0$ and is translated only along the x axis by the linear stages. Since we are dealing with a relatively short magnet, we can ignore the impact of gravity-induced sag. The ends of the wire and the return wire must be sufficiently far from the magnet to lie in a field-free region, where stray and environmental fields can be ignored. As a result, the flux linked through the wire loop at any time can be expressed as follows:

$$\Phi(X(t), X_R, I) = \int_{-\frac{\ell}{2}}^{+\frac{\ell}{2}} dz \int_{X(t)}^{X_R} B_y(x, 0, z, I)\, dx \tag{4}$$

where $X(t)$ and X_R are the transverse positions of the stretched and return wire, respectively. Under the assumptions made, the flux does not depend either on the precise length of the stretched wire, or on the exact path followed by the return wire [14].

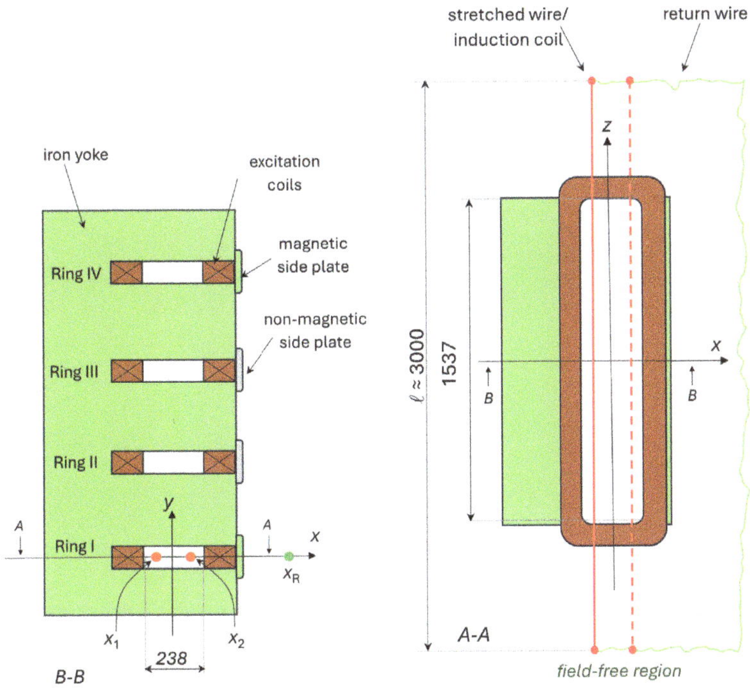

(a) (b)

Figure 3. Schematic representation of the SSW system mounted in the four-gap PSB bending dipole magnet, where each gap corresponds to a different accelerator ring and all lengths are given in millimeters. Red lines denote the stretched wire, while green shows the return wire closing the flux loop. The induction coil can be placed along the same path followed by the stretched wire, being only slightly shorter (although both its ends still reach into the field-free region). (**a**) Front view. (**b**) Top view.

A static SSW measurement consists of the wire moving from X_1 to X_2 at an approximately constant speed dX/dt. According to Faraday's law, the induced loop emf V_s depends only on position and velocity:

$$V_s(t) = -\frac{\partial \Phi}{\partial t} = \frac{dX}{dt} \int_{-\frac{\ell}{2}}^{+\frac{\ell}{2}} B_y(X(t), 0, z, I)\, dz \tag{5}$$

In a good field region of a highly uniform dipole, the integral term is essentially constant, and the voltage is simply proportional to the stretched wire velocity, which, therefore, should be as high as possible to improve the signal-to-noise ratio (SNR). By digitally acquiring Equation (5) and then integrating it numerically with respect to time, we can derive the expression of the flux $\Delta\Phi_s$ linked through the rectangular surface swept by the stretched wire:

$$\Delta\Phi_s(X_1, X_2, I) = -\int_{t_1}^{t_2} V_s(u)\, du = \Phi(X_2, X_R, I) - \Phi(X_1, X_R, I) = \int_{-\frac{\ell}{2}}^{+\frac{\ell}{2}} dz \int_{X_1}^{X_2} B_y(x, 0, z, I)\, dx \tag{6}$$

where u is the integration variable, $X_1 = X(t_1)$, and $X_2 = X(t_2)$. In principle, this result does not depend upon the exact law of motion $X = X(t)$ followed by the wire. In practice, however, the wire will vibrate every time it is accelerated, and the integration interval must be adjusted to allow the vibrations to damp out. The need to curb the velocity and acceleration of the wire may represent a limitation of the method, especially in the case of fast-cycled magnets such as the PSB dipole, where the duration of current plateaus is severely constrained.

Finally, the average field can be obtained according to Equation (2) as

$$\bar{B}_s(I) = \frac{1}{l_m d} \lim_{d \to 0} \Delta\Phi_s(-\frac{d}{2}, \frac{d}{2}, I) \tag{7}$$

where $d = X_2 - X_1$ represents the stroke of the wire's movement, which, in this paper, we will always take to be symmetric with respect to $x = 0$. However, the magnitude of the integrated flux and, hence, its accuracy tend to vanish as $d \to 0$. In practice, an optimal wire stroke d^* can be defined as a compromise between measurement accuracy and the additional error due to field nonuniformity, which must be found experimentally as discussed in Section 4.1, to obtain

$$\bar{B}_s(I) = \frac{1}{l_m d^*} \Phi_s(-\frac{d^*}{2}, \frac{d^*}{2}, I) \tag{8}$$

2.2. Pulsed SSW

The pulsed SSW method subtracts two dynamic measurements at the positions of the wire X_1 and X_2. The voltage induced in the wire loop when the magnet excitation current is changing while the wire position is kept fixed is given by

$$V_p(t) = -\frac{\partial \Phi}{\partial t} = \frac{dI}{dt} \int_{-\frac{\ell}{2}}^{+\frac{\ell}{2}} dz \int_{X}^{X_R} \frac{\partial}{\partial I} B_y(X, 0, z, I(t))\, dx \tag{9}$$

By acquiring and integrating Equation (9) we can obtain the change between the initial and final linked fluxes:

$$\Delta\Phi_p(X, X_R, I_0, I(t)) = -\int_{t_0}^{t} V_p(u)\, du = \Phi(X, X_R, I(t)) - \Phi(X, X_R, I_0) \tag{10}$$

where $I_0 = I(t_0)$. In principle, the flux change depends only upon the initial and final states, much like in the case of static wire. However, the nonlinear relationship between current and field often makes it necessary to implement suitable measures to improve the reproducibility of results. Most importantly, the change between minimum and maximum current should always be monotonic to avoid switching to a different branch of the magnetic hysteresis loop. For the same reason, the rate of change of the current should also be constant to ensure reproducible losses. As a result, it is strongly recommended that a fixed current cycle $I(t)$ be repeatedly applied at X_1 and X_2. The results shall then be subtracted numerically to finally obtain

$$\begin{aligned} \Delta^2\Phi_p(X_1, X_2, I_0, I) = \Delta\Phi_p(X_2, X_R, I_0, I) - \Delta\Phi_p(X_1, X_R, I_0, I) = \\ \int_{-\frac{\ell}{2}}^{+\frac{\ell}{2}} dz \int_{X_1}^{X_2} (B_y(X, 0, z, I(t)) - B_y(X, 0, z, I_0))\, dx \end{aligned} \tag{11}$$

where the squared delta symbol represents the difference with respect to the position of a difference with respect to time. In other words, Equation (11) represents the change of the flux linked through a rectangular area equivalent to a virtual, single-turn fixed coil of width $d = X_2 - X_1$. Similarly to Equation (8), the average field can be derived by moving the wire by the optimal stroke d^*, thus obtaining

$$\Delta\bar{B}_p(I_0, I(t)) = \frac{1}{l_m d^*}\Delta^2\Phi_p(-\frac{d^*}{2}, \frac{d^*}{2}, I_0, I(t)) \quad (12)$$

where the Δ symbol emphasizes the fact that Equation (12) represents a difference between two excitation current levels. The accuracy of this result depends crucially on the reproducibility of the magnetic field, which, unfortunately, is not guaranteed, even when identical $I(t)$ cycles are applied in succession. This is mainly due to the possible impact of the previous excitation history and, to a lesser degree, to common imperfections of the excitation current, such as high-frequency ripple or uncontrolled transients (glitches or overshoots). When a test campaign includes a variety of different magnetic cycles, as is the case in the present work, two well-known strategies can be applied to improve reproducibility: (1) systematic execution of one or more normalization precycles; and (2) forcing each current cycle to swing monotonically between fixed minimum and maximum values to remain always on the same hysteresis loop. Both techniques are used in the present work, as discussed in Section 3.

Additional considerations must be made about the optimal position of the return wire, which may be placed externally or internally to the magnet. Let us consider two common cases, as shown in Figure 4: a window-frame-type dipole (top), where the return flux splits evenly between the two vertical legs of the iron yoke, and a C-shaped dipole (bottom), where one side is left open for easier access. For simplicity, we shall assume that all the magnetic flux generated by the excitation coils Φ_e is captured as the wire moves from X_1 to X_2. We find the following:

- **Window-frame dipole** (inset a, b): When the return wire is external to the magnet (X_R), the measured flux changes antisymmetrically with respect to the origin, reaching, at both ends, the same absolute value $|\Phi(X_1, X_R, I)| = |\Phi(X_2, X_R, I)| \approx \frac{1}{2}\Phi_e$. If, instead, the return wire is inside the magnet gap at either position $X_{R'}$ or $X_{R''}$ due to the symmetry of the yoke, the linked flux will be $\sim$0 at one end and reach the maximum value $\sim|\Phi_e|$ at the other.
- **C-shaped dipole** (inset c, d): If the return wire is placed on the side where the yoke is open, either externally (X_R) or internally ($X_{R'}$) to the gap, the measured flux will behave identically i.e., it will swing from the maximum value $\sim|\Phi_e|$ to $\sim$0; if, instead, the return is moved on the closed side of the yoke ($X_{R''}$), the sign of flux will be opposite as the curve $\Phi(X)$ is shifted by $-\Phi_e$.

For a static SSW measurement, these differences are irrelevant since the results depend only on the initial and final positions of the wire. Instead, in the case of pulsed SSW, it is important to maximize the peak value of the measured flux to gain dynamic range, which may improve the final accuracy, provided the acquisition chain can be adapted accordingly. For a C-shaped dipole, the maximum dynamic range is always guaranteed, while for a window-frame type, either internal position leads to a factor $\sim$2 improvement. The case of the PSB dipole corresponds to an intermediate situation closer to the C-shaped design since the flux returns on the $x > 0$ side through a relatively thin plate. For the present case study, the return wire was fixed externally, and the resulting flux curves, which retain the same sign throughout the wire movement, are shown in Figure 4. Running the return wire internally may provide an additional advantage linked to lower EM interference pick-up due to the smaller loop area (see also the discussion in Section 4.3). Should one choose this configuration, fixing the mechanically stable wire is paramount since even very small movements in the high-field region may perturb the voltage readout.

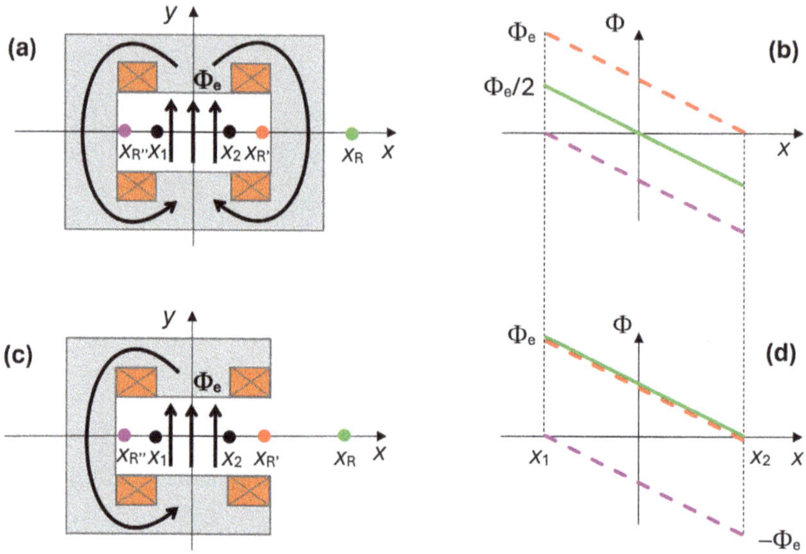

Figure 4. Impact of the position of the return wire, according to the type of magnet. (**a**): Magnetic flux in a symmetric window-frame dipole. (**b**): corresponding qualitative representation of the measured flux as the wire moves from X_1 to X_2. (**c**): Flux in an asymmetric C-shaped dipole. (**d**): corresponding measured flux. Green: external return wire (X_R). Purple: internal return wire ($X_{R'}$).

2.3. Combined SSW

Embodiment in the same physical setup gives the unique opportunity to combine static and pulsed SSW measurement procedures to derive the absolute field, even under dynamic excitation conditions (c-SSW method). Assuming that DC excitation at I_0 is possible, we can simply add Equations (6) and (11) to define the combined flux as follows:

$$\begin{aligned}\Phi_c(X_1, X_2, I(t)) &= \Delta\Phi_s(X_1, X_2, I_0) + \Delta^2\Phi_p(X_1, X_2, I_0, I(t)) \\ &= \int_{-\frac{\ell}{2}}^{+\frac{\ell}{2}} dz \int_{X_1}^{X_2} B_y(x, 0, z, I(t))\, dx\end{aligned} \tag{13}$$

The average field can then be expressed as

$$\bar{B}_c(I(t)) = \bar{B}_s(I_0) + \Delta\bar{B}_p(I_0, I(t)) = \frac{1}{l_m d^*}\Phi_c(-\frac{d^*}{2}, \frac{d^*}{2}, I(t)) \tag{14}$$

As it incorporates the residual field, Equation (14) captures, in principle, the current magnetic state irrespective of the previous excitation history. However, as discussed in the previous section, the reproducibility of the two consecutive cycles necessary for the pulsed SSW is generally guaranteed only if the magnet is already on a stable hysteresis loop. A suitable number of normalization precycles, to be established experimentally for each magnet type, should be run before any measurement to ensure that is the case.

2.4. Induction Coil

Ideally, an induction coil can be represented by N_T infinitely thin identical rectangular turns of length l_c and width w_c. In practice, the width of each winding is affected by the pile-up of manufacturing tolerances, which increase with the total length and the number of turns. As a result, the average width of the windings should be considered instead of an unknown function $w_c(z)$ of the longitudinal position [15]. In general, the flux linked by the coil can be expressed as

$$\Phi_{\text{coil}}(I) = N_T \int_{-\frac{l_c}{2}}^{+\frac{l_c}{2}} dz \int_{-\frac{w_c(z)}{2}}^{+\frac{w_c(z)}{2}} B_y(x,0,z,I)dx \tag{15}$$

In the following, we shall assume that the length of the coil is sufficient to reach the field-free region at both ends, allowing us to use the same longitudinal integration limits as in the case of the stretched wire. In addition, we shall ignore the nonuniformity of the field across the coil, which is usually only a few millimeters wide. Under these assumptions, we can express the total flux linked through the coil as

$$\Phi_{\text{coil}}(I) = N_T \int_{-\frac{\ell}{2}}^{+\frac{\ell}{2}} w_c(z) B_y(0,0,z,I)\, dz \tag{16}$$

The flux at any given current level can be measured by integrating the voltage V_c induced according to Faraday's law in different ways. The following two methods were used in this work:

- **Flip-coil method**: The coil is first positioned flat in its central rest position inside the magnet's gap (roll angle $\theta = 0$), then it is flipped upside down ($\theta = \pi$) while measuring the induced voltage $V_c(t)$. The final flux linkage will be equal and opposite to the initial one, and we can derive the following:

$$\Phi_{\text{coil}}(I) = \frac{1}{2} \int_{t(\pi)}^{t(0)} V_c(u)\, du \tag{17}$$

  where the integration bounds correspond to mechanically stable coil configurations. The result, in principle, does not depend upon the precise law of motion $\theta(t)$ followed, even if some translation is unwittingly superposed to the rotation. The main practical difficulty is turning the coil quickly enough to avoid the build-up of integrator drift while at the same time ensuring that the initial and final configurations are not offset in any direction. Mainly for this reason, a suitable nonmagnetic, nonconducting, rotating, mechanical support should be preferred to manual operation.

- **Fixed coil method**: We keep the coil fixed and instead ramp the current starting from zero up to its desired value, to obtain the flux change:

$$\Delta\Phi_{\text{coil}}(I_0, I) = \int_{t(I_0)}^{t(I)} V_c(u)\, du \tag{18}$$

  This method is the most rapid and practical since it does not involve any coil handling; however, the measurement is blind to the initial flux $\Phi_0 = \Phi(0)$. This is typically associated with a remanent field in the iron poles, which depends upon the previous magnetization history and can be of the order of a few milliteslas. Ideally, a demagnetization cycle should be applied before the measurement, but doing so requires bipolar power converters, which are not always available. In the alternative, Φ_0 must be measured independently, for example, by the flip-coil method. Another independent measurement of Φ_0 could be obtained with the static SSW method by scaling the results with the appropriate number of induction coil turns. However, the measured flux will not be entirely consistent with Equation (16), since the wire is much more straight than the coil windings, at least horizontally. Fortunately, the resulting error is often negligible since the integrated remanent field is very small, typically at least two orders of magnitude below the lowest level of interest (corresponding to beam injection). The value thus obtained is then added to the pulsed measurement to derive the absolute flux:

$$\Phi_{\text{coil}}(I) = \Phi_0 + \Delta\Phi_{\text{coil}}(0, I) \tag{19}$$

The same Φ_0 can be reused multiple times for fixed coil measurements at different current levels. Doing so, however, requires that a stable hysteresis cycle be followed to ensure repeatability, as we have performed in the present work.

Regardless of the method followed, a straightforward and accurate derivation of the average field, such as in Equation (8) or Equation (12), is not possible due to the coil width being affected by a large uncertainty. We shall therefore define a suitable calibration parameter and the effective coil width as follows [16]:

$$w_{\text{eff}}(I) = \frac{\Phi_{\text{coil}}(I)}{l_{\text{m}}\bar{B}(I)} = \frac{N_{\text{T}}\int_{-\frac{\ell}{2}}^{+\frac{\ell}{2}} w_{\text{c}}(z)B_y(0,0,z,I)\,dz}{\int_{-\frac{\ell}{2}}^{+\frac{\ell}{2}} B_y(0,0,z,I)\,dz} \tag{20}$$

which can be interpreted as the average coil turn width, weighed with the local magnetic field, additionally incorporating the number of turns to simplify practical usage as a single calibration coefficient. The calibration of the coil depends, at the same time, upon its geometry and the longitudinal field profile of the magnet being measured. In iron-dominated magnets, the most common choice for particle accelerators is the shape of the longitudinal field profile, which is mainly a function of the excitation current. At high fields, saturation increases the flux leaking out of the iron yoke, leading to a relative increase in the field strength of the fringe field region and a subsequent flattening of the profile. Calibration of the effective width from Equation (20) requires an independent measurement of $\bar{B}$ in the same magnet with a reference method, such as any form of SSW. For example, omitting for simplicity the incorporation of the residual field in both coil and SSW measurements, the pulsed SSW method can be used to derive the effective width as

$$w_{\text{eff}}(I) = \frac{\Delta\Phi_{\text{coil}}(0,I)}{l_{\text{m}}\Delta\bar{B}_{\text{p}}(0,I)} = d^* \frac{\Delta\Phi_{\text{coil}}(0,I)}{\Delta^2\Phi_{\text{p}}(-\frac{d^*}{2},\frac{d^*}{2},0,I)} \tag{21}$$

We note that any nominally identical magnets belonging to a series production run can be used as a reference, with their relative differences usually being very small. Once the calibration has been obtained, the average field can finally be derived from Equation (20) as follows:

$$\bar{B}_{\text{coil}}(I) = \frac{\Phi_{\text{coil}}(I)}{l_{\text{m}}w_{\text{eff}}(I)} \tag{22}$$

3. Test Setup

The measurements discussed in Section 4 were obtained with the setup based on the PSB dipole magnet shown in Figures 5 and 6 and represented schematically in Figure 7. The main design parameters and field quality requirements are listed in Table 1. The design of the PSB magnet is unique in that it consists of four vertically stacked apertures powered in series and sharing the same iron yoke, allowing for a four-fold increase in the number of protons that can be accelerated in parallel. In the context of this work, all apertures are functionally equivalent to a standalone magnet, and the measurements were performed in the bottom aperture.

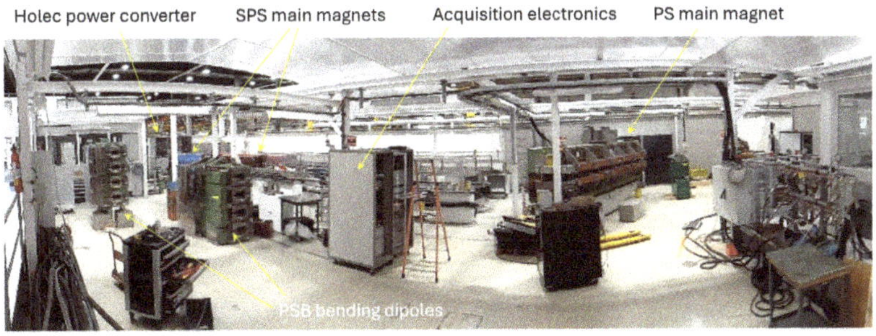

Figure 5. CERN test hall 867-R-29, showing two PSB bending dipoles being prepared for magnetic measurements. Thanks to the availability of the Holec power converter, this hall is a unique asset dedicated to testing potentially activated, high-pulsed-current magnets such as the other PS and SPS units shown.

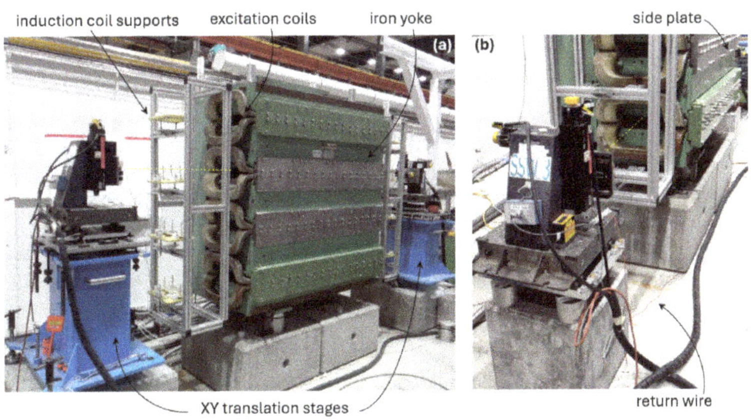

Figure 6. Test setup including the four-aperture PSB magnet and the SSW stages. (**a**) Illustration of the wire (dotted yellow line) stretched through the magnet between the two XY translation stages. (**b**) Detail of one stage placed in front of the lowest of the magnet's apertures, i.e. the one used for the present work.

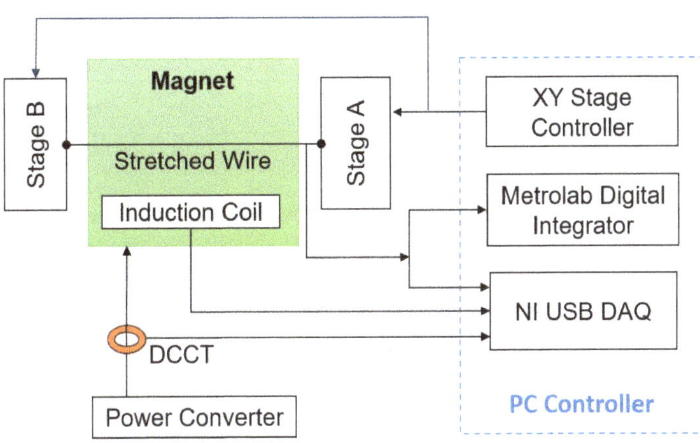

Figure 7. Schematic block diagram of the test setup.

Table 1. Main geometrical and electrical design parameters of the PSB bending dipole, along with the principal magnetic field quality requirements [17].

Parameter	Value	Unit
Iron core length	1537	mm
Gap height	70	mm
Gap width	238	mm
Good field region width	160	mm
Excitation turns/gap	12	-
Total excitation resistance	9.9	mΩ
Total excitation inductance	41.5	mH
Peak excitation current	5400	A
Integrated dipole tolerance $\frac{B}{B_{max}} - 1$	$\pm 5 \cdot 10^{-4}$	-
Field uniformity in the good field region $\frac{\Delta B}{B}$ (23)	$\pm 4 \cdot 10^{-4}$	-

3.1. SSW Setup

The setup used for this work is based on the systems originally developed at Fermilab to measure series LHC magnets [11,12]. The PC-controlled system consists of a 3 m long, ⌀125 µm beryllium copper wire stretched through the aperture and supported on either side by a Newport ESP 7000 XY translation stage, having a maximum ±75 mm stroke on both axes and mounted on a granite support. An additional motor keeps the wire under constant tension 8 N to limit the gravity-induced sagitta to about 0.1 mm [13]. The acquisition chain includes a Metrolab PDI-5035 voltage integrator [18] to measure the flux change induced by the displacement of the wire in static mode, as well as a National Instruments USB-6366 16-bit DAQ module to acquire the voltage induced in pulsed mode in both the wire loop and the coil up to 1 MS/s, as well as the DCCT output measuring the excitation current. We point out that the large inductance of accelerator magnets usually limits the effective bandwidth well below 1 kHz; however, oversampling at high frequency is often very useful for estimating and then subtracting the low-frequency $1/f$ noise components leading to integration drift.

3.2. Induction Coil

The integral induction coil used for this work is part of a batch of nominally identical units developed for the PSB dipoles. It consists of $N_T = 70$ turns of copper wire wound around a $l_c = 2.75$ m long, $w_c = 10$ mm wide fiberglass core. For series measurements, these coils are installed inside grooves milled in G10 supports, held in place by an extruded aluminum profile structure, as shown in Figure 6. The coil used for this work was calibrated at $I = 1000$ A by combining in (20) the flip-coil method (17) with a static SSW measurement, resulting in a calibration coefficient $w_{1000} = 0.7083$ m.

3.3. Magnet Powering

The PSB dipole was powered with a custom 6500 A_{peak}/3200 A_{RMS}, ±200 V converter developed in 1997 by Holec Projects BV, NL, especially for testing CERN pulsed magnets [19]. This is a two-quadrant, current-controlled converter based on an 18 kV transformer and two parallel 6-pulse bridge thyristor rectifiers, preloaded with an additional 12-pulse thyristor rectifier at 150 A so as to be able to output stable controlled current down to zero. The combination of a passive LC filter having an 80 Hz cut-off with an active filter injection choke and amplifier enables suppression of the inherent 300 Hz ripple, while ensuring a good match to rapid transients in the input reference, up to a bandwidth of 10 kHz. In particular, the active filter is designed to always approach the steady-state step response from below, i.e., without any overshoot that may switch the magnet's response onto a different branch of the hysteresis loop. As a result, the overall accuracy of the output current is better than 23 ppm of the full scale for loads up to 200 mH and 120 mΩ.

The excitation current waveforms are plotted in Figure 8. The current was measured with a 10 V/6000 A TOPACC 1.0 Zero-Flux DCCT, also developed by Holec and implemented as an integral part of the current control loop of the converter [19]. The DCCT working principle is based on a high-turn-ratio secondary winding, powered by a feedback-controlled amplifier in order to cancel out the flux generated by the primary, represented by the current to be measured. In this way, the secondary current is used to derive the primary with combined DC offset and linearity errors below 5 ppm with respect to the full scale, in addition to a long-term stability of 5 ppm/year and a thermal stability of 0.25 ppm/°C. The dynamic performance of the DCCT can be expressed in terms of the maximum output slew rate, i.e., 1.5 V μs^{-1}, which corresponds to 900 A μs^{-1}. This is four orders of magnitude higher than the fixed ramp rate of the excitation cycles, that is, 10 kA s^{-1}, resulting in a completely negligible delay. The rated noise level of the DCCT from DC up to 10 kHz is as low as 1.5 ppm of full scale, i.e., about 10 mA. The level of random noise measured in the acquired signal is about 100 mA, i.e., about $3 \cdot 10^{-5}$ of the nominal peak value $I_{\mathrm{max}} = 5400$ A. This can be largely attributed to the cabling and acquisition system, and it remains well below the required accuracy.

The waveforms consist of a sequence of two cycles that start at $I = 0$, reach I_{max}, and then return to $I = 0$. The first is a normalization precycle that has the function of improving the reproducibility of the magnet's response, with a flat-top duration of 1 s. As a rule, normalization precycles should reach at least as high as the highest level of the cycles being run, due to the well-known wiping out property [20]. Typically, the number of normalization cycles necessary to ensure repeatability of the flat-top field within a given tolerance tends to decrease with the level of saturation, which asymptotically represents a uniquely defined reference state. However, based on our own experience, no general quantitative prescription can be formulated; instead, the number of repetitions should be established experimentally. In this case study, a single precycle was enough, plausibly due to the high level of saturation reached by the PSB magnet in operation.

After the precycles, the proper test cycle begins, introducing an intermediate plateau at a level ranging from 500 A to 5400 A. For measurements with pulsed SSW and fixed coil methods, the duration of the plateau was 2 s, which is enough to allow the eddy currents to decay (as discussed in Section 4.4). For measurements using the static SSW method, the duration of the plateau was extended to 120 s. This long duration is needed for at least three back-and-forth wire sweeps, which are averaged to improve the SNR. However, extended DC powering of the PSB dipole is possible only up to a maximum of 3000 A to keep the temperature of the excitation coils below a safe level, typically 60 °C. In all cases, the absolute value of the ramp rate was fixed to ensure stable power converter operation.The measured stability and reproducibility of the current plateaus is about $\sim 10^{-5}$. The strategy of combining a precycle with a fixed maximum current level proved very effective in controlling the effects of hysteresis, and the residual field measured with the static SSW method, $l_{\mathrm{m}} \bar{B}_0 = 0.54$ mTm (i.e., $3 \cdot 10^{-4}$ with respect to the nominal field), was found to be reproducible throughout the test campaign.

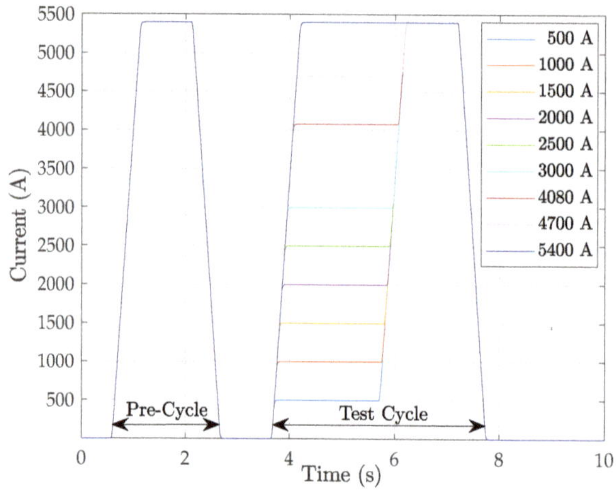

Figure 8. Excitation current waveforms used to power the PSB magnet.

4. Measurement Results and Analysis

In this section, we analyze and compare the results obtained with the three measurement methods. A detailed summary of the main results is given in Table 5.

4.1. Field Uniformity

The uniformity of the transverse field was measured with the fixed coil method as a preliminary step in order to determine the optimal range of the SSW stroke. Measurements were taken at $x = 0, \pm 25, \pm 50, \pm 75$, and ± 90 mm for $I = 1000, 3000$, and 5000 A, and the results normalized to the central values, expressed in terms of

$$\frac{\Delta B}{B}(x, I) = \frac{\int_{-\infty}^{\infty} B_y(x, 0, z, I) dz}{\int_{-\infty}^{\infty} B_y(0, 0, z, I) dz} - 1, \tag{23}$$

are plotted in Figure 9 and listed in Table 2. The table also provides the lowest-order normal field harmonics b_n [5], obtained by performing a least-squares fit of the measured field profiles over the interval ± 90 mm. The uniformity at all current levels is equal to or better than $3.4 \cdot 10^{-4}$, which is within the tolerance given in Table 1. The field error is dominated by the sextupole and decapole components, which are allowed by the nominal symmetry of the main dipole component. However, because of the actual asymmetry of the construction, smaller quadrupole and octupole components are also present, as can be observed in the vicinity of the axis. These and all other even-order harmonic field components do not affect the measurement of the dipole, as long as the SSW stroke remains symmetric with respect to the origin.

In order to obtain a high level of integrated signal, it is always preferable to sweep the stretched wire across the widest possible range, provided the nonuniformity of the field does not impact the result. In our case, we find that the stroke $d^* = 60$ mm, corresponding to the extreme positions of the wire ± 30 mm, gives a maximum relative error with respect to the central value equal to $0.2 \cdot 10^{-4}$, which represents a reasonable compromise. In the following sections, a detailed comparison of the results obtained with shorter strokes of $\pm 10, \pm 20$, and ± 5 mm is given.

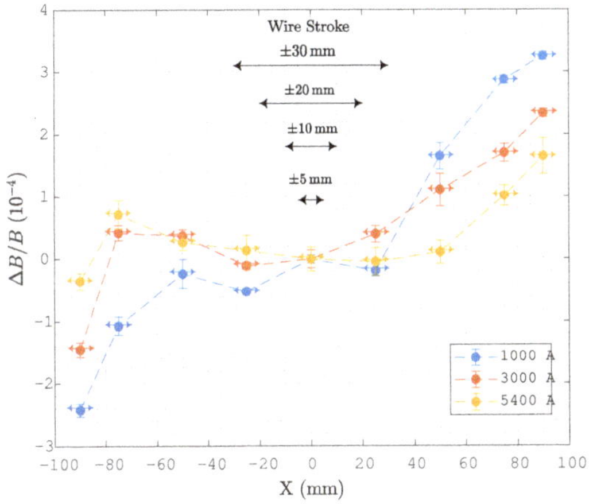

Figure 9. Field uniformity relative to the value at 0 mm, as measured using the fixed coil method. Vertical error bars represent the standard deviation over three measurements, while the horizontal arrows represent the induction coil's width of 10 mm. Dashed lines denote positions of X_1 and X_2 for the different SSW configurations under test.

Table 2. Integrated normal field error harmonics and field uniformity, corresponding to the curves plotted in Figure 9. The harmonics are numbered according to the convention n = 2 quadrupole, n = 3 sextupole, etc., refer to the radius r_0 = 80 mm, and are normalized with respect to the integrated dipole field at each respective current level.

Current	b_2	b_3	b_4	b_5	$\frac{\Delta B}{B}$ (23)	Unit
1000 A	1.00	4.28	−1.87	−3.47	3.4	10^{-4}
3000 A	−0.02	3.90	1.70	−3.34	2.5	10^{-4}
5400 A	−0.77	1.89	1.68	−1.06	3.0	10^{-4}

4.2. Pulsed SSW Results

Let us analyze in detail an example of the flux measured with the pulsed SSW method. As illustrated in Figure 10, we focus on a test cycle with a 1000 A plateau and dynamic measurements taken plotted in the positions $X = \pm 5, \pm 10, \pm 20$, and ± 30 mm. In particular, measurements at $X_1 = +5$ mm, $X_2 = -5$ mm correspond to the nominal width of the induction coil and can be used to make a direct comparison, as discussed in Section 4.6.

Inset (a) shows an example of the voltage output of the wire loop. Thanks to the high field ramp rate, even a single-turn loop provides a peak voltage of the order of one volt, which, combined with the 16-bit acquisition, is enough to guarantee accurate and drift-free integration of the flux. The flux $\Delta\Phi_p$ integrated at all wire positions is plotted vs. time in the inset (b). Due to the high uniformity of the field in the interval scanned by the wire, the curves scale with a good approximation linearly with respect to the wire position. The flux at the end of the cycle flat-top, plotted in inset (c), varies linearly from about 240 to 130 mVs, with a relative residual RMS of the order of 10^{-5}. This should be compared with the ideal expectation illustrated in Figure 4d, considering that the measurement range ± 30 mm captures only a fraction of the total flux in the 180 mm gap width.

The curves $\Delta\Phi_p(t)$ exhibit a peculiar kink at high field, both at the end of the ramp-up and at the beginning of the ramp-down. By instead plotting the flux as a function of the excitation current, as shown in inset (d), it appears that the kink is due to the saturation of the yoke. When the wire is in the rightmost position, $X = +30$ mm, the fall-off due to saturation above ~3500 A is about 40 % with respect to the approximately linear behavior at low current. This level of saturation is much higher than what is observed in the center of the gap (see Section 4.6) because, in this case, the flux measured by the wire loop is dominated by the contribution of the thin side plates. Finally, according to (11), the measured flux curves were subtracted numerically pairwise to obtain the virtual-coil equivalent $\Delta^2\Phi_\mathrm{p}$. Two examples of the results for the cases $X = \pm5$ and ±30 mm are plotted in the insets (e) and (f), respectively. The flux difference curves resemble the current waveform much more closely than the $\Delta\Phi_\mathrm{p}$ curves since the kinks cancel out, as expected. The peak values of $\Delta^2\Phi_\mathrm{p}$ are reported in Table 5 for the case $X = \pm30$ mm.

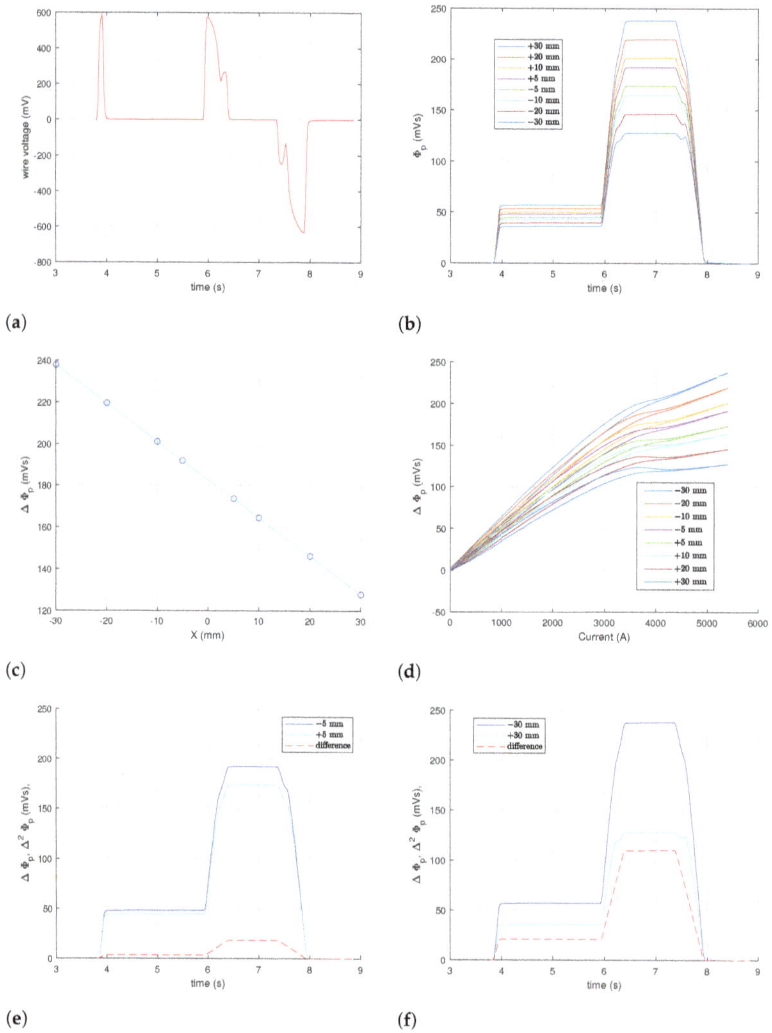

Figure 10. Examples of pulsed SSW measurements of a 1000 A plateau test cycle, taken at eight wire positions between $X = \pm30$ mm. (**a**) Example of raw output voltage at $X = +5$ mm; (**b**) Integrated flux $\Delta\Phi_\mathrm{p}$. (**c**) Peak flux at $t = 7.2$ s vs. wire position. (**d**) Magnetization curves of $\Delta\Phi_\mathrm{p}$ vs. current. (**e**) Difference of $\Delta\Phi_\mathrm{p}$ at ±5 mm. (**f**) Difference of $\Delta\Phi_\mathrm{p}$ at ±30 mm.

4.3. Noise Analysis

The noise of the fixed coil and pulsed SSW was analyzed in the frequency domain as a function of the stroke width d and acquisition sampling rate for a magnetic cycle with a 5400 A plateau. The results, summarized in Table 3, are expressed in terms of the fraction η of the power spectrum that can be attributed to extraneous noise, defined as

$$\eta = \sqrt{\frac{\int_{f_m}^{\infty} |\Psi(f)|^2 df}{\int_{0}^{\infty} |\Psi(f)|^2 df}} \tag{24}$$

where $\Psi(f) = \mathfrak{F}(\bar{B}(t))$ is the Fourier transform of the measured field, its modulus squared provides the power spectrum, and $f_m \approx 80$ Hz is the highest appreciable frequency content of the magnetic cycle waveforms. In practice, (24) is evaluated via the FFT of the discrete field signal, and integration is truncated at the Nyquist frequency.

Overall, the noise level is very low, being on the order of a few 10^{-5} for the fixed coil and the pulsed SSW. The results obtained at the sampling rate of 1 MS/s are plotted as a function of the wire stroke in Figure 11. The pulsed SSW data points fit roughly a relationship of inverse proportionality with respect to d, as one might expect, assuming that the noise level in the raw acquisition remains constant, while the measured flux $\Delta\Phi_p$ is proportional to d. The noise content of the fixed coil measurement is $\eta = 3.6 \cdot 10^{-6}$, which is equivalent to an ideal wire stroke as wide as ~39 mm. The pulsed SSW results are clearly penalized by the intrinsic limitations of the method, which are (a) lower signal levels due to the single-turn nature of the wire loop, and (b) the difference between two separate measurements adds noise components (including quantization noise and excitation current ripple) and common modes quadratically, instead of canceling them out.

Further insight into the relative performance of the two methods can be gained from the amplitude spectra plotted in Figure 12, which include the fixed coil acquired at 1 MS/s and the ±30 mm pulsed SSW acquired at sampling rates down to 10 kS/s. All measurements were carried out with the same anti-aliasing analog filter set at the maximum Nyquist frequency, that is, 500 kHz. For comparison, pulsed SSW signals were also downsampled by the decimation of the 1 MS/s acquisition in the post-processing phase. The spectrum of the induction coil (in black) is characterized by an initial steep slope, up to ~200 Hz, associated with the shape of the magnetic cycle, followed by a roll-off slope about −20 dB/decade, indicating the effect of the coil as a first-order RL low-pass filter. The slope approximately doubles above ~ 80 kHz, marking the lowest-order resonance associated with the self-capacitance of the coil. At the same sampling frequency of 1 MS/s, the pulsed SSW spectrum (in green) exhibits a first-order slope similar to that of the coil but shifted about 15 dB higher, consistently with the lower L/R of the wire, leading to a higher cutoff frequency. By lowering the sampling rate down to 10 kHz while intentionally maintaining the same anti-aliasing filter, we observe that the noise of the pulsed SSW increases by as much as ~50 dB. This result does not depend appreciably on the way downsampling is achieved, either by lowering the ADC settings or by decimation of the signal acquired at the highest rate in post-processing. This increase can be entirely ascribed to aliasing and demonstrates how the single-turn wire loop behaves as a very effective antenna picking up high-frequency disturbances, which, instead, are filtered out by the induction coil. This result highlights the importance of reducing the setup susceptibility to broadband noise, not only, of course, by including an appropriate anti-aliasing filter, but also by minimizing the surface area of the return loop.

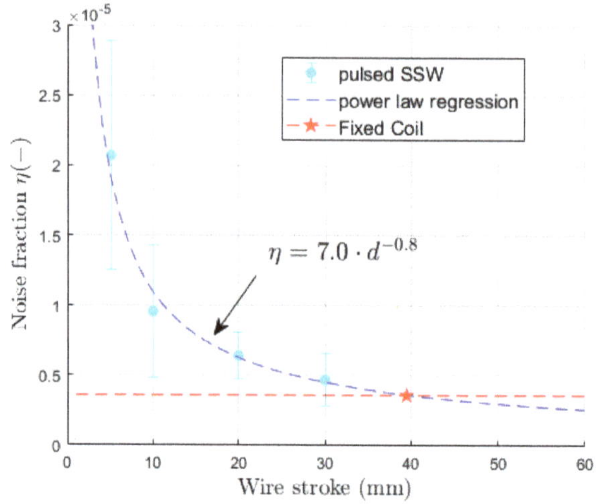

Figure 11. Effect of displacement width on the p-SSW setup's noise content for a 5400 A plateau cycle. Vertical error bars represent the standard deviation over 3 repetitions.

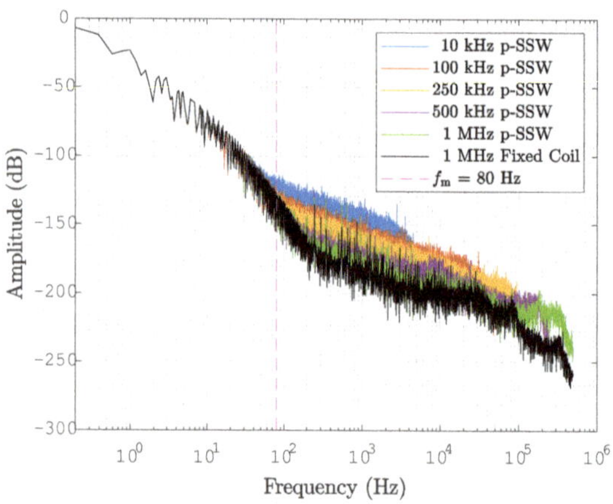

Figure 12. Amplitude spectrum of p-SSW and fixed coil, acquired at different sampling rates. The part of the spectrum above f_m is attributed to noise only.

Table 3. Noise content η (24) for the fixed coil and the ±30 mm pulsed SSW. The quoted uncertainty is the standard deviation over three consecutive repetitions.

Sampling Rate	η Fixed Coil (1 MS/s Decimated)	η Pulsed SSW (1 MS/s Decimated)	η Pulsed SSW (Directly Sampled)	Unit
10 kS/s	5.8 ± 2.6	58 ± 11	56 ± 11	10^{-6}
100 kS/s	3.7 ± 2.0	17 ± 3.5	17 ± 3.5	10^{-6}
250 kS/s	3.6 ± 1.7	10 ± 1.7	9.6 ± 1.7	10^{-6}
500 kS/s	3.6 ± 1.7	5.0 ± 2.0	4.9 ± 1.4	10^{-6}
1 MS/s	3.6 ± 1.7	4.7 ± 1.4	4.7 ± 1.4	10^{-6}

4.4. Eddy Current Decay Transient

The measurement of the amplitude and time constant of eddy current decay transients is necessary to characterize magnets for operation, as well as to establish a time interval on the cycle plateaus where the comparison between static and dynamic measurements is meaningful. The magnetic cycles used for our tests, as those generally used for synchrotron operation, are composed of successions of linear current ramps and plateaus where eddy currents develop at the start of ramps, and decay at their end. As an example, we focus on the transient at the end of a ramp-up to 5400 A, depicted in Figure 13. The normalized current $\tilde{I}$, normalized field $\tilde{B}$, and their difference δ are defined as follows:

$$\tilde{I}(t) = \frac{I(t)}{I(t_e)}, \quad \tilde{B}(t) = \frac{\bar{B}(t)}{\bar{B}(t_e)}, \quad \delta(t) = \tilde{I}(t) - \tilde{B}(t) \tag{25}$$

where the average field is measured with the fixed coil, and the time t_e represents steady-state conditions, when the eddy current can be assumed to have fully decayed. The difference δ represents the relative impact of nonlinear effects, and it can be clearly seen to undergo an exponential decay according to the following expression:

$$\delta(t) = -A_0 e^{-\frac{t-t_0}{\tau}} \tag{26}$$

where $t_0 = 4.2$ s is the end of the ramp-up, A_0 represents the peak relative field error, and τ is the time constant of the decay [14]. Equation (26) has been fitted with least squares to measurements taken with the induction coil and the pulsed SSW method with the wire stroke range $\pm 5, \pm 10, \pm 20$, and ± 30 mm, and the results are reported in Table 4. The variance of the fitted parameters is lowest for the induction coil and decreases as the SSW wire stroke increases, which is consistent with the levels of signal noise appearing in the example shown in Figure 14a. This result is also in agreement with the findings reported in Figure 11, which shows the signal noise decreasing with d.

Measurements were repeated for cycles with plateaus at different levels, and the results are also listed in Table 4. The fitted parameters are reported in Table 5, and the RMSE residuals of the exponential fitting are plotted in Figure 13 as a function of the integrated field. The accuracy of the fitting improves with increasing d and, even more substantially, with the inverse of the field level.

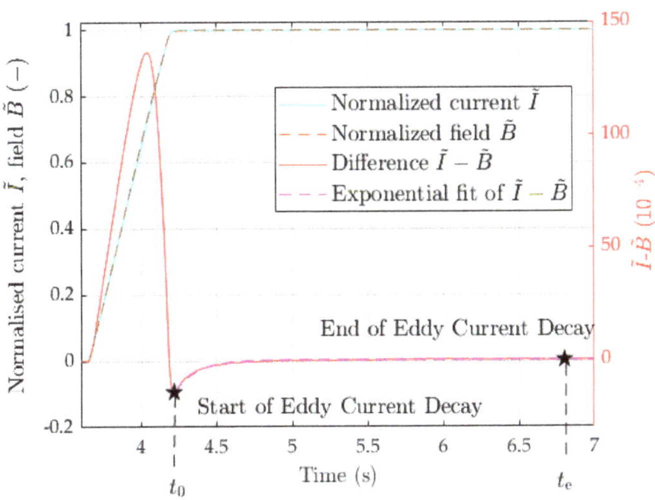

Figure 13. Normalized current and field showing the eddy current decay.

Table 4. Dynamic performance of the p-SSW and fixed coil setups. Values were calculated from the results across 81 measurements. Relative values are presented with respect to the magnetic field at 5400 A (1.835 Tm).

Measurement Method	Current (A)	τ (ms) Avg. $\pm\ \sigma$	A_0 (10^{-4}) Avg. $\pm\ \sigma$	Fit Error (10^{-4}) RMS
Fixed Coil	500	48 ± 17	3.0 ± 0.2	0.2
Fixed Coil	1000	93 ± 15	4.4 ± 0.3	0.2
Fixed Coil	1500	121 ± 14	4.9 ± 0.2	0.2
Fixed Coil	2000	85 ± 18	5.1 ± 0.2	0.2
Fixed Coil	2500	99 ± 11	5.5 ± 0.3	0.2
Fixed Coil	3000	116 ± 7	6.2 ± 0.2	0.2
Fixed Coil	4080	125 ± 7	6.6 ± 0.3	0.2
Fixed Coil	4700	142 ± 3	7.1 ± 0.2	0.2
Fixed Coil	5400	146 ± 3	14.9 ± 0.3	0.2
$\pm$5 mm p-SSW	5400	137 ± 42	15.2 ± 1.4	0.8
$\pm$10 mm p-SSW	5400	143 ± 21	14.7 ± 0.9	0.4
$\pm$20 mm p-SSW	5400	149 ± 14	14.6 ± 0.6	0.3
$\pm$30 mm p-SSW	5400	144 ± 9	14.5 ± 0.4	0.3

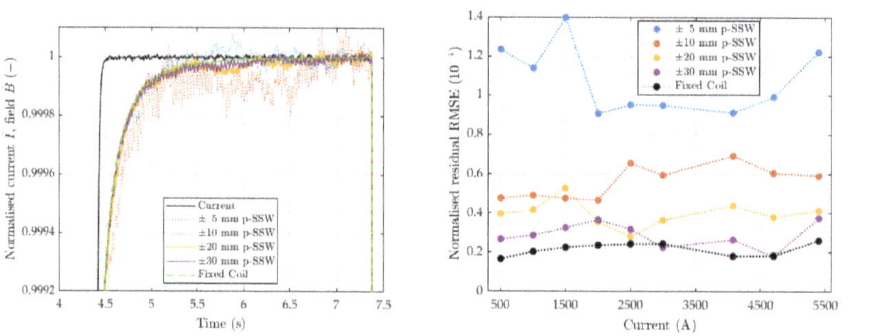

(**a**) (**b**)

Figure 14. (**a**) Normalized magnetic field measured on the 5400 A test cycle. (**b**) Normalized RMS residual calculated across all measurement plateaus. Values are relative to corresponding magnetic field.

4.5. Comparison with Static SSW

The performance of the combined SSW method was compared with that of the static SSW, which is the reference method for DC conditions. For both methods, the integrated field measured with a stroke of $\pm$30 mm is reported in Table 5 for each plateau level from 500 to 3000 A, along with the standard deviation over three consecutive repetitions. The relative difference over all current levels has a systematic (average) value of $3.8 \cdot 10^{-5}$, while the standard deviation is $1.2 \cdot 10^{-4}$. These results confirm that the performance of the combined SSW method meets operational requirements, which are typically of the order of 10^{-4}.

The comparison is also made in terms of the integral transfer function, illustrated in Figure 15 and defined as the ratio between the integral field and the current:

$$TF(I) = l_m \frac{\bar{B}(I)}{I} \tag{27}$$

The transfer function is a very useful tool, not only because it allows machine operators to easily set the current necessary to achieve a given field, but also because it helps to visualize even minor deviations from the desired linear behavior, which ideally corresponds to a simple flat line. Moreover, by normalizing with respect to the current, the confounding impact of current reproducibility is eliminated inherently. This is reflected in the systematic relative difference with respect to the static SSW, which reduces to $2.1 \cdot 10^{-5}$, while the standard deviation is $1.3 \cdot 10^{-4}$.

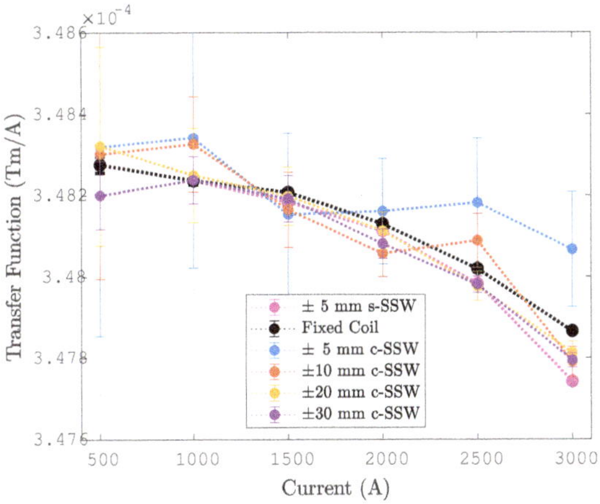

Figure 15. Transfer function measured with the fixed coil, static, and combined SSW in the current range allowing for DC powering. Error bars represent twice the standard deviation over three consecutive repetitions.

Table 5. Summary of the main results obtained with the different measurement methods. All SSW measurements were performed with a ±30 mm stroke.

Current (A)	Integral Field $l_m\bar{B}$ (Tm)		Integral Transfer Function $\frac{l_m\bar{B}}{I}$ ($10^{-4}\frac{Tm}{A}$)		Induction Coil Calibration w_{eff} (mm)	
	s-SSW	c-SSW	s-SSW	c-SSW	s-SSW	c-SSW
0	0.00054	-	-	-	-	-
500	0.17414 ± 0.00001	0.1741 ± 0.00005	3.4828 ± 0.0002	3.4820 ± 0.0008	-	708.27 ± 0.24
1000	0.34824 ± 0.00001	0.34823 ± 0.00006	3.4824 ± 0.0001	3.4824 ± 0.0006	708.30 ± 0.03	708.26 ± 0.17
1500	0.52228 ± 0.00001	0.52228 ± 0.00008	3.4818 ± 0.0000	3.4819 ± 0.0006	-	708.27 ± 0.17
2000	0.69623 ± 0.00001	0.69616 ± 0.00007	3.4811 ± 0.0000	3.4808 ± 0.0004	-	708.29 ± 0.10
2500	0.86996 ± 0.00001	0.86995 ± 0.00005	3.4798 ± 0.0000	3.4798 ± 0.0002	-	708.29 ± 0.06
3000	1.04323 ± 0.00001	1.04338 ± 0.00005	3.4774 ± 0.0000	3.4779 ± 0.0002	-	708.34 ± 0.05
4080	-	1.41406 ± 0.00006	-	3.4659 ± 0.0001	-	708.43 ± 0.03
4700	-	1.61743 ± 0.00004	-	3.4413 ± 0.0001	-	708.50 ± 0.03
5400	-	1.83548 ± 0.00006	-	3.3990 ± 0.0001	-	708.56 ± 0.03

4.6. Comparison with the Fixed Coil Setup

The comparison between pulsed SSW and fixed coil is analyzed in detail with the aim of deriving the coil calibration across the whole range of currents. We remark that adding or not the residual field to the pulsed SSW results has no impact, provided, of course, that the same is applied to the coil results. The transfer functions measured over the entire set of test cycles with the fixed coil and the pulsed SSW are plotted in Figure 16, separately, for the four-stroke lengths ±5, ±10, ±20, and ±30 mm. The calibration used for the fixed coil is that obtained with the flip-coil method.

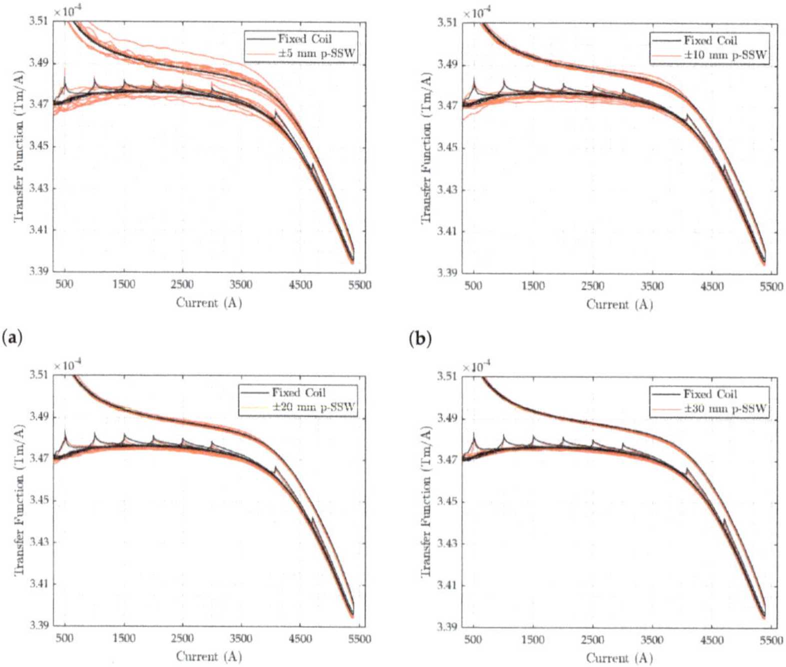

(a) (b) (c) (d)

Figure 16. Transfer function of cycles with plateaus from 500 to 5400 A measured with the p-SSW and the fixed coil. **(a)** $w_d = \pm 5\,\text{mm}$. **(b)** $w_d = \pm 10\,\text{mm}$. **(c)** $w_d = \pm 20\,\text{mm}$. **(d)** $w_d = \pm 30\,\text{mm}$.

As a general characteristic, the transfer function is approximately flat on the ramp-up between 1000 and 3500 A, which is indicative of the desired linear response. At higher currents, the curve drops by about 2.3% due to iron saturation. This drop, measured in the magnet gap, depends on the average level of saturation of the whole iron yoke and is, therefore, one order of magnitude below the drop registered by individual $\Delta\Phi_p$ measurements, which capture the much higher level of saturation of the side plates. On the intermediate plateaus and on the flat-top of the excitation cycles, we observe an increase in the field at constant current due to the decay of the eddy currents that screen the field in the gap. As discussed in Section 4.4, steady-state conditions apply at the end of the decay transient and the corresponding values are listed in Table 5. The ramp-down of the hysteresis loop is not relevant for synchrotrons such as the PSB, as no beam is circulating. Both branches of the loop diverge as the current approaches zero, due to the presence of a residual field. All these features are captured equally well by both measurement methods. Consistent with the results reported in the previous sections, the reproducibility of the pulsed SSW improves dramatically as the stroke lengthens and the field increases. At injection ($I \approx 1100$ A), the relative standard deviation changes from 0.6% at ±5 mm (Figure 16a) to 0.1% at ±30 mm (Figure 16d), further dropping to 0.03% at

5400 A. This result confirms the choice of the wire stroke $d^* = 60$ mm as the optimal one, which we also used as the basis to derive the coil calibration.

The comparison between the two methods was carried out using the ratio ρ between the respective transfer functions, which can be expressed as follows:

$$\rho(I) = \frac{TF_{\text{coil}}(I)}{TF_{\text{p}}(I)} = \frac{\bar{B}_{\text{coil}}(I)}{\bar{B}_{\text{p}}(I)} = \frac{d^*}{w_{1000}} \frac{\Delta\Phi_{\text{coil}}(0, I)}{\Delta^2\Phi_{\text{p}}(-\frac{d^*}{2}, \frac{d^*}{2}, 0, I)} \tag{28}$$

The ratio corresponding to the ±30 mm pulsed SSW is plotted as a function of time in Figure 17a and of current in Figure 17b, where only the 1000 A plateau cycle is shown. As expected, at low current, $\rho \approx 1$, except for a discontinuity corresponding to the decay of the eddy currents on the plateau. However, as the current increases above 2000 A, the ratio increases almost linearly by as much as 0.04% at 5400 A, which denotes a substantial change in coil calibration with field level. The effective coil width can be derived by combining (21) and (29) to obtain

$$w_{\text{eff}}(I) = \rho(I) w_{1000} \tag{29}$$

The results are listed in Table 5 and plotted in Figure 18a, where the error bars represent the standard deviation over the set of nine test cycles. As a first approximation for practical use, the effective width can be reasonably well fitted by the following linear expression:

$$w_{\text{eff}}(I) = w_0 + w' \frac{I}{I_{\text{max}}} \tag{30}$$

where $w_0 = 0.7082$ m and $w' = 4.0 \cdot 10^{-4}$. The initial value is very close to the one obtained by the flip-coil calibration method, i.e., 0.7083 m. This could be expected to apply throughout the linear part of the range. However, an accurate evaluation in this region is impeded by the high standard deviations due to low signal levels. The increase in the effective width at a high field can be explained by one, or both, coil ends being marginally wider than the central part, thus being able to capture more of the high-field leakage due to saturation. Such a geometrical imperfection is actually often observed in coils wound with multiconductor cables due to their stiffness, which makes it relatively difficult to bend the, around the 90° corners at the ends of the winding form. Regardlesss of the physical origin, using the calibration (30) allows the elimination of any systematic current-dependent discrepancy with respect to pulsed SSW measurements, as shown in Figure 18b.

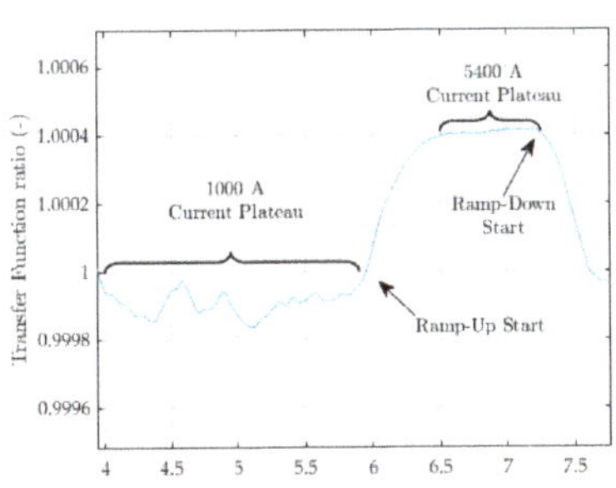

(a)

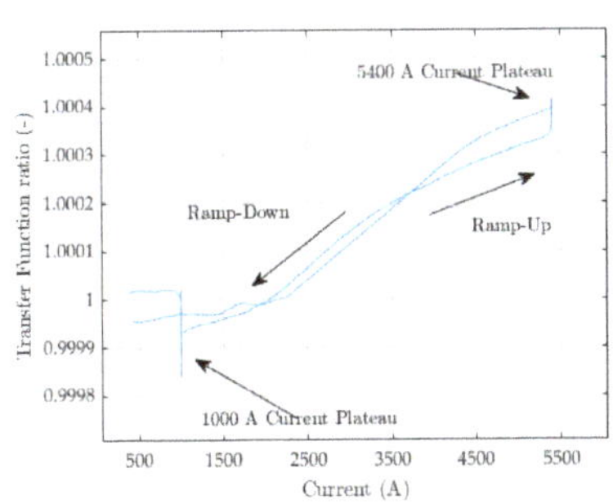

(b)

Figure 17. Ratio between the transfer functions measured with fixed coil and the ±30 mm pulsed SSW over a 1000 A test cycle. (**a**) As a function of time. (**b**) As a function of current.

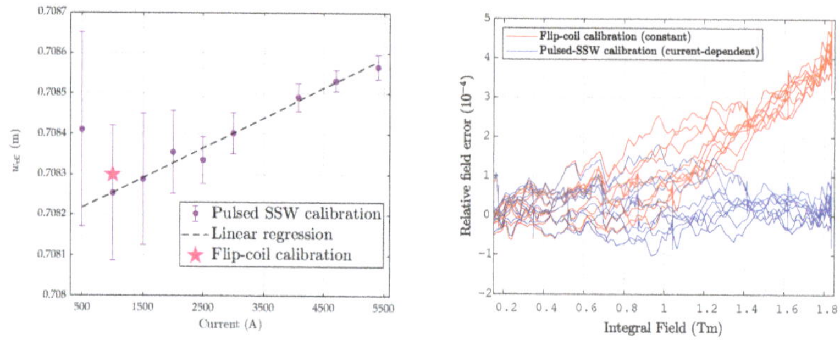

(a) (b)

Figure 18. (**a**) Induction coil effective width derived from the difference with respect to ±30 mm combined SSW measurements. Vertical error bars represent 2-sigma standard deviation of over 9 measurements. (**b**) Relative difference between the test cycles measured with the fixed coil and ±30 mm c-SSW setups. The plot shows the improvement in accuracy due to the current-dependent $w_{\text{eff}}(I)$ calibration.

5. Conclusions and Future Work

In this paper, we presented a novel method for measuring the dynamic integral field of an accelerator magnet using the SSW measurement bench. The pulsed SSW method was shown to accurately characterize the integral field of the CERN PSB bending dipole over its entire range of operation, which was not possible using the previous static implementation. Similar to static SSW, the measurement uncertainty was shown to depend strongly on the wire stroke. Therefore, we suggest identifying an optimal stroke width that meets the measurement precision without compromising the uncertainty from the field uniformity. The pulsed SSW shows no significant systematic difference compared to the static SSW, with a relative RMS difference of less than $2 \cdot 10^{-4}$ across their shared field range. This level of performance is closely comparable to that of other instrumentation and generally up to beamline requirements.

No implementation of the SSW method can clearly compete with a multiturn induction coil in terms of SNR, immunity to high-frequency external perturbations, and practicality of use in a context requiring multiple dynamic measurements of different magnetic cycles. However, the pulsed SSW method offers crucial advantages with its extreme geometrical adaptability and accuracy of the width over the flux-capturing area. Moreover, the dynamic performance was shown to improve with the rate of change of the magnetic field under conditions that make static SSW measurements more difficult. Overall, this method represents a valid general alternative to the static SSW, demonstrating its viability as a metrological reference when calibrating an induction coil as a function of the iron core's saturation level. Future planned developments include widening the wire stroke into the nonuniform field region and the estimation of the field harmonic errors; upgrading the signal conditioning electronics to improve EMI robustness and SNR; and extending the method to quadrupole and higher-order multipole magnets.

Author Contributions: Conceptualization, M.B.; methodology, C.P.; investigation, J.V.W. and A.P.; data curation, J.V.W.; writing—original draft preparation, J.V.W.; writing—review and editing, M.B.; supervision, N.S. All authors have read and agreed to the published version of the manuscript.

Funding: This research received no external funding.

Data Availability Statement: The datasets presented in this article are not readily available because of time limitations. Requests to access the datasets should be directed to the corresponding author.

Acknowledgments: The authors wish to thank Antony Newborough (CERN) for his support of the measurement campaign and for his comments on the paper.

Conflicts of Interest: The authors declare no conflicts of interest.

References

1. Bryant, P.; Johnsen, K. *The Principles of Circular Accelerators and Storage Rings*; Cambridge University Press: Cambridge, UK , 2005. [CrossRef]
2. Zickler, T. Basic design and engineering of normal-conducting, iron-dominated electromagnets. In Proceedings of the 2009 CAS-CERN Accelerator School: Specialised Course on Magnets, Bruges, Belgium, 16–25 June 2009; CERN: Geneva, Switzerland, 2011. [CrossRef]
3. Henrichsen, K.N. Overview of magnet measurement methods. In Proceedings of the 1997 CAS-CERN Accelerator School: Measurement and Alignment of Accelerator Magnets, Anacapri, Italy, 11–17 April 1997; CERN: Geneva, Switzerland, 1998; pp. 128–142. [CrossRef]
4. Tumanski, S. Induction coil sensors—A review. *Meas. Sci. Technol.* **2007**, *18*, R31. [CrossRef]
5. Walckiers, L. Magnetic measurement with coils and wires. In Proceedings of the 2009 CAS-CERN Accelerator School: Specialised Course on Magnets, Bruges, Belgium, 16–25 June 2009; CERN: Geneva, Switzerland, 2009; pp. 357–386 . [CrossRef]
6. Sanfilippo, S. Hall probes: Physics and application to magnetometry. In Proceedings of the CERN Accelerator School CAS 2009: Specialised Course, Magnets, Bruges, Belgium, 16–25 June 2009; CERN-2010-004; CERN: Geneva, Switzerland, 2010; pp. 423–462. [CrossRef]
7. Albright, S. *PS Booster Magnetic Cycles to 1.4 and 2 GeV after LS2*; Techreport EDMS: 1770413; PSB-OP-ES-0001; CERN: Geneva, Switherland, 2018.
8. Newborough, A.; Buzio, M.; Chritin, R. Upgrade of the CERN Proton Synchrotron Booster Bending Magnets for 2 GeV Operation. *IEEE Trans. Appl. Supercond.* **2014**, *24*, 0500304 . [CrossRef]
9. Moritz, G. Eddy currents in accelerator magnets. In Proceedings of the 2009 CAS-CERN Accelerator School: Specialised Course on Magnets, Bruges, Belgium, 16–25 June 2009; CERN: Geneva, Switzerland, 2011; pp. 103–140. [CrossRef]
10. Zangrando, D.; Walker, R.P. A stretched wire system for accurate integrated magnetic field measurements in insertion devices. *Nucl. Instrum. Methods Phys. Res. Sect. Accel. Spectrometers Detect. Assoc. Equip.* **1996**, *376*, 275–282. [CrossRef]
11. DiMarco, J. *MTF Single Stretched Wire System*; Technical Report; Fermilab: Batavia, IL, USA, 1996.
12. DiMarco, J.; Glass, H.; Lamm, M.J.; Schlabach, P.; Sylvester, C.; Tompkins, J.C.; Krzywinski, J. Field alignment of quadrupole magnets for the LHC interaction regions. *IEEE Trans. Appl. Supercond.* **2000**, *10*, 127–130. [CrossRef]
13. Petrone, C.; Russenschuck, S. Wire Methods for Measuring Field Harmonics, Gradients and Magnetic Axes in Accelerator Magnets. Ph.D. Thesis, University of Sannio, Benevento, Italy, 2013.
14. Moritz, G. Mechanical Equipment. In Proceedings of the 1997 CAS-CERN Accelerator School: Measurement and Alignment of Accelerator Magnets, Anacapri, Italy, 11–17 April 1997; CERN: Geneva, Switzerland, 1998; pp. 250–272. Available online: http://cds.cern.ch/record/1246521 (accessed on 19 June 2024).
15. Bottura, L.; Henrichsen, K.N. Field Measurements. In *CAS—CERN Accelerator School on Superconductivity and Cryogenics for Accelerators and Detectors*; CERN: Geneva, Switzerland, 2002; pp. 118–148. [CrossRef]
16. Buzio, M. Fabrication and calibration of search coils. In Proceedings of the 2009 CAS-CERN Accelerator School: Specialised Course on Magnets, Bruges, Belgium, 16–25 June 2009; CERN: Geneva, Switzerland, 2009; pp. 387–421. [CrossRef]
17. Asner, A.; Brianti, G.; Giesch, M.; Lohmann, K.D. The PS Booster Main Bending Magnets and Quadrupole Lenses. In Proceedings of the 3rd AnnualInternational Conference on Magnet Technology MT-3, DESY, Hamburg, Germany, 19–22 May 1970.
18. METROLAB. *PDI 5025 User Manual*, Version 2.2 ; METROLAB Instruments SA: Geneva, Switzerland, 2000.
19. Lieferink, J.G. *User Manual—Test Magnet CERN Power Converter with High Precision, Pulsed Current Controlled Output*, Version 0.0; HOLEC Projects BV: Hengelo, The Netherlands, 1997.
20. Mayergoyz, I.D. Mathematical Models of Hysteresis. *IEEE Trans. Magn.* **1986**, *22*, 603–608. [CrossRef]

sensors

MDPI

Article

Measurement of Hydraulic Fracture Aperture by Electromagnetic Induction

Mohsen Talebkeikhah [1], Alireza Moradi [2] and Brice Lecampion [1,*]

1 Institute of Civil Engineering, Gaznat Chair on Geo-Energy, École Polytechnique Fédérale de Lausanne (EPFL), 1015 Lausanne, Switzerland; m.talebkeikhah@gmail.com
2 Faculty of Engineering, Tarbiat Modares University (TMU), Tehran 14115-111, Iran; amoradi@blueheartenergy.com
* Correspondence: brice.lecampion@epfl.ch

Abstract: We present a new method for accurately measuring the aperture of a fluid-driven fracture. This method uses an eddy current probe located within a completion tool specifically designed to obtain the fracture aperture in the wellbore at the location where the fluid is injected into the fracture. The probe induces an eddy current in a target object, producing a magnetic field that affects the overall magnetic field. It does not have any limitations with respect to fluid pressure and temperature within a large range, making it unlike other methods. We demonstrate the accuracy and performance of the sensor under laboratory conditions. A hydraulic fracture experiment in a porous sandstone is conducted and discussed. The obtained measurement of the evolution of the fracture inlet aperture by the eddy current probe during the multiple injection cycles performed provided robust information. The residual fracture aperture (after the test) measured by the probe is in line with estimations from image processing of X-ray CT scan images as well as a thin-section analysis of sub-parts of the fractured specimen. The robustness and accuracy of this electromagnetic induction probe demonstrated herein under laboratory conditions indicate an interesting potential for field deployment.

Keywords: fracture mechanics; hydraulic fracturing; aperture measurement; porous sandstone; eddy current probe; eddy current; fracture opening; fracture width

1. Introduction

Hydraulic fracturing is widely used to enhance production from oil and gas wells. It consists of creating fractures within a reservoir, which are then propped open by the injection of sand to improve the overall hydraulic transmissivity of the reservoir in the area surrounding the well. Fractures primarily govern fluid flow within tight rock formations, essential in various applications like oil and gas extraction, geothermal energy production, CO_2 sequestration, and underground nuclear waste storage. A precise understanding of how fractures propagate under fluid injection is critical for the efficiency and safety of these geo-energy applications. Understanding fracture aperture and its variability is essential for validating hydraulic fracture models and comprehending the extent of fracturing and the characteristics of the proppant used to keep the fracture open.

Records of the injection fluid pressure are the simplest measurements available to understand fracture propagation. It, however, provides only partial information and is often insufficient to properly assess the growth of hydraulic fractures. Active and passive seismic monitoring are often used to estimate the geometrical extent of the fracture [1–4] with more or less accuracy depending on the measurement location. The measurement of the induced elastic strain or rotation associated with hydraulic fracture propagation at distant locations from the created fracture provides a robust estimation of fracture volume and orientation. However, it lacks sensitivity to fracture shapes if measurements are not located in the near field [5,6]. Direct and accurate measurements of fracture aperture are particularly challenging in non-transparent materials, like rocks, but would provide important additional information to better monitor fracture propagation.

Some measurement methods can measure the fracture shape and aperture profile. For instance, Bunger [7] developed a photometry method for measuring fracture aperture in transparent materials. Groenenboom et al. [8–10] observed the dispersion of compressional waves across hydraulic fractures, which is indicative of fracture width and suitable for thin fluid-filled layers but necessitates both compressional and shear wave measurements. Furthermore, geophysical imaging techniques, such as seismic reflection, electrical methods, and ground-penetrating radars, have been utilized to detect subsurface features [11].

Distributed strain measurements from fiber optics placed in wellbores have recently been deployed in hydraulic fracturing operations. Chen et al. [12] developed a fiber optic sensing scheme to image 3D strain fields in concrete during hydraulic fracturing. Leggett et al. [13] and Yang et al. [14] employed hydraulic fracture experiments with embedded optical strain sensors and fiber Bragg grating strain sensors, respectively. Qasim et al. [15] conducted a field data analysis using distributed fiber optic technology in various international basins. Additionally, precise point-wise measurement methods exist but can only measure the aperture at specific points, such as injection points. Kakurina et al. [16] used a specific fiber-optic probe (the SIMFIP probe) to measure 3D displacement at a given location in boreholes. Additionally, De Pater et al. [17] used a Linear Voltage Differential Transformer (LVDT) to measure fracture opening during propagation and closure in the laboratory. A similar technique was used by Warpinski [18], who measured fracture width during hydraulic fracture propagation using LVDTs in boreholes at a U.S. Nevada test site. Furthermore, Overbey et al. [19], Palmer and Sparks [20], Darilek [21] utilized borehole video cameras or television systems to "estimate" fracture openings. Smith et al. [22] conducted measurements using a downhole closed-circuit television camera during hydraulic fracturing propagation stimulation in an oil-bearing sandstone formation. Moreover, Medlin and Masse [23] utilized a capacitance bridge to continuously measure fracture aperture in a cubic sandstone specimen in the laboratory.

In this study, we propose a robust method for directly measuring the aperture at the fracture inlet. We present its mechanism and calibration procedure under various conditions. Subsequently, we conduct a laboratory-scale hydraulic fracture experiment on a cubic sandstone block. Throughout these experiments, we monitor both pressure and aperture at the inlet of the fracture within the rock sample. The obtained results will be validated by comparing the probe measurement with the sample's 3D X-ray computer tomography (CT) scan and thin-section analysis, demonstrating the method accuracy and reliability.

2. Problem Description

The main goal of this research is to conduct a laboratory-scale hydraulic fracturing experiment and evaluate the practicality of a new method for measuring fracture aperture at the fracture inlet. Our approach involves placing a probe directly at the point where the fluid is injected to measure the fracture aperture. This injection point is positioned at the center of a cubic sandstone block in the middle of a drilled well within the specimen. However, there are several challenges in this process. The drilled well has a diameter of approximately 20 mm, requiring the completion tool, which contains the probe as well as the fluid injection line, to be compact yet accessible at a depth of 125 mm.

Additionally, the measurement process must be sufficiently fast to capture the real-time evolution of fracture aperture during an experiment. It is essential to maintain a high level of accuracy and acquisition frequency. The probe must precisely measure fracture apertures on the order of micrometers (μm), with an accuracy below 1 μm, and the measurement frequency must be sufficient to record all events. Moreover, the probe must withstand high pressures, as injection fluid pressures can exceed 60 MPa, be sealed to the injection fluid, and resist any chemical reactions from the components. Lastly, our design must allow for the probe's reusability by making it easy to remove from the completion tool after the experiment.

3. Gap Measurement via Electromagnetic Induction

3.1. Probe Description

An electromagnetic inductive probe called an Eddy Current (EC) probe is a type of contact or non-contact probe that utilizes electromagnetic induction to measure a target object's position, displacement, or distance. This probe possesses several advantages, such as high sensitivity, strong anti-interference ability, and resistance to oil, water, and other media, ensuring reliable operation over extended periods. The probe's non-contact measurement and rapid response make it suitable for long-term, real-time monitoring. It is based on the principle that when a magnetic field is introduced into a conductive material, it will induce eddy currents (according to Faraday's law of electromagnetic induction) that generate their magnetic field, which, according to Lenz's law, opposes the original magnetic field. As a result, there is a change in the overall magnetic field and, accordingly, the impedance value of the coil in the probe. The magnitude of this change is directly proportional to the distance between the probe and the target object [24]. The eddy current probe has numerous applications in various industries, such as industrial automation, aerospace, and automotive. For instance, it can be used to inspect metal parts, measure the thickness of metal plates, and track the position of moving machinery.

The mechanism employed by the eddy current probe to measure distance, as illustrated in Figure 1, involves producing an alternating current (AC) in a coil of wire inside the probe through an oscillator. This current generates a magnetic field (blue lines), which induces eddy currents (black lines) in the conductive plate. The eddy currents then produce their magnetic field (red lines), which superimposes the original magnetic field produced by the coil [25]. The resulting interaction changes the coil's impedance, which can be measured and utilized to determine the distance between the probe and the plate. The probe's electronics convert this measurement into an analog or digital signal that a computer or other control system can read. In this study, a Miran eddy current probe, provided by Miran Technology Co., Ltd. (Shenzhen, China) (http://www.miransensor.com/ accessed on 1 September 2022), was utilized. Table 1 summarizes the main properties of this EC probe.

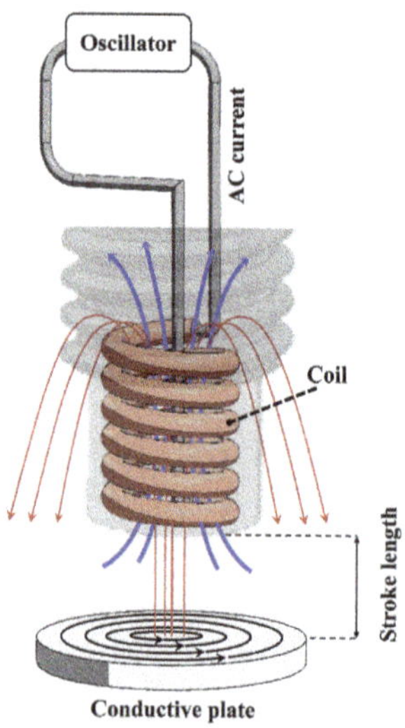

Figure 1. Schematic representation of the operational mechanism of an eddy current (EC) probe.

Table 1. Main specifications of the EC probe used in this study.

Property	Value	Unit	Property	Value	Unit
Stroke length	1	mm	Probe diameter	6	mm
Probe length	5	mm	Temperature range	−30∼150	C
Resolution	0.05	µm	Output signal range	0∼5	V

3.2. Calibration

3.2.1. Calibration Setup

Large fluid pressures may affect the behavior of the EC probe during hydraulic fracturing experiments. Additionally, the choice of the conductive plate material, whether steel, iron, or aluminum, also impacts the EC probe readings. In all cases, it is essential to calibrate the EC probe's linear output by measuring the slope of the linear function between the EC probe value and the corresponding physical distance. To do so, a calibration device has been designed and built as shown in Figure 2. This calibration device positions the EC probe in front of an aluminum conductive plate that simulates the conditions of the primary experiment. The conductive plate can be moved vertically by rotating the bottom part of the device along its thread, creating a controlled movement in front of the EC probe tip that can be compared to the EC probe reading. The system is connected to an injection line, an interface vessel containing injection fluid (glycerol), and an air riser line that removes air during fluid pressurization. A specialized micrometer is attached to the calibration device, which can accurately measure the exact distance variation in the system. This length difference corresponds to the vertical movement caused by the bottom rotation and conductive parts around the thread. The volume between the EC probe and the conductive plate is pressurized by fluid injection during the calibration test. The injection fluid is sourced through a $1/8''$ (in) line tube connected to an interface vessel. This interface vessel is then connected to a high-pressure pump that injects water at a constant rate/pressure. The interface vessel conveys the water pressure from the pump to the injection fluid, which can be glycerol, glucose, silicon oil, or any other suitable fluid. At various constant pressures, the manual rotation of the calibration device's bottom part creates a controlled displacement between the EC probe and the conductive plate. The EC probe output value is recorded alongside the corresponding displacement value measured by the micrometer, providing a correlation between the EC probe readings and the actual displacement for each pressure level.

3.2.2. Calibration Results

The calibration device filled with glycerol is intended to be used during the hydraulic fracturing test. The pressure was kept constant during each calibration phase, during which the bottom part of the calibration device was manually rotated. The corresponding values from the micrometer and EC probe readings are recorded while rotating the bottom part until the EC probe reading changes from 0 to 1. This procedure is repeated several times for different fluid pressure values from 0.1 to 20 MPa.

The results of this calibration are reported in Figure 3. Each set of constant pressure data was fitted with a line, revealing that all lines share the same constant slope. The difference in the intercept value (b) between the lines for different pressures can be attributed to the calibration device's deformation, especially at higher pressure levels.

This is inconsequential, as the difference between the current and the initial EC probe reading (prior to the start of fluid injection and fracturing) is taken to determine the aperture of a hydraulic fracture at the wellbore during the hydraulic fracturing experiment. Therefore, the required calibration is solely given by the slope value of the EC probe–micrometer value $m = 0.604$. Afterward, the EC probe is placed within the completion tool at the injection point inside the rock, as detailed in the next section. After conducting the experiment, the obtained EC probe values are multiplied by the slope value of the line to obtain the actual fracture aperture values.

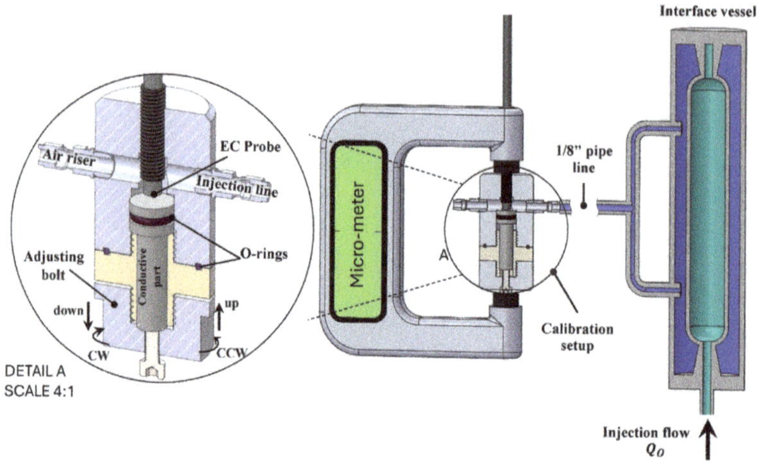

Figure 2. Schematic of the setup designed for calibrating the EC probe with the condition of the hydraulic fracturing experiment.

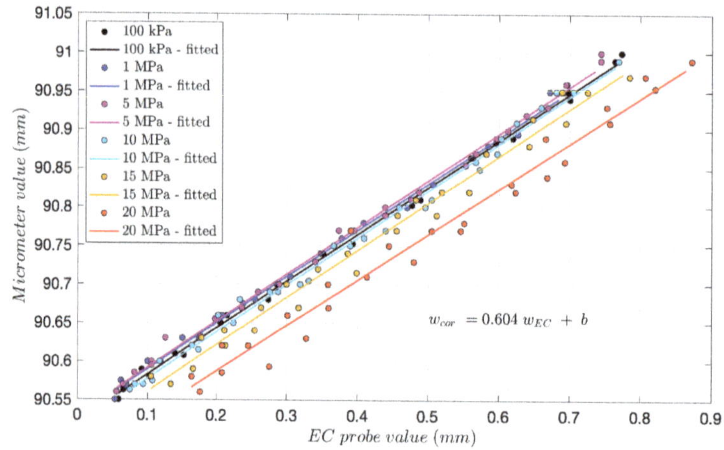

Figure 3. EC probe readings and corresponding micrometer results obtained in the calibration device at various fluid pressures of 0.1, 1, 5, 10, 15, and 20 MPa.

4. Hydraulic Fracture Experimental Setup

4.1. Materials

A cubic block of Molasse sandstone was selected to conduct a hydraulic fracturing experiment on a laboratory scale. The sandstone was sourced from the Molasse de Villarlod quarry in Switzerland and had a nominal length of 250 mm. This sandstone sample is a porous and permeable rock, which makes it an ideal choice for the experiment. By selecting a porous and permeable sample, the experiment aimed to simulate the opening and closure of a fracture, as the fluid inside the fracture would diffuse into the surrounding porous media.

4.2. Experimental Setup Description

The experimental setup consists of a polyaxial frame designed to investigate hydraulic fracture propagation under a true-triaxial state of stress. A schematic of the setup is given in Figure 4. Triaxial confining stresses are applied via three independent symmetric pairs of flat-jack (σ_3) and pistons (σ_1, σ_2). The fracturing fluid is injected through a high-pressure line 1/8″ in diameter, cemented with epoxy in a centrally drilled hole within the block. An axisymmetric notch is created at the middle of the well to facilitate fracture initiation and enhance fracture planarity. A planar hydraulic fracture is then propagated by injecting high-pressure viscous fluid from the notch to the sample's boundary. Throughout the experiment, injection fluid rates and pressures are continuously measured. We refer to [3,26–29] for a detailed description of the experimental setup. The main addition discussed here relates to the addition of an EC probe inside the central wellbore.

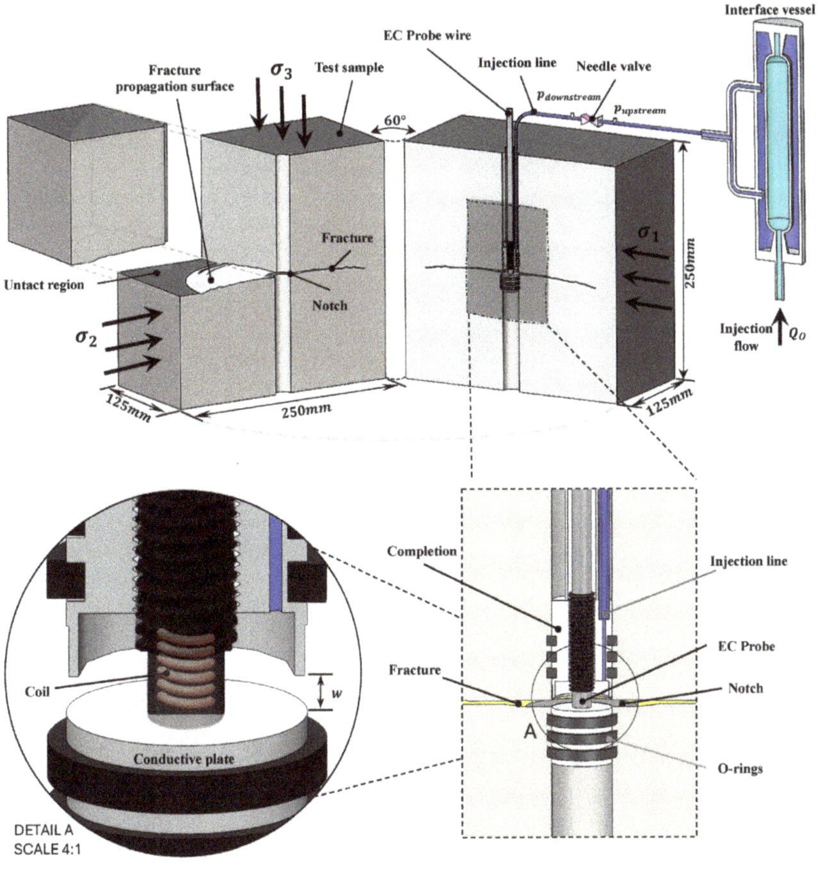

Figure 4. Schematic of the experimental setup for the propagation of hydraulic fracture under confinement, with details of the completion and the EC probe to measure fracture aperture at the fracture inlet.

4.3. *Installation of the EC Probe in the Wellbore*

The EC probe introduced in Section 3.1 is positioned in the central wellbore right at the position of the notch, precisely where the fluid is injected into the rock. The EC probe, therefore, shall provide a precise measurement of the fracture aperture at the injection point. A specialized completion tool aligns the probe's tip with the notch, simultaneously accommodating the injection fluid line. This specifically designed completion tool is depicted in Figure 4. It consists of an upper and a lower part. The upper part, made from stainless steel, centrally holds the probe while the injection line is positioned eccentrically. The lower part, made of aluminum, is the EC probe's conductive plate and target object. As the fracture initiates and its aperture expands, the distance between the EC probe and the conductive plate increases. Consequently, the induced eddy currents in the conductive plate decrease and the probe measures this distance, which corresponds to the aperture of fracture, in real time throughout the experiment.

5. Results and Discussion

5.1. Overview of the Experiment

As an illustrative example, we report a hydraulic fracture experiment performed in a porous sandstone rock from the Molasse de Villarlod quarry in Switzerland. This experiment uses glycerol as a fracturing fluid with a viscosity of $\mu = 0.57$ Pa·s measured by a rotational cone and a plate viscometer. Fluorescence additives were dissolved into the glycerol fluid to visualize the fracture path under UV light, facilitating easy observation of the fracture once the sample was cut after the experiment.

After the preparation of the sandstone block for the experiment and the placement of the EC probe inside the completion tool, we initially subject the block to lateral confining pressures of $\sigma_1 = \sigma_2 = 15$ MPa and a vertical confining pressure of $\sigma_3 = 5$ MPa. This ensures that the created fracture will propagate horizontally in a radial manner from the notch. Additionally, before starting the injection of the glycerol, we saturate all fluid lines with it to avoid air bubbles. The records of the fluid injection pressure and rates, as well as of the measurement of the wellbore fracture aperture w, are reported in Figure 5. In Figure 5a, the fracture aperture over time at the injection point is plotted in black, while Figure 5b depicts the fluid pressure in blue and the injection rate in red over time. It is important to note that the confining stresses were applied early on (at t = 0 s), and the reading of the EC probe is "zeroed" just after.

As reported in Figure 5, glycerol was injected into the central wellbore at a rate of 5 mL/min starting at time t = 0 s. However, due to the large pressurization rate, the fluid injection rate was reduced to 3 mL/min to avoid reaching the maximum pressure that the injection pump could handle. The fluid pressure largely exceeded the minimum applied confinement, and subsequently dropped while the fluid was still injected, clearly indicating the propagation of a fracture [30].

The fracturing continued until 462 s, when a needle valve in the injection was closed, suddenly stopping the fluid injection. As a result, fracture propagation stopped, and the hydraulic fracture began to recess and close. The sandstone is porous and permeable, allowing the fluid within the fracture to diffuse into the surrounding porous medium. This results in a decrease in fluid pressure in the injection line and fracture, thus decreasing the wellbore fracture aperture.

Subsequently, after some time, the fluid injection was re-started at 937 s with a rate of 3 mL/min, then stopped. This was repeated once more, resulting in three complete cycles of fluid injection, fracture creation and propagation, closure of the needle valve, cessation of fluid injection, and fracture closure. In Figure 5a, the highlighted green sections correspond to the fluid injection and fracture propagation periods. In contrast, the highlighted red sections indicate the periods without any injection where the fracture recesses and closes. Finally, at 2135 s, the needle valve is opened, but no further injection is performed. The experiment concludes. From 2944 to 3267 s, the applied confining pressures of 15 MPa and 5 MPa are released (time period highlighted in yellow in Figure 5a). This results in an elastic rebound of the measured fracture aperture in the wellbore, as measured by the EC probe (see Figure 5a).

Once the experiment is completed, the sandstone sample is cut in order to visually inspect the created fracture. The sample is sliced into two vertical cuts parallel to the wellbore axis, resulting in four parts. Figure 6 displays one of these parts, including the wellbore and the completion tool, in Figure 6b, a closer view of the area surrounding the wellbore inlet is provided, where the EC probe is located. The upper and lower completion and the EC probe are visible. In Figure 6c, the same section is exposed to UV light, revealing the fracture path and the leakoff area where the fracturing fluid has diffused inside the porous rock.

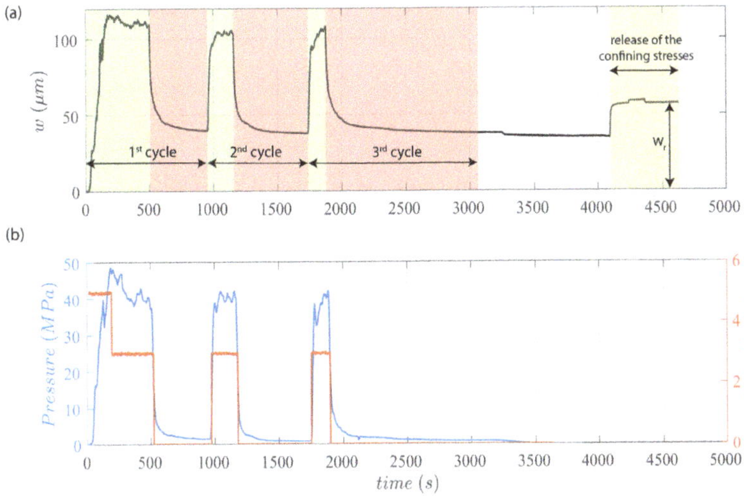

Figure 5. Hydraulic fracturing experiment in a porous sandstone. (**a**) Time evolution of the fracture aperture at the wellbore measured via the EC probe (in black), (**b**) time evolution of the fluid pressure in the well (in blue) and the injection rate (in red). Three injection cycles were performed. After a hold period following the last cycle, the applied confining stresses were released.

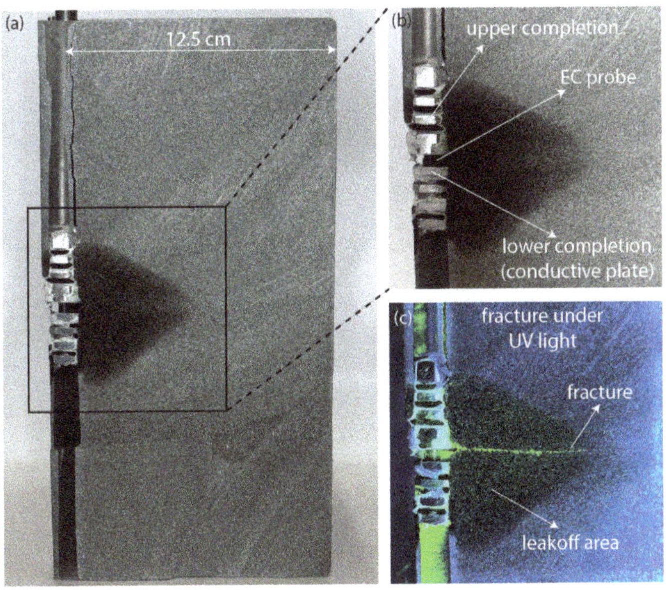

Figure 6. (**a**) Sandstone sample sliced into four parts after the experiment, displaying the wellbore, completion tool, and the EC probe. (**b**) A closer view of the area surrounding the fracture inlet, where the EC probe is located. (**c**) The same section is exposed to UV light, revealing the fracture path and the leakoff area.

5.2. Discussion

It is important to note that even after completely depressurizing the sample, the measured fracture aperture at the wellbore did not return to zero. This means that the initiation, propagation, and closure of the hydraulic fracture leaves a residual aperture in the rock. Such a residual fracture aperture has been reported in numerous experiments [8,9,18,23]. In the experiment reported herein, as highlighted in the yellow section of Figure 5a, the residual fracture aperture w_r after all the stresses applied to the sample were released was about 57 μm according to the EC probe reading. This residual aperture could be related to the characteristic grain size of the rock. During the fracturing process, grains induce a deviation from planarity, and a slight misalignment of the fracture surfaces necessarily occurs upon closure. A thin section of the Molasse sandstone can be seen in Figure 7a. An image processing code has been developed to obtain the particle's size distribution in 2D (see panel b in Figure 7). It follows a classical log-normal distribution, with a mean particle size of 48.8 μm, on par with the measured residual aperture w_r from the EC probe (57 μm).

We also performed an additional inspection of the fractured specimen after the test. A core was drilled in the sandstone specimen within a zone containing the created fracture. This core spans a radial zone from the central notch to the boundary of the sample (see Figure 8 for a schematic). An X-ray microcomputer tomography (CT) scan of this core was performed at the PIXE facility at EPFL. This CT scan process involved capturing 2D images of the core in a plane perpendicular to the fracture surface. To do so, CT scans were taken every 10 microns (the adjusted resolution of the CT scan instrument) in the mentioned direction, from the notch to the boundary of the sample. Figure 8 depicts the rock sample along with the core extracted schematically. Furthermore, three example images are reported at the locations of the fracture inlet (near the notch), fracture middle part, and fracture tip. In all three CT scan images depicted in Figure 8, a noticeable dark and rough area can be observed. This area, in each CT scan image, corresponds to the fracture voids, thus allowing us to estimate the fracture aperture. This aperture is the residual fracture aperture because it remains in the specimen after the fracturing experiment with all loadings removed. Based on the information obtained from the CT scan instrument, it has been determined that the size of each pixel in the CT scan image is 10 microns. This detail can be used to estimate the size of the fracture aperture by counting the number of pixels that correspond to the dark area. It is important to note that due to the rough surface of the fracture, the aperture normal to the mid-fracture plane is used to measure the average fracture aperture in each CT scan image, rather than the vertical extent of the fracture void (referred to as the "vertical" aperture here). The difference between the normal aperture and vertical aperture is depicted in Figure 9.

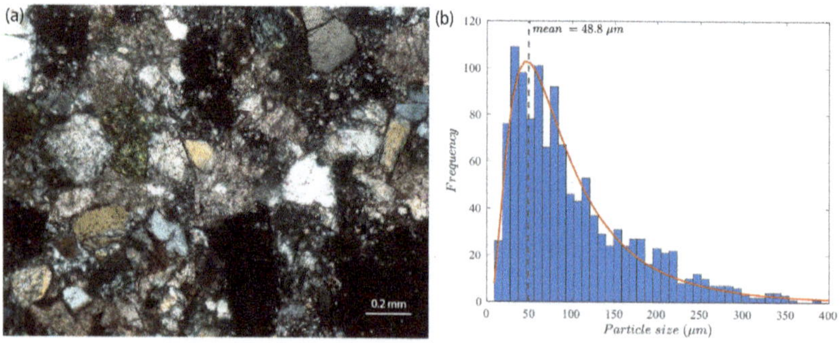

Figure 7. (**a**) Thin-section image depicting the microstructure of the Molasse sandstone. (**b**) The 2D particle size distribution of the sandstone obtained from image processing of the thin section.

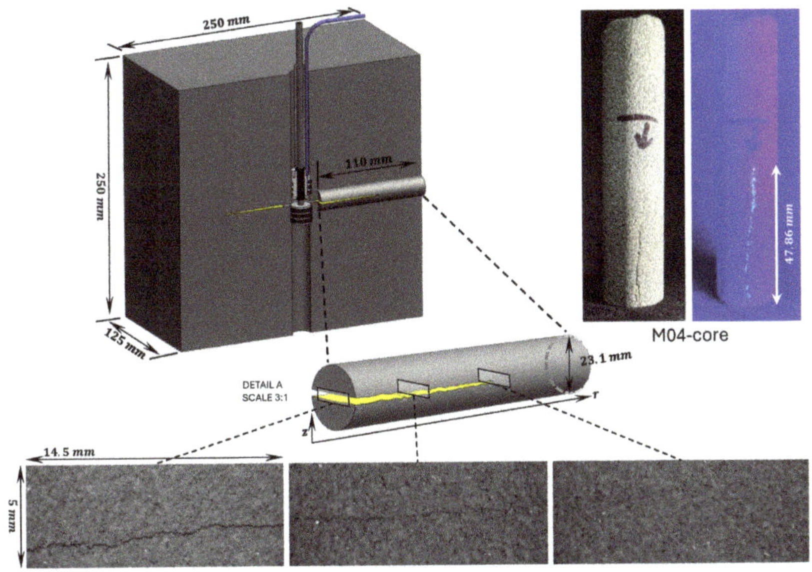

Figure 8. Schematic diagram of the rock sample with core extraction and location of the corresponding CT scan images taken near the fracture inlet, in the middle part of the fracture, and near the fracture front. The corresponding core sample images are in the right section.

Manually estimating the average perpendicular aperture for each CT scan image is impractical due to the large number of images—over 1000 per centimeter. Automatic image processing was, therefore, developed to do so. Its algorithm uses image processing techniques to calculate the average perpendicular aperture for all CT scan images. The process begins with reading TIFF images provided by the CT scan instrument, whose pixel values are then normalized between 0 and 1. A median filter is then applied to eliminate noise, and a contrast-limited adaptive histogram equalization is utilized to enhance the contrast of the grayscale image. Following this, a threshold is applied to remove all connected components that have fewer pixels than a specific number from the binary image (area opening operation). Finally, the Canny edge detection method is applied to identify the darker area edges and find the row and column indices of the non-zero elements (edges). These CT scan image datasets are available for download at M04 core CT scan images (https://doi.org/10.5281/zenodo.13357511 accessed on 1 September 2024), and the code for processing these CT scan images is available for access at FractureSurf (https://github.com/GeoEnergyLab-EPFL/FractureSurf/ accessed on 1 September 2024).

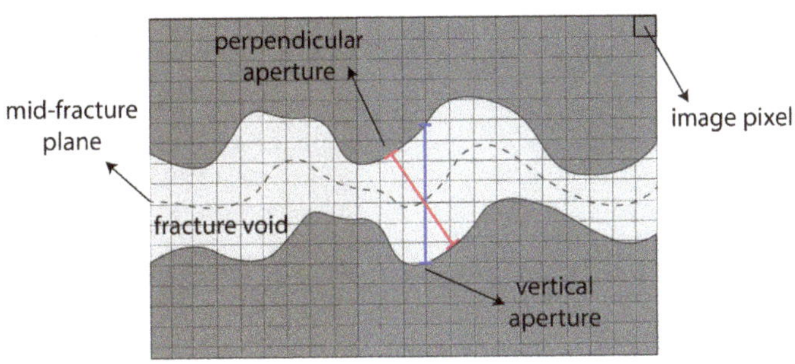

Figure 9. Comparison of the aperture of the normal fracture to the mid-plane fracture with the vertical aperture of the fracture.

In this way, an average value for the perpendicular aperture can be estimated for each CT scan image. By plotting the average aperture value of each CT scan image as a function of its position from the central wellbore, it is possible to obtain a comprehensive profile of the average aperture of the fracture along the length of the core. This profile extends from the notch to the fracture front and is shown in Figure 10. The gray area in this figure represents the error bars associated with the adjusted resolution of the CT scan. The residual aperture at the notch measured by the EC probe at the end of the experiment (in Figure 5a) of 57 microns falls within the range of 53–73 microns of the CT scan estimations (Figure 10). Both estimations correspond to similar conditions, without fluid and confinement pressures, which indicates that the EC probe and CT scans provide similar measurements of the fracture aperture.

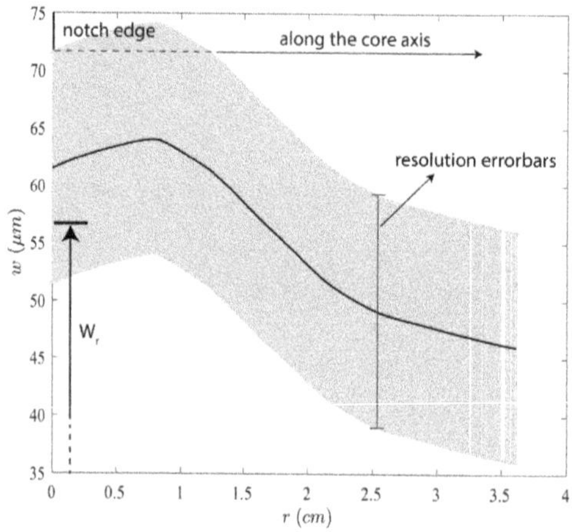

Figure 10. Profile of the residual aperture of the fracture along the length of the core sample.

5.3. Additional Requirements for Field Applications

The research conducted in this study demonstrated the very good performance of the EC (eddy current) probe, which was able to produce accurate results. The characteristics of the sensor, which are presented in Table 1, include high pressure and high temperature tolerance, as well as immunity to all types of fluids. Therefore, it could be implemented at field scale in harsh downhole environments at a depth typical of geo-energy applications (1 to 5 km). Specific arrangements must be made to deploy such a measurement at field scale. For instance, it will be necessary to clamp the target conductive plate (lower completion) and the EC probe holder (upper completion) to the walls of the well to place them on both sides of the fracture while allowing vertical relative movement. This would enable measuring the gap changes between the probe and the target plate, which corresponds to the aperture of the fracture. In addition, a stronger EC probe with a larger stroke length should be selected in order to measure a larger fracture aperture. If running a wire connection between the bottom and top of the well is not feasible, a downhole electronic system for the probe will have to be developed. This system should not only store the measured signals but also provide the energy required for the probe's operation.

The erosion of the probe and conductive plates by, for example, proppant used in hydraulic fracturing operations could be mitigated by coating the probe and conductive plates with more durable materials, such as diamond-like carbon. Alternatively, a layer of compacted protective material can be placed around them. Additionally, a stiffer metal could be used for the target plate to further enhance durability. In any case, proper probe calibration according to the material of the target conductive plate and the type of injected fluid, as conducted in this study, will always be required.

6. Conclusions

In this study, we successfully integrated an eddy current (EC) probe into a specialized completion tool positioned in a wellbore to accurately measure the fracture aperture during high-pressure fluid injection. Our findings demonstrate the reliability and applicability of this method to monitor the evolution of fracture aperture throughout hydraulic fracture initiation, propagation, and closure (once the injection has stopped). The residual fracture width measurements obtained from the EC probe at the conclusion of the fluid injection, after the stresses were removed, align with those derived from X-ray CT scan images and thin-section analysis. We have observed that the EC probe performs exceptionally well in laboratory settings. It is, therefore, a promising technique with potential applications in larger-scale field conditions. However, achieving this will require the development of a specifically designed EC probe and a hardware mounting tool to enable placement and retrieval in deep wellbores.

Author Contributions: M.T.: conceptualization, data curation, formal analysis, funding acquisition, investigation, methodology, resources, software, validation, visualization, writing—original draft. A.M.: conceptualization, investigation, methodology, visualization, writing—review and editing. B.L.: conceptualization, methodology, funding acquisition, resources, writing—review and editing. All authors have read and agreed to the published version of the manuscript.

Funding: This work was partly funded by the Swiss National Science Foundation (grant number: 192237). Mohsen Talebkeikhah also received partial funding from the Swiss Government Excellence Scholarship, administered by the Federal Commission for Scholarships for Foreign Students.

Institutional Review Board Statement: Not applicable.

Informed Consent Statement: Not applicable.

Data Availability Statement: The dataset, which includes detailed measurements from a radial hydraulic fracturing experiment conducted on a cubic M04 sample, is available for download at M04 Radial Hydraulic Fracturing (https://doi.org/10.5281/zenodo.13402755 accessed on 1 September 2024). The CT scan images of the present study (M04 core sample) are available for download at M04 core CT scan images https://doi.org/10.5281/zenodo.13357511 accessed on 1 September 2024. The code for processing low-rate data, including opening and pressure measurements, is available for access at FracLowRate https://github.com/GeoEnergyLab-EPFL/FracLowRate accessed on 1 September 2024. The code for processing the CT scan images is available for access at FractureSurf https://github.com/GeoEnergyLab-EPFL/FractureSurf/ accessed on 1 September 2024.

Acknowledgments: We want to express our sincere gratitude to Gary Perrenoud, Albert Taureg, and Lionel Pittet, the technical experts of the PIXE platform at École Polytechnique Fédérale de Lausanne (EPFL). Their proficiency and assistance in running the CT scan device were pivotal to the success of this study. Their commitment and the exceptional quality of their contributions are deeply valued.

Conflicts of Interest: The authors state that they do not have any financial interests or personal relationships that could have influenced the work reported in this publication. They declare that they have no known competing interests that could have affected the objectivity and integrity of their research.

References

1. Cipolla, C.L.; Maxwell, S.C.; Mack, M.G. Engineering guide to the application of microseismic interpretations. In Proceedings of the SPE Hydraulic Fracturing Technology Conference, The Woodlands, TX, USA, 6–8 February 2012; p. SPE-152165.
2. Butt, A.; Hedayat, A.; Moradian, O. Laboratory investigation of hydraulic fracturing in granitic rocks using active and passive seismic monitoring. *Geophys. J. Int.* **2023**, *234*, 1752–1770. [CrossRef]
3. Liu, D.; Lecampion, B.; Blum, T. Time-lapse reconstruction of the fracture front from diffracted waves arrivals in laboratory hydraulic fracture experiments. *Geophys. J. Int.* **2020**, *223*, 180–196. [CrossRef]
4. Lu, G.; Momeni, S.; Peruzzo, C.; Moukhtari, F.E.; Lecampion, B. Rock anisotropy promotes hydraulic fracture containment at depth. *J. Geophys. Res. Solid Earth* **2024**, *129*, e2023JB028011. [CrossRef]
5. Lecampion, B.; Jeffrey, R.G.; Detournay, E. Resolving the geometry of hydraulic fractures from tilt measurements. *Pure Appl. Geophys.* **2005**, *162*, 2433–2452. [CrossRef]

6. Lecampion, B.; Gunning, J. Model Selection in Fracture Mapping from Elastostatic data. *Int. J. Solids Struct.* **2007**, *44*, 1391–1408. [CrossRef]
7. Bunger, A. A photometry method for measuring the opening of fluid-filled fractures. *Meas. Sci. Technol.* **2006**, *17*, 3237. [CrossRef]
8. Groenenboom, J.; Fokkema, J.T. Monitoring the width of hydraulic fractures with acoustic waves. *Geophysics* **1998**, *63*, 139–148. [CrossRef]
9. Groenenboom, J.; Van Dam, D.; De Pater, C. Time-lapse ultrasonic measurements of laboratory hydraulic-fracture growth: Tip behavior and width profile. *SPE J.* **2001**, *6*, 14–24. [CrossRef]
10. Groenenboom, J.; van Dam, D.B. Monitoring hydraulic fracture growth: Laboratory experiments. *Geophysics* **2000**, *65*, 603–611. [CrossRef]
11. National Research Council, Division on Earth, Life Studies, Commission on Geosciences, Committee on Fracture Characterization and Fluid Flow. *Rock Fractures and Fluid Flow: Contemporary Understanding and Applications*; National Academies Press: Washington, DC, USA, 1996.
12. Chen, R.; Zaghloul, M.A.; Yan, A.; Li, S.; Lu, G.; Ames, B.C.; Zolfaghari, N.; Bunger, A.P.; Li, M.J.; Chen, K.P. High resolution monitoring of strain fields in concrete during hydraulic fracturing processes. *Opt. Express* **2016**, *24*, 3894–3902. [CrossRef] [PubMed]
13. Leggett, S.; Reid, T.; Zhu, D.; Hill, A. Experimental investigation of low-frequency distributed acoustic strain-rate responses to propagating fractures. *SPE J.* **2022**, *27*, 3814–3828. [CrossRef]
14. Yang, X.; Zhang, G.; Du, X.; Liu, Z.; Dong, H.; Wang, Y.; Nie, Y. Measurement and implications of the dynamic fracture width in hydraulic fracturing using FBG strain sensors. In Proceedings of the 51st U.S. Rock Mechanics/Geomechanics Symposium, San Francisco, CA, USA, 25–28 June 2017; p. ARMA-2017.
15. Qasim, M.; Rehman, Z.U.; Ding, L.; Tanoli, J.I.; Abbas, W.; Jamil, M.; Bhatti, Z.I.; Umar, M. Foreland basin unconformity, Western Himalaya, Pakistan: Timing gap, regional correlation and tectonic implications. *Prog. Earth Planet. Sci.* **2023**, *10*, 51. [CrossRef]
16. Kakurina, M.; Guglielmi, Y.; Nussbaum, C.; Valley, B. In situ direct displacement information on fault reactivation during fluid injection. *Rock Mech. Rock Eng.* **2020**, *53*, 4313–4328. [CrossRef]
17. De Pater, C.; Desroches, J.; Groenenboom, J.; Weijers, L. Physical and numerical modeling of hydraulic fracture closure. *SPE Prod. Facil.* **1996**, *11*, 122–128. [CrossRef]
18. Warpinski, N. Measurement of width and pressure in a propagating hydraulic fracture. *Soc. Pet. Eng. J.* **1985**, *25*, 46–54. [CrossRef]
19. Overbey, W.; Yost, L.; Yost, A. Analysis of natural fractures observed by borehole video camera in a horizontal well. In Proceedings of the SPE Unconventional Resources Conference/Gas Technology Symposium, Dallas, TX, USA, 8–11 June 1988; p. SPE-17760.
20. Palmer, I.; Sparks, D. Measurement of induced fractures by downhole TV camera in black warrior basin coalbeds. *J. Pet. Technol.* **1991**, *43*, 270–328. [CrossRef]
21. Darilek, G. A color borehole television system for developing gas production from Devonian shales. In Proceedings of the SPE Unconventional Gas Technology Symposium, Louisville, Kentucky, 18–21 May 1986; p. SPE-15219.
22. Smith, M.; Rosenberg, R.; Bowen, J. Fracture Width–Design vs. Measurement. In Proceedings of the SPE Annual Technical Conference and Exhibition, New Orleans, LA, USA, 26–29 September 1982; p. SPE-10965.
23. Medlin, W.; Masse, L. Laboratory experiments in fracture propagation. *Soc. Pet. Eng. J.* **1984**, *24*, 256–268. [CrossRef]
24. Nabavi, M.R.; Nihtianov, S.N. Design strategies for eddy-current displacement sensor systems: Review and recommendations. *IEEE Sens. J.* **2012**, *12*, 3346–3355. [CrossRef]
25. Webster, J.G.; Eren, H. *Measurement, Instrumentation, and Sensors Handbook: Two-Volume Set*; CRC Press: Boca Raton, FL, USA, 2018.
26. Liu, D.; Lecampion, B. Laboratory investigation of hydraulic fracture growth in Zimbabwe gabbro. *J. Geophys. Res. Solid Earth* **2022**, *127*, e2022JB025678. [CrossRef]
27. Talebkeikhah, M.; Momeni, S.; Lu, G.; Lecampion, B. Experimental investigation of hydraulic fracture closure in a porous sandstone. In Proceedings of the ARMA/DGS/SEG International Geomechanics Symposium, Abu Dhabi, United Arab Emirates, 7–9 November 2022; p. ARMA-IGS.
28. Talebkeikhah, M.; Momeni, S.; Lecampion, B. Measuring hydraulic fracture propagation and closure in sandstone. In Proceedings of the 58th U.S. Rock Mechanics/Geomechanics Symposium, Golden, CO, USA, 23–26 June 2024.
29. Talebkeikhah, M.; Lecampion, B. Hydraulic Fracture Closure Revealed Under Controlled Laboratory Condition. In Proceedings of the SPE/AAPG/SEG Unconventional Resources Technology Conference, Houston, TX, USA, 17–19 June 2024; p. D011S010R002.
30. Nolte, K.G. Determination of fracture parameters from fracturing pressure decline. In Proceedings of the SPE Annual Technical Conference and Exhibition, Las Vegas, NV, USA, 23–26 September 1979; p. SPE-8341.

sensors MDPI

Article

Convolutional Neural Network-Based Electromagnetic Imaging of Uniaxial Objects in a Half-Space

Chien-Ching Chiu [1,*], Jen-Shiun Chiang [1], Po-Hsiang Chen [1] and Hao Jiang [2]

[1] Department of Electrical and Computer Engineering, Tamkang University, Tamsui 251301, Taiwan; chiang@mail.tku.edu.tw (J.-S.C.); 810440031@gms.tku.edu.tw (P.-H.C.)
[2] School of Engineering, San Francisco State University, San Francisco, CA 94117-1080, USA; jianghao@sfsu.edu
* Correspondence: chiu@mail.tku.edu.tw

Abstract: In this paper, we adopt artificial intelligence (AI) technology for the electromagnetic imaging of uniaxial objects buried in a half-space environment. The limited measurement angle inherent to half-space configurations significantly increases the difficulty of data collection. This paper discusses the simultaneous emission of Transverse Magnetic (TM) and Transverse Electric (TE) electromagnetic waves to illuminate a uniaxial object embedded in a half-space. The dominant current scheme (DCS) and the backpropagation scheme (BPS) are subsequently employed to compute the initial permittivity distribution, which is then used as a dataset for training Convolutional Neural Networks (CNNs). The numerical results compare the reconstruction capabilities of both methods under identical conditions, demonstrating that the DCS exhibits superior generalization and noise immunity compared to the BPS. These findings confirm the effectiveness of both schemes in reconstructing the dielectric constant distribution of uniaxial objects buried in a half-space.

Keywords: artificial intelligence; electromagnetic imaging; uniaxial objects; half-space; convolutional neural network; dominant current scheme; backpropagation scheme

1. Introduction

Electromagnetic imaging, being an inverse scattering problem, faces challenges such as nonlinearity and non-uniqueness. Electromagnetic sensing technology serves as a powerful tool for identifying the shapes and material of unknown objects by analyzing the scattering of the electromagnetic waves. When combined with the Internet of Things (IoT), it enhances capabilities in sensing, communication systems, and data collection. This synergy drives the development of intelligent and automated systems with applications in diverse domains, such as industrial automation and smart cities [1–5]. Traditional methods like the Born Approximation [6] and its variants [7] provide solutions based on simplified linear models. However, these methods are highly dependent on prior information and battle to surmount inherent nonlinearity and ill-posedness problems [8]. To overcome these challenges, researchers have explored more robust iterative algorithms tailored to full-wave models. Prominent examples include the Distorted Born Iterative Method [9], Source Inversion [10], and the Subspace-based Optimization Method [11]. These advanced techniques deliver high-quality reconstructions but often increase computational complexity and time demands. While iterative algorithms have been employed to address these issues, they are computationally expensive and time-consuming.

Inverse scattering problem (ISP) algorithms are primarily developed to establish nonlinear relationships between the scattered field and the unknown material properties of scatterers. Neural networks have shown considerable promise in building such relationships, although these methods usually rely on pre-existing information about the scatterers. Recently, the exceptional representational power of Deep Neural Networks (DNNs) has spurred their uptake for solving ISPs, with Convolutional Neural Networks (CNNs) playing a particularly prominent role [12–27]. Deep learning-based methods for ISPs can generally be divided into two main approaches. The first approach seeks to substitute the computationally demanding aspects of traditional nonlinear iterative techniques with trained neural models, thereby simplifying the process [12–14]. The second approach frames the ISP as an image-to-image transformation task [15–27]. A non-iterative physical method is used to generate an initial guess based on the scattered field, and then a CNN is used to refine the generated initial guess into a high-resolution electromagnetic image. This paper emphasizes the second approach, analyzing its potential for delivering superior reconstructions. The second approach showcases the capability of neural networks to generate high-resolution reconstructions, although it remains in its nascent stages. In recent years, significant efforts have been directed toward enhancing imaging quality, with researchers concentrating on optimizing the objective function and integrating regularization techniques to improve both the precision and resilience of the reconstruction process. In 2019, Wei proposed a CNN to tackle the ISP more efficiently. Three training schemes were compared: direct inversion, the BPS, and the DCS. The numerical results showed that the DCS demonstrated superior accuracy and was capable of solving typical ISPs in about 1 s and that the CNN was a promising approach for real-time quantitative imaging [15]. In 2020, Zhang utilized a CNN to reconstruct high-resolution dielectric targets. This approach combined a qualitative Direct Sampling Method (DSM) with a quantitative backpropagation (BP) technique to capture spatial information for preliminary parameter estimation. The proposed method improved the reconstruction quality of the U-Net CNN without increasing computational complexity [16]. In 2021, Zhou introduced the Modified Contrast Scheme (MCS) to address the nonlinear ISP. This scheme leveraged local wave amplifier coefficients to create a modified contrast ratio, which helped to reduce the overall nonlinearity while maintaining accuracy. The numerical results indicated that after offline training, the MCS was effective in both two-dimensional and three-dimensional environments, particularly showing significant improvements in reconstructing high-contrast scatterers [17]. In 2022, Song proposed a unified learning-based approach to solve the ISP involving mixed boundary conditions. He first used a T-matrix method to simulate the scattering behavior of mixed dielectrics and Perfect Electric Conductors (PECs). Next, he reconstructed a rough image of the unknown scatterer using the backpropagation method, which was then enhanced by an attention-assisted pix2pix Generative Adversarial Network (GAN). Numerical tests validated the effectiveness of this method in reconstructing scatterers without any prior knowledge of their boundary conditions [18]. In 2023, Wang introduced a novel Early Fusion Framework (EFF) designed to address the ISP. The EFF architecture integrated a Scattered Field Encoder (SFE) and a Backpropagation Encoder (BPE). The SFE primarily resolved the nonlinear relationships arising from multiple scattering, while the BPE offered prior information based on the pre-reconstructed backpropagated distribution. The numerical results demonstrated that the EFF exhibited excellent accuracy, stability, and generalization capability [19]. In 2024, Wu introduced a method called Contrastive Learning-Based Subspace Optimization and Semantic Segmentation Assisted Reconstruction (CLSO-SSR). This network improved the initial estimation of contrasting sources, which helped reduce the likelihood of reaching local optima and decreased the time needed for inversion. It also incorporated a semantic segmentation network with an attention mechanism to address the nonlinearities arising from multiple scattering effects. Numerical experiments demonstrated the effectiveness of CLSO-SSR in estimating dielectric constants [20].

Recent studies have made significant advancements in electromagnetic imaging technology within half-space scenarios. These research studies are focused on enhanced methodologies and applications that leverage the unique challenges posed by the half-space environment. In 2019, Chiu used Transverse Magnetic (TM) polarized waves to illuminate a periodical rough surface and applied the Self-Adaptive Dynamical Differential Evolution (SADDE) algorithm to tackle the inverse scattering problem. The results indicated that SADDE effectively converged to the global extrema. Additionally, the period length exhibited a significantly faster convergence speed compared to the shape function and dielectric constant [21]. In 2023, Topbaş developed a hybrid method for locating cylinders buried in a half-space with a known orientation. This approach combined the classical moment method applied to the vertical coordinate with a spectral expansion for the horizontal coordinate. It utilized Green's function and Tikhonov regularization to address the problem's ill-posedness. An example was provided to demonstrate the method's applicability and accuracy [22]. In the same year, Chiu proposed a Deep Convolutional Neural Network to reconstruct the shape of the conductor. The measured scattered field was used as the direct input of the neural network to reconstruct the shape of the conductor in a half-space [23]. In 2024, Chiu introduced a novel artificial intelligence technique that incorporated an attention mechanism for half-space electromagnetic imaging. The BPS generated an initial guess of the image for the scatterer located in the lower half-space based on the measured scattered field. These guessed images were then input into a Self-Attention Generative Adversarial Network (SAGAN) for reconstruction. The numerical results demonstrated that SAGAN had effectively reconstructed the electromagnetic images with high accuracy [24].

The architecture of our system is depicted in Figure 1. In this setup, we strategically place transmitters within the simulation environment to illuminate the target object while receivers capture the scattered field generated. We then utilize the gathered scattered field data to produce an initial image using the BPS and the DCS. These initial images are subsequently processed by a CNN for image reconstruction, transforming the estimated images into high-resolution electromagnetic images. This research makes several significant contributions:

1. To date, there are no existing publications addressing the electromagnetic imaging of buried uniaxial objects in a half-space using artificial intelligence technology. We introduce a novel approach that integrates the DCS with a CNN to tackle the complex nonlinear inverse scattering problem.
2. Given that measurements are restricted to the upper space, the range of measuring angles is inherently limited. Our numerical simulations demonstrate that the proposed method effectively images highly nonlinear scatterers with both speed and accuracy while exhibiting robust noise immunity.
3. The dielectric constant vector sum derived from the electric field of Transverse Electric (TE) polarized waves manifests as a tensor across the cross-section, presenting greater complexity compared to the dielectric constant vector sum associated with Transverse Magnetic (TM) polarized waves. The interaction between the dielectric constant tensor and the electric field results in strong directional dependence in the TE polarized waves, leading to difficult reconstruction.
4. A uniaxial scatterer possesses dielectric constant components that vary with direction. The nonlinear characteristics of the TE polarized waves present considerably more challenges than their TM counterparts, complicating the reconstruction process using the scattered fields.

5. We successfully amalgamate the DCS with CNN methodologies to reconstruct electromagnetic images of buried objects in a half-space environment. Our numerical analyses reveal the reconstruction efficacy of this combined approach. To validate the robustness of our method, we employ a pretrained model to reconstruct scenarios involving high dielectric constant distributions. The results confirm that our technique maintains high reliability, even in half-space settings.

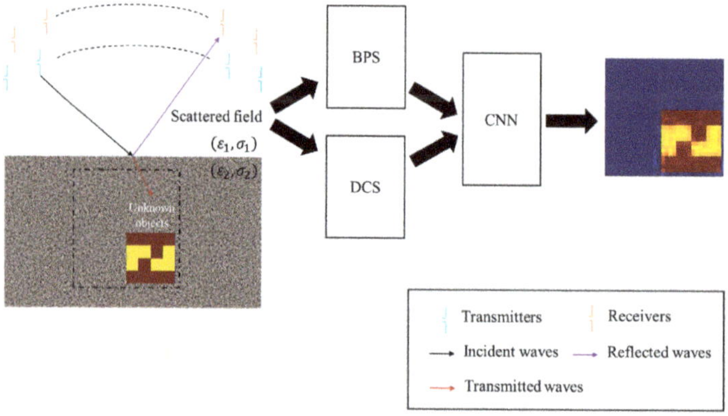

Figure 1. Electromagnetic sensing architecture.

In Section 2, we provide a theoretical derivation of the TM and TE modes. Section 3 focuses on utilizing the U-Net architecture to reconstruct images from uniaxial objects. Section 4 presents numerical results and simulation analyses, comparing the outcomes from the U-Net reconstruction with the initial estimates obtained from the DCS and BPS techniques. This comparison is made under various noise conditions while maintaining consistent deep learning training parameters and network architecture. Finally, Section 5 summarizes the findings and conclusions drawn from our research.

2. Theory

2.1. Direct Problems

Assume a uniaxial object $\overline{\overline{\varepsilon}}_r$ is buried in a lossy half-space, as shown in Figure 2. ε_1 and ε_2 represent the dielectric constants of Region 1 and Region 2, respectively. Similarly, σ_1 and σ_2 signify the conductivity of Region 1 and Region 2, respectively. Let μ_0 be the permeability of free space in each region. In other words, this article focuses exclusively on scatterers as nonmagnetic materials. We begin by introducing a dielectric object in which the scatterer is infinitely extended along the z-axis. Specifically, we consider an incident wave with TM polarization. We define the relationship between the incident wave and its time variation as $e^{j\omega t}$, and we denote its angle of incidence as $\varnothing_1$.

In the context of the Cartesian coordinate system, which consists of the axes x, y, and z, we can define a diagonal matrix for permittivities, as illustrated in Equation (1). $\varepsilon_x(x,y)$, $\varepsilon_y(x,y)$, and $\varepsilon_z(x,y)$ are typically represented as complex values.

$$\overline{\overline{\varepsilon}} = \begin{bmatrix} \varepsilon_x(x,y) & 0 & 0 \\ 0 & \varepsilon_y(x,y) & 0 \\ 0 & 0 & \varepsilon_z(x,y) \end{bmatrix}_{xyz} \quad (1)$$

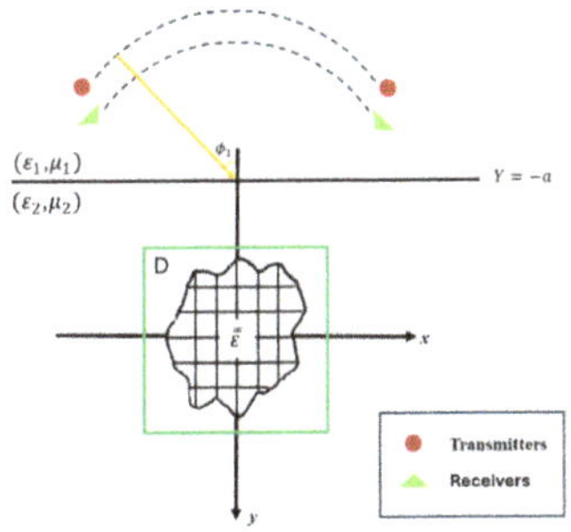

Figure 2. Schematic diagram of a two-dimensional object located in a lossy half-space.

To make our analysis clearer and more accurate, we investigate the TM wave first, which is aligned parallel to the z-axis. We designate the electric field in a medium without any scatterers as E_z^i. We can describe this electric field with Equation (2), which shows how it behaves and distributes under the given conditions. This method helps us focus on the key parts of wave propagation and interaction, without being distracted by the effects of scatterers.

$$E_z^i(x,y) = \begin{cases} E_1(x,y) = e^{-jk_1[xsin\varnothing_1+(y+a)cos\varnothing_1]} + \frac{1-n^{TM}}{1+n^{TM}}e^{-jk_1[xsin\varnothing_1-(y+a)cos\varnothing_1]} & , y \le -a \\ E_2(x,y) = \frac{2}{1+n^{TM}}e^{-jk_2[xsin\varnothing_2+(y+a)cos\varnothing_2]} & , y \ge -a \end{cases} \quad (2)$$

where $n^{TM} = \frac{cos\varnothing_2}{cos\varnothing_1}\sqrt{\frac{\varepsilon_0\varepsilon_2 - j\sigma_2/\omega}{\varepsilon_0\varepsilon_1 - j\sigma_1/\omega}}$ and $k_1 sin\varnothing_1 = k_2 sin\varnothing_2$.

$$k_i^2 = \omega^2\varepsilon_0\varepsilon_i\mu_0 - j\omega\mu_0\sigma_i \, , \, i = 1,2 \; Im(k_i) \le 0 \quad (3)$$

In cases where Regions 1 and 2 consist of lossless materials that do not absorb energy as electromagnetic waves pass through them, $\varnothing_1$ and $\varnothing_2$ represent the angles of incidence and refraction, respectively. However, when Regions 1 and 2 contain lossy materials that dissipate energy as heat or through other mechanisms, the behavior of the angles $\varnothing_1$ and $\varnothing_2$ becomes more complex. In these situations, the interaction of waves with the lossy materials may introduce additional factors, such as attenuation and phase shifts. This complexity necessitates the implementation of more advanced models to accurately describe the angles and the associated wave propagation. Therefore, analyzing angles in lossy media requires a deeper understanding of how energy loss influences the overall behavior of the incident and refracted waves. Based on the principles of induced current and Maxwell's equations, we can derive the following relationships:

$$\nabla \times \vec{E} = -j\omega\mu_0\vec{H} \quad (4)$$

$$\nabla \times \vec{H} = j\omega\varepsilon_0\varepsilon_2\vec{E} + \vec{j}_{eq} \quad (5)$$

where $\vec{j}_{eq} = j\omega\varepsilon_0[\varepsilon_z(x,y) - \varepsilon_2]E\hat{z}$ is the equivalent current density of the dielectric object.

In this context, $\vec{j}_{eq} = j\omega\varepsilon_0[\varepsilon_z(x,y) - \varepsilon_2]E^t\hat{z}$ represents the equivalent current density of the dielectric object. The total electric field within the object, where $\vec{E}_z^t(x,y) = E_z^t(x,y)\hat{z} = [E_z^i(x,y) + E_z^s(x,y)]\hat{z}$ can be expressed using the two-dimensional Green's function as:

$$E_z^i(\bar{r}) = \int_s G(r,r')k_2^2[\varepsilon_r(r') - \varepsilon_2]E_z^t(r')ds', \; y > -a \quad (6)$$

The scattered field can be written as:

$$E_z^s(\bar{r}) = -\int_s G(r,r')k_2^2[\varepsilon_r(r') - \varepsilon_2]E_z^t(r')ds' \tag{7}$$

To solve the half-space issue, we need to solve for the Green's function first, denoted as $G(x,\ y;\ x\prime,\ y\prime)$. This requires using a line current source located at a specific point and determining the scattered field at another point. By employing the Fourier transform technique, the half-space Green's function can be expressed as:

$$G(x,y;x',y') = \begin{cases} G_1(x,y;x',y') & ,y \le -a \\ G_2(x,y;x',y') = G_f(x,y;x',y') + G_s(x,y;x',y') & ,y > -a \end{cases} \tag{8}$$

$$G_1(x,y;x',y') = \frac{1}{2\pi}\int_{-\infty}^{\infty} \frac{j}{\gamma_1+\gamma_2} e^{j\gamma_1(y+a)} e^{-j\gamma_2(y\prime+a)} e^{-j\alpha(x-x')} d\alpha \tag{9}$$

$$G_f(x,y;x',y') = \frac{j}{4}H_0^{(2)}\left[k_2\sqrt{(x-x\prime)^2+(y-y\prime)^2}\right] \tag{10a}$$

$$G_s(x,y;x',y') = \frac{1}{2\pi}\int_{-\infty}^{\infty} \frac{j}{2\gamma_2}\left(\frac{\gamma_2-\gamma_1}{\gamma_2+\gamma_1}\right) e^{-j\gamma_2(y+2a+y')} e^{-j\alpha(x-x')} d\alpha \tag{10b}$$

$$\gamma_i^2 = k_i^2 - \alpha^2,\ i = 1,2,\ Im(\gamma_i) \le 0 \tag{11}$$

In this context, k_i represents the wave number for the i-th region, while $G(x,\ y;\ x\prime,\ y\prime)$ denotes the half-space Green's function obtained through the Fourier transform. Here, H_0^2 refers to the second-order zero Hankel function. Accurately calculating the Green's function in Equation (8) is essential for the numerical solutions of Equations (9) and (10). However, when the points $(x,\ y)$ and $(x\prime,\ y\prime)$ are near the interface between the two regions at $y = -a$, the integration process will converge slowly. This leads to a significant computational burden when evaluating the half-space Green's function in such scenarios.

In this paper, the electric field for TE waves propagates along the x-axis and y-axis in a half-space. In a homogeneous medium without scatterers, we can use Equation (12) to represent the electric field associated with these TE waves. The component of the electric field along the x-axis is denoted as E_x^i, while the component along the y-axis is represented as E_y^i. By expressing the overall electric field in terms of these components, we can analyze the behavior of the fields in relation to one another.

$$E_{xy}^i(x,y) = \begin{cases} E_3(x,y) = [-\hat{x}cos\phi_1 + \hat{y}sin\phi_1]e^{-jk_1[xsin\varnothing_1+(y+a)cos\varnothing_1]} + [\hat{x}cos\phi_1 + \hat{y}sin\phi_1]\frac{1-n^{TE}}{1+n^{TE}}e^{-jk_1[xsin\varnothing_1-(y+a)cos\varnothing_1]} & ,y \le -a \\ E_4(x,y) = (-\hat{x}cos\phi_2 + \hat{y}sin\phi_2)\frac{\eta_2}{\eta_1}\frac{2}{1+n^{TE}}e^{-jk_2[xsin\varnothing_2+(y+a)cos\varnothing_2]} & ,y > -a \end{cases} \tag{12}$$

$$n^{TE} = \frac{cos\varnothing_2}{cos\varnothing_1}\sqrt{\frac{\varepsilon_0\varepsilon_1 - j\sigma_1/\omega}{\varepsilon_0\varepsilon_2 - j\sigma_2/\omega}},\eta_1 = \sqrt{\frac{\mu_0}{\varepsilon_0\varepsilon_1 - j\sigma_1/\omega}},\ \eta_2 = \sqrt{\frac{\mu_0}{\varepsilon_0\varepsilon_2 - j\sigma_2/\omega}} \tag{13}$$

where E_3 represents the direct wave of the electromagnetic wave irradiating Region 1, as well as the reflected wave from the uniform medium. E_4 represents the refracted wave of the electromagnetic wave incident on Region 2 through Region 1. The integral equation of the total electric field is more complex to analyze due to the coupling effect between E_x and E_y, as described by Maxwell's equations.

$$\nabla \times \vec{E} = -j\omega\mu_0\vec{H} \tag{14}$$

$$\nabla \times \vec{H} = j\omega\varepsilon_0\varepsilon_2\vec{E} + \vec{J}_{eq} \tag{15}$$

where $\vec{J}_{eq} = j\omega\varepsilon_0[\varepsilon_x(x,y) - \epsilon_2]E_x\hat{x} + j\omega\varepsilon_0\left[\varepsilon_y(x,y) - \epsilon_2\right]E_y\hat{y}$ denotes the equivalent current density in the medium. The total field and scattered field are given by the following Equations (16)–(19):

$$\begin{aligned} E_x(x,y) = &-\left(\frac{\partial^2}{\partial x^2} + k_2^2\right)\left[\int_S G(x,y;x',y')\left(\frac{\varepsilon_x(x',y')}{\epsilon_2} - 1\right)E_x(x',y')ds'\right] \\ &-\frac{\partial^2}{\partial x\partial y}\left[\int_S G(x,y;x',y')\left(\frac{\varepsilon_y(x',y')}{\epsilon_2} - 1\right)E_y(x',y')ds\prime\right] + E_x^i(x,y) \end{aligned} \tag{16}$$

$$\begin{aligned} E_y(x,y) = &-\frac{\partial^2}{\partial x\partial y}\left[\int_S G(x,y;x',y')\left(\frac{\varepsilon_x(x',y')}{\epsilon_2} - 1\right)E_x(x',y')ds'\right] \\ &-\left(\frac{\partial^2}{\partial y^2} + k_2^2\right)\left[\int_S G(x,y;x',y')\left(\frac{\varepsilon_y(x',y')}{\epsilon_2} - 1\right)E_y(x',y')ds\prime\right] + E_y^i(x,y) \end{aligned} \tag{17}$$

$$\begin{aligned} E_x^s(x,y) = &-\left(\frac{\partial^2}{\partial x^2} + k_2^2\right)\left[\int_S G(x,y;x',y')\left(\frac{\varepsilon_x(x',y')}{\epsilon_2} - 1\right)E_x(x',y')ds'\right] \\ &-\frac{\partial^2}{\partial x\partial y}\left[\int_S G(x,y;x',y')\left(\frac{\varepsilon_y(x',y')}{\epsilon_2} - 1\right)E_y(x',y')ds\prime\right] \end{aligned} \tag{18}$$

$$\begin{aligned} E_y^s(x,y) = &-\frac{\partial^2}{\partial x\partial y}\left[\int_S G(x,y;x',y')\left(\frac{\varepsilon_x(x',y')}{\epsilon_2} - 1\right)E_x(x',y')ds'\right] \\ &-\left(\frac{\partial^2}{\partial y^2} + k_2^2\right)\left[\int_S G(x,y;x',y')\left(\frac{\varepsilon_y(x',y')}{\epsilon_2} - 1\right)E_y(x',y')ds\prime\right] \end{aligned} \tag{19}$$

The Green's functions for the TE wave in a half-space are expressed by the following equations:

$$(G_3)_{mn} = \begin{cases} \left.\frac{\partial^2 G_s(x,y;x_n,y_n)}{\partial x^2}\right|_{\substack{x=x_m \\ y=y_m}} \cdot \Delta S_n + G_s(x_m,y_m;x_n,y_n)\cdot k_2^2 \cdot \Delta S_n + \frac{j\pi a_n J_1(k_2 a_n)}{2\rho_{mn}^3}\left[k_2\rho_{mn}(y_m-y_n)^2 H_0^{(2)}(k_2\rho_{mn}) + \left((x_m-x_n)^2-(y_m-y_n)^2\right)H_1^{(2)}(k_2\rho_{mn})\right] & , m \neq n \\ \left.\frac{\partial^2 G_s(x,y;x_n,y_n)}{\partial x^2}\right|_{\substack{x=x_m \\ y=y_m}} \cdot \Delta S_n + G_s(x_m,y_m;x_n,y_n)\cdot k_2^2 \cdot \Delta S_n + \frac{j}{4}\left[\pi k_2 a_n H_1^{(2)}(k_2 a_n) - 4j\right] & , m = n \end{cases} \tag{20}$$

$$(G_4)_{mn} = \begin{cases} \left.\frac{\partial^2 G_s(x,y;x_n,y_n)}{\partial x\partial y}\right|_{\substack{x=x_m \\ y=y_m}} \cdot \Delta S_n + \frac{j\pi a_n J_1(k_2 a_n)}{2\rho_{mn}^3}(x_m-x_n)(y_m-y_n)\left[2H_1^{(2)}(k_2\rho_{mn}) - k_2\rho_{mn}H_0^{(2)}(k_2\rho_{mn})\right] & , m \neq n \\ 0 & , m = n \end{cases} \tag{21}$$

$$(G_5)_{mn} = \begin{cases} \left.\frac{\partial^2 G_s(x,y;x_n,y_n)}{\partial y^2}\right|_{\substack{x=x_m \\ y=y_m}} \cdot \Delta S_n + G_s(x_m,y_m;x_n,y_n)\cdot k_2^2 \cdot \Delta S_n + \frac{j\pi a_n J_1(k_2 a_n)}{2\rho_{mn}^3}\left[k_2\rho_{mn}(x_m-x_n)^2 H_0^{(2)}(k_2\rho_{mn}) + \left((y_m-y_n)^2-(x_m-x_n)^2\right)H_1^{(2)}(k_2\rho_{mn})\right] & , m \neq n \\ \left.\frac{\partial^2 G_s(x,y;x_n,y_n)}{\partial y^2}\right|_{\substack{x=x_m \\ y=y_m}} \cdot \Delta S_n + G_s(x_m,y_m;x_n,y_n)\cdot k_2^2 \cdot \Delta S_n + \frac{j}{4}\left[\pi k_2 a_n H_1^{(2)}(k_2 a_n) - 4j\right] & , m = n \end{cases} \tag{22}$$

$$(G_6)_{mn} = \left.\frac{\partial^2 G_1(x,y;x_n,y_n)}{\partial x^2}\right|_{\substack{x=x_m \\ y=y_m}} \cdot \Delta S_n + G_1(x_m,y_m;x_n,y_n)\cdot k_2^2 \cdot \Delta S_n \tag{23}$$

$$(G_7)_{mn} = \left.\frac{\partial^2 G_1(x,y;x_n,y_n)}{\partial x\partial y}\right|_{\substack{x=x_m \\ y=y_m}} \cdot \Delta S_n \tag{24}$$

$$(G_8)_{mn} = \left.\frac{\partial^2 G_1(x,y;x_n,y_n)}{\partial y^2}\right|_{\substack{x=x_m \\ y=y_m}} \cdot \Delta S_n + G_1(x_m,y_m;x_n,y_n)\cdot k_2^2 \cdot \Delta S_n \tag{25}$$

where $\rho_{mn} = \sqrt{(x_m - x_n)^2 + (y_m - y_n)^2}$ and $H_0^{(2)}$ is the zero-order Hankel function of the second kind, $H_1{}^{(2)}$ is the first-order Hankel function of the second kind, J_1 is the first-order Bessel function, (x_m, y_m) is the m-th observation point, and (x_n, y_n) is the n-th source point. In the next subsection, we introduce an imaging method that solves the inverse scattering problem. The measured scattered field is used to generate initial guesses for the BPS and the DCS. This process of generating initial guesses can greatly reduce the difficulty of CNN training.

2.2. Inverse Problem

Within the framework of the inverse scattering problem, we focus on measuring the scattered field in the external region surrounding the scatterer. Notably, in the TM case, the sole unknown dielectric parameter is ε_z, while in the TE case, the unknowns involve the dielectric constants ε_x and ε_y. Next, we proceed by applying the method of moments to reformulate the resulting integral equations into a matrix representation. Subsequently, we employ the BPS and the DCS to generate the initial dielectric constant distribution. As a final step, we leverage a U-Net architecture for deep learning to accurately reconstruct the dielectric constant tensor.

2.2.1. BPS

In this section, we leverage the measured scattered field data to derive an initial permittivity distribution via the BPS, aiming to streamline the training process of the U-Net model. Previous studies have shown that the BPS is particularly good at reconstructing weak scatterers. We further postulate that the backpropagation field is directly proportional to the induced currents I_z^b, I_x^b, and I_y^b.

$$\left(I_z^b\right) = Y_m \cdot [G_2]^H \left(E_z^s\right) \tag{26}$$

$$\begin{pmatrix} I_x^b \\ I_y^b \end{pmatrix} = Y_e \cdot \begin{bmatrix} [G_6] & [G_7] \\ [G_7] & [G_8] \end{bmatrix}^H \begin{pmatrix} E_x^s \\ E_y^s \end{pmatrix} \tag{27}$$

where H represents the conjugate transpose. The loss function is defined as follows:

$$L_m^b(Y) = \left\| \left(E_z^s\right) - [G_2] \cdot Y_m \cdot [G_2]^H \left(E_z^s\right) \right\|^2 \tag{28}$$

$$L_e^b(Y) = \left\| \begin{pmatrix} E_x^s \\ E_y^s \end{pmatrix} - \begin{bmatrix} [G_6] & [G_7] \\ [G_7] & [G_8] \end{bmatrix} \cdot Y_e \cdot \begin{bmatrix} [G_6] & [G_7] \\ [G_7] & [G_8] \end{bmatrix}^H \begin{pmatrix} E_x^s \\ E_y^s \end{pmatrix} \right\|^2 \tag{29}$$

To determine the minimum of the loss function, we set its derivative to zero. The resulting analytical expressions for Y_m and Y_e can be derived as follows:

$$Y_m = \frac{\left(E_z^s\right)^T \cdot \left([G_2]\left([G_2]^H \cdot \left(E_z^s\right)\right)\right)^*}{\left\| [G_2]\left([G_2]^H \cdot \left(E_z^s\right)\right) \right\|^2} \tag{30}$$

$$Y_e = \frac{\begin{pmatrix} E_x^s \\ E_y^s \end{pmatrix}^T \cdot \left(\begin{bmatrix} [G_6] & [G_7] \\ [G_7] & [G_8] \end{bmatrix} \left(\begin{bmatrix} [G_6] & [G_7] \\ [G_7] & [G_8] \end{bmatrix}^H \cdot \begin{pmatrix} E_x^s \\ E_y^s \end{pmatrix} \right) \right)^*}{\left\| \begin{bmatrix} [G_6] & [G_7] \\ [G_7] & [G_8] \end{bmatrix} \left(\begin{bmatrix} [G_6] & [G_7] \\ [G_7] & [G_8] \end{bmatrix}^H \cdot \begin{pmatrix} E_x^s \\ E_y^s \end{pmatrix} \right) \right\|^2} \tag{31}$$

where T and * represent the transpose and complex conjugate, respectively.

The induced current can be derived from Y, and the total field of BPS can also be determined accordingly. The total electric field E_z^b is defined as follows:

$$\left(E_z^b\right) = \left(E_z^i\right) + [G_1]\left(I_z^b\right) \tag{32}$$

$$\begin{pmatrix} E_x^b \\ E_y^b \end{pmatrix} = \begin{pmatrix} E_x^i \\ E_y^i \end{pmatrix} + \begin{bmatrix} [G_3] & [G_4] \\ [G_4] & [G_5] \end{bmatrix} \begin{pmatrix} I_x^b \\ I_y^b \end{pmatrix} \tag{33}$$

The correlation between the induced current and contrast can be expressed as:

$$\left(I_{z,p}^b\right) = diag\left(\left[\tau_z^b\right]\right)\left(E_z^b\right) \tag{34}$$

$$\begin{pmatrix} I_{x,p}^b \\ I_{y,p}^b \end{pmatrix} = diag\left(\begin{bmatrix} \left[\tau_x^b\right] & 0 \\ 0 & \left[\tau_y^b\right] \end{bmatrix}\right) \begin{pmatrix} E_x^b \\ E_y^b \end{pmatrix} \tag{35}$$

where p represents the number of incidences and τ^b represents the distribution of dielectric coefficients during BP. By aggregating all instances of Equations (34) and (35) and applying the least squares method, we can derive an analytical expression for the n-th element of the contrast τ^b as follows:

$$\left[\tau_z^b\right] = \frac{\sum_{p=1}^{Ni} I_{z,\,p}^b(n) \cdot \left[E_z^b(n)\right]^*}{\sum_{p=1}^{Ni} \left\| E_z^b(n) \right\|^2} \tag{36}$$

$$\begin{bmatrix} \left[\tau_x^b\right] & 0 \\ 0 & \left[\tau_y^b\right] \end{bmatrix} = \frac{\sum_{p=1}^{Ni} \begin{matrix} I_{x,p}^b(n) \\ I_{y,p}^b(n) \end{matrix} \cdot \left[\begin{pmatrix} E_x^b \\ E_y^b \end{pmatrix}(n)\right]^*}{\sum_{p=1}^{Ni} \left\| \begin{matrix} E_x^b(n) \\ E_y^b(n) \end{matrix} \right\|^2} \tag{37}$$

where Ni is the number of incidences.

2.2.2. DCS

In this section, we utilize the scattered field data to obtain the initial distribution of the dielectric constant through the DCS. This approach streamlines the CNN training process and is robust against noise interference, ensuring a more accurate estimation.

In the first step, we apply Singular Value Decomposition (SVD) to decompose the matrices $[G_2]$ and $\begin{bmatrix} [G_6] & [G_7] \\ [G_7] & [G_8] \end{bmatrix}$

$$[G_2] = [U_m][W_m][V_m]^* \tag{38}$$

$$\begin{bmatrix} [G_6] & [G_7] \\ [G_7] & [G_8] \end{bmatrix} = [U_e][W_e][V_e]^* \tag{39}$$

where U_m and U_e are the left singular vectors, while W_m and W_e are the diagonal matrices containing the singular values. Additionally, V_m and V_e are the right singular vectors. From Equations (8) and (9), it is evident that the dominant current I_z^s is associated with the first L_m row vectors of the right matrix V_m, where L_m serves as the adjustment parameter.

Similarly, from Equation (37), the dominant current $\begin{pmatrix} I_x^s \\ I_y^s \end{pmatrix}$ is associated with the first L_e row vectors of the right matrix V_e, where L_e serves as the adjustment parameter.

$$I_z^s = \sum_{i=1}^{L_m} \frac{(U_m)_i^* \cdot E_z^s}{(W_m)_{i,\,i}} (V_m)_i,\ i = 1 \ldots L_m \tag{40}$$

$$\begin{pmatrix} I_x^s \\ I_y^s \end{pmatrix} = \sum_{i=1}^{L_e} \frac{(U_e)_i^* \cdot \begin{pmatrix} E_x^s \\ E_y^s \end{pmatrix}}{(W_e)_{i,\,i}} (V_e)_i,\ i = 1 \ldots L_e \tag{41}$$

In Equations (40) and (41), the first L singular values are considerably larger than the remaining singular values. As a result, when the measured scattered field experiences noise disturbances, the induced current remains relatively stable. The induced current can be mathematically defined as follows:

$$I_z^d = I_z^s + I_z^l \tag{42}$$

$$\begin{pmatrix} I_x^d \\ I_y^d \end{pmatrix} = \begin{pmatrix} I_x^s \\ I_y^s \end{pmatrix} + \begin{pmatrix} I_x^l \\ I_y^l \end{pmatrix} \tag{43}$$

The low-frequency induced current is defined as follows:

$$I_z^l = \overset{=l}{F}_z \cdot \psi_z^l \tag{44}$$

$$\begin{pmatrix} I_x^l \\ I_y^l \end{pmatrix} = \overset{=l}{F}_{xy} \cdot \begin{pmatrix} \psi_x^l \\ \psi_y^l \end{pmatrix} \tag{45}$$

where $\overset{=l}{F}_z$ and $\overset{=l}{F}_{xy}$ denote the low-frequency Fourier transforms, while ψ_z^l, ψ_x^l and ψ_y^l represent the associated coefficients. Section 4 elaborates on the criteria for selecting the parameters L_m and L_e.

In the second step, we apply the operator $diag(\tau)$ to both sides of Equations (32) and (33) rearrange the resulting terms to derive the equation that describes the induced current as follows:

$$\left(I_z^d\right) = diag[\tau_z] \cdot \left[\left(E_z^i\right) + [G_1] \cdot \left(I_z^d\right)\right] \tag{46}$$

$$\begin{pmatrix} I_x^d \\ I_y^d \end{pmatrix} = diag\begin{bmatrix} [\tau_x] & 0 \\ 0 & [\tau_y] \end{bmatrix} \cdot \left[\begin{pmatrix} E_x^i \\ E_y^i \end{pmatrix} + \begin{bmatrix} [G_3] & [G_4] \\ [G_4] & [G_5] \end{bmatrix} \cdot \begin{pmatrix} I_x^d \\ I_y^d \end{pmatrix}\right] \tag{47}$$

By standardizing the state and data equations, we can obtain:

$$F\left(\psi_z^l\right) = \left[\frac{\left\| \left([G_2]\overset{=l}{F}_z\right)\psi_z^l + \left([G_2](I_z^s) - (E_z^s)\right) \right\|^2}{\|(E_z^s)\|^2} + \frac{\left\| \left(\overset{=l}{F}_z - [\tau]\left([G_1]\overset{=l}{F}_z\right)\right)\psi_z^l - \left([\tau]\left((E_z^i) + [G_1](I_z^s)\right) - (I_z^s)\right) \right\|^2}{\|(I_z^s)\|^2}\right] \tag{48}$$

$$F\begin{pmatrix} \psi_x^l \\ \psi_y^l \end{pmatrix} = \left[\frac{\left\| \left(\begin{bmatrix} [G_6] & [G_7] \\ [G_7] & [G_8] \end{bmatrix} \overset{=l}{F}_{xy}\right)\begin{pmatrix} \psi_x^l \\ \psi_y^l \end{pmatrix} + \left(\begin{bmatrix} [G_6] & [G_7] \\ [G_7] & [G_8] \end{bmatrix}\begin{pmatrix} I_x^s \\ I_y^s \end{pmatrix} - \begin{pmatrix} E_x^s \\ E_y^s \end{pmatrix}\right) \right\|^2}{\left\| \begin{pmatrix} E_x^s \\ E_y^s \end{pmatrix} \right\|^2} + \right.$$

$$\frac{\left\|\left(\overline{\overline{F}}^{l}_{xy}-\begin{bmatrix}[\tau_x]&0\\0&[\tau_y]\end{bmatrix}\left(\begin{bmatrix}[G_3]&[G_4]\\{[G_4]}&[G_5]\end{bmatrix}\overline{\overline{F}}^{l}_{xy}\right)\right)\begin{pmatrix}\psi^l_x\\\psi^l_y\end{pmatrix}-\left(\begin{bmatrix}[\tau_x]&0\\0&[\tau_y]\end{bmatrix}\left(\begin{pmatrix}E^i_x\\E^i_y\end{pmatrix}+\begin{bmatrix}[G_3]&[G_4]\\{[G_4]}&[G_5]\end{bmatrix}\begin{pmatrix}I^s_x\\I^s_y\end{pmatrix}\right)-\begin{pmatrix}I^s_x\\I^s_y\end{pmatrix}\right)\right\|^2}{\left\|\begin{pmatrix}I^s_x\\I^s_y\end{pmatrix}\right\|^2} \tag{49}$$

To avoid the need for matrix inversion, we propose an approximate analytical solution for ψ^l. This method is designed to minimize the cost functions (8) and (9) in the direction of the gradient $\overline{\rho}$.

$$\psi^l_z = \beta_z\overline{\rho}_z \tag{50}$$

$$\begin{pmatrix}\psi^l_x\\\psi^l_y\end{pmatrix}=\begin{pmatrix}\beta_x\overline{\rho}_x\\\beta_y\overline{\rho}_y\end{pmatrix} \tag{51}$$

The gradient direction function is defined as follows:

$$F'\left(\psi^l_z=0\right)=\overline{\rho}_z=\frac{\left([G_2]\overline{\overline{F}}^{l}\right)^H\left([G_2](I^s_z)-(E^s_z)\right)}{\|(E^s_z)\|^2}+\frac{-\left(\overline{\overline{F}}^{l}-[\tau_z]\left([G_1]\overline{\overline{F}}^{l}\right)\right)^H\left([\tau_z]\left((E^i_z)+[G_1](I^s_z)\right)-(I^s_z)\right)}{\|I^s_z\|^2} \tag{52}$$

$$F'\begin{pmatrix}\psi^l_x\\\psi^l_y\end{pmatrix}=0\Bigg)=\begin{pmatrix}\overline{\rho}_x\\\overline{\rho}_y\end{pmatrix}=\left[\frac{\left(\begin{bmatrix}[G_6]&[G_7]\\{[G_7]}&[G_8]\end{bmatrix}\overline{\overline{F}}^{l}_{xy}\right)^H\left(\left(\begin{bmatrix}[G_6]&[G_7]\\{[G_7]}&[G_8]\end{bmatrix}\begin{pmatrix}I^s_x\\I^s_y\end{pmatrix}\right)-\begin{pmatrix}E^s_x\\E^s_y\end{pmatrix}\right)}{\left\|\begin{pmatrix}E^s_x\\E^s_y\end{pmatrix}\right\|^2}+\right.$$
$$\left.\frac{-\left(\overline{\overline{F}}^{l}_{xy}-\begin{bmatrix}[\tau_x]&0\\0&[\tau_y]\end{bmatrix}\left(\begin{bmatrix}[G_3]&[G_4]\\{[G_4]}&[G_5]\end{bmatrix}\overline{\overline{F}}^{l}_{xy}\right)\right)^H\left(\begin{bmatrix}[\tau_x]&0\\0&[\tau_y]\end{bmatrix}\left(\begin{pmatrix}E^i_x\\E^i_y\end{pmatrix}+\begin{bmatrix}[G_3]&[G_4]\\{[G_4]}&[G_5]\end{bmatrix}\begin{pmatrix}I^s_x\\I^s_y\end{pmatrix}\right)-\begin{pmatrix}I^s_x\\I^s_y\end{pmatrix}\right)}{\left\|\begin{pmatrix}I^s_x\\I^s_y\end{pmatrix}\right\|^2}\right] \tag{53}$$

The minimizer function is defined as follows:

$$\beta_z=\frac{-\dfrac{\left([G_2]\overline{\overline{F}}^{l}_z\overline{\rho}_z\right)^H\left([G_2](I^s_z)-(E^s_z)\right)}{\|(E^s_z)\|^2}+\dfrac{\left(\left(\overline{\overline{F}}^{l}_z-[\tau]\left([G_1]\overline{\overline{F}}^{l}_z\right)\right)\overline{\rho}_z\right)^H\left([\tau]\left((E^i_z)+[G_1](I^s_z)\right)-(I^s_z)\right)}{\|(I^s_z)\|^2}}{\dfrac{\left\|[G_2]\overline{\overline{F}}^{l}_z\overline{\rho}_z\right\|^2}{\|(E^s_z)\|^2}+\dfrac{\left\|\left(\overline{\overline{F}}^{l}_z-[\tau]\left([G_1]\overline{\overline{F}}^{l}_z\right)\right)\overline{\rho}_z\right\|^2}{\|(I^s_z)\|^2}} \tag{54}$$

$$\begin{pmatrix}\beta_x\\\beta_y\end{pmatrix}=\frac{-\dfrac{\left(\begin{bmatrix}[G_6]&[G_7]\\{[G_7]}&[G_8]\end{bmatrix}\overline{\overline{F}}^{l}_{xy}\begin{pmatrix}\overline{\rho}_x\\\overline{\rho}_y\end{pmatrix}\right)^H\left(\begin{bmatrix}[G_6]&[G_7]\\{[G_7]}&[G_8]\end{bmatrix}\begin{pmatrix}I^s_x\\I^s_y\end{pmatrix}-\begin{pmatrix}\overline{E}^s_x\\\overline{E}^s_y\end{pmatrix}\right)}{\left\|\begin{pmatrix}E^s_x\\E^s_y\end{pmatrix}\right\|^2}}{\dfrac{\left\|\begin{bmatrix}[G_6]&[G_7]\\{[G_7]}&[G_8]\end{bmatrix}\overline{\overline{F}}^{l}_{xy}\begin{pmatrix}\overline{\rho}_x\\\overline{\rho}_y\end{pmatrix}\right\|^2}{\left\|\begin{pmatrix}E^s_x\\E^s_y\end{pmatrix}\right\|^2}}+$$

$$\frac{\dfrac{\left(\left(\overline{\overline{F}}^{l}_{xy}-\begin{bmatrix}[\tau_x]&0\\0&[\tau_y]\end{bmatrix}\left(\begin{bmatrix}[G_3]&[G_4]\\{[G_4]}&[G_5]\end{bmatrix}\overline{\overline{F}}^{l}_{xy}\right)\right)\begin{pmatrix}\overline{\rho}_x\\\overline{\rho}_y\end{pmatrix}\right)^H\left(\begin{bmatrix}[\tau_x]&0\\0&[\tau_y]\end{bmatrix}\left(\begin{pmatrix}E^i_x\\E^i_y\end{pmatrix}+\begin{bmatrix}[G_3]&[G_4]\\{[G_4]}&[G_5]\end{bmatrix}\begin{pmatrix}I^s_x\\I^s_y\end{pmatrix}\right)-\begin{pmatrix}I^s_x\\I^s_y\end{pmatrix}\right)^H}{\left\|\begin{pmatrix}I^s_x\\I^s_y\end{pmatrix}\right\|^2}}{\dfrac{\left\|\left(\overline{\overline{F}}^{l}_{xy}-\begin{bmatrix}[\tau_x]&0\\0&[\tau_y]\end{bmatrix}\left(\begin{bmatrix}[G_3]&[G_4]\\{[G_4]}&[G_5]\end{bmatrix}\overline{\overline{F}}^{l}_{xy}\right)\right)\begin{pmatrix}\overline{\rho}_x\\\overline{\rho}_y\end{pmatrix}\right\|^2}{\left\|\begin{pmatrix}I^s_x\\I^s_y\end{pmatrix}\right\|^2}} \tag{55}$$

The final step involves calculating the contrast $\left[\tau_p^d\right]$ of the DCS using I^d. We update the total electric field function E_p^d for the p-th incidence DCS as follows:

$$\left(E_{z,p}^d\right) = \left(E_{z,p}^i\right) + [G_1]\left(I_{z,p}^d\right) \tag{56}$$

$$\begin{pmatrix} E_{x,p}^d \\ E_{y,p}^d \end{pmatrix} = \begin{pmatrix} E_x^i \\ E_y^i \end{pmatrix} + \begin{bmatrix} [G_3] & [G_4] \\ [G_4] & [G_5] \end{bmatrix} \begin{pmatrix} I_{x,p}^d \\ I_{y,p}^d \end{pmatrix} \tag{57}$$

Based on the definitions of (53) and (54), the n-th component of the contrast τ_p^d for the p-th incidence

$$I_{z,p}^d = diag\left(\tau_{zp}^{\ d}\right) \cdot \left(E_{z,p}^d\right) \tag{58}$$

$$\begin{pmatrix} I_{x,p}^d \\ I_{y,p}^d \end{pmatrix} = diag \begin{bmatrix} \left[\tau_{xp}^{\ d}\right] & 0 \\ 0 & \left[\tau_{yp}^{\ d}\right] \end{bmatrix} \cdot \begin{pmatrix} E_{x,p}^d \\ E_{y,p}^d \end{pmatrix} \tag{59}$$

can be calculated:

$$\tau_{zp}^{\ d}(n) = \frac{I_{z,p}^d(n) \cdot \left(E_{z,p}^d(n)\right)^*}{\left\| E_{z,p}^d(n) \right\|^2} \tag{60}$$

$$\begin{bmatrix} \left[\tau_{xp}^{\ d}\right] & 0 \\ 0 & \left[\tau_{yp}^{\ d}\right] \end{bmatrix}(n) = \frac{\begin{pmatrix} I_{x,p}^d \\ I_{y,p}^d \end{pmatrix}(n) \cdot \left(\begin{pmatrix} E_{x,p}^d \\ E_{y,p}^d \end{pmatrix}(n)\right)^*}{\left\| \begin{pmatrix} E_{x,p}^d \\ E_{y,p}^d \end{pmatrix}(n) \right\|^2} \tag{61}$$

3. Convolutional Neural Network

AI technology is evolving at an impressive pace, demonstrating broad applicability across various domains such as speech recognition, image processing, and autonomous vehicles. In addition to these fields, the integration of AI methodologies into electromagnetic imaging techniques has become increasingly prominent, particularly in conjunction with CNNs, Deep Convolutional Neural Networks (DCNNs), and Generative Adversarial Networks (GANs). This paper implements a CNN architecture specifically tailored to tackle the inverse scattering problem in a half-space, as depicted in Figure 3. The architecture is composed of two main components: a contraction network on the left and an expansion network on the right. On the contraction side, we sequentially arrange layers that include 3×3 convolutional layers, normalization layers, ReLU (Rectified Linear Unit) activations, and 2×2 pooling layers. The expansion network is structured to mirror the contraction network, utilizing 3×3 deconvolutional layers, normalization layers, and a ReLU activation layer. The architecture culminates in a 1×1 convolutional layer that acts as a fully connected layer. The resultant averaged output is then forwarded to a regression layer to evaluate the error associated with the dielectric coefficient distribution. The cost function can be defined as follows:

$$\underset{A_{i1},i}{argmin} : \sum_{N=1}^{N_t} f(A_{i1}(\varepsilon_z^\alpha), \varepsilon_z) + Q_1(i) \tag{62}$$

$$\underset{A_{i2},i}{argmin} : \sum_{N=1}^{N_t} f\left(A_{i2}\begin{pmatrix} \varepsilon_x^\alpha \\ \varepsilon_y^\alpha \end{pmatrix}, \begin{pmatrix} \varepsilon_x \\ \varepsilon_y \end{pmatrix}\right) + Q_2(i) \tag{63}$$

where A_{i1} and A_{i2} indicate the parameters of the neural network architecture. f signifies the difference between the image generated by the CNN and the ground truth. ε^{α} refers to the approximate permittivity coefficient used in the model. $Q_{1(i)}$ and $Q_{2(i)}$ are regularization functions applied to mitigate overfitting and enhance model generalization, where $Q_1(i)$ is an L2-norm regularization function of z-axis permittivity and $Q_2(i)$ is an L2-norm regularization function of x-axis permittivity, respectively. In addition to helping mitigate overfitting, it also improves the convergence of the optimization process.

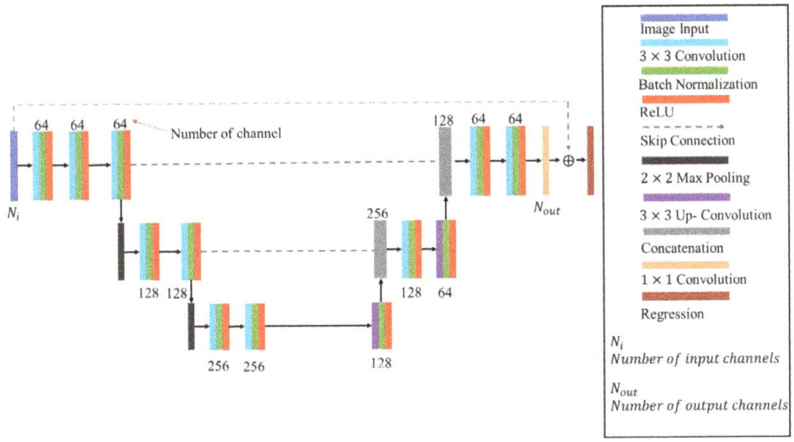

Figure 3. Convolutional Neural Network architecture.

As CNNs are widely recognized for their superior performance in image processing tasks, we opted to employ a CNN architecture to reconstruct electromagnetic images in this paper. The rationale behind selecting CNNs is multi-faceted, involving the following factors:

1. Skip connections between the input and output layers of a CNN play a critical role in addressing the vanishing gradient problem, ensuring a more consistent gradient flow during backpropagation.
2. Down-sampling in the contraction pathway of a CNN expands the receptive data, which improves the network's ability to make accurate pixel-level predictions in the output.
3. Batch normalization, an integral component of the CNN architecture, mitigates internal covariate shift, accelerates convergence, and reduces sensitivity to parameter initialization and gradient instability.

In the next section, we explore the reconstruction performance of the BPS and the DCS in a half-space environment using a CNN. Our analysis emphasizes the noise resilience of both methods and examines how the material properties of buried objects affect reconstruction accuracy, providing insights into their respective strengths and limitations.

4. Numerical Result

In this study, we explore the dielectric constant distributions of uniaxial objects buried in soil at a depth of 2 m. In other words, we assume that ε_x is equivalent to ε_y. A configuration comprising 32 receivers spanning from $\theta = 195°$ to $350°$, with a radius of distance set at 3 m and 32 transmitters ranging from $\varnothing_1 = -80°$ to $80°$ at $5°$ intervals, are employed in our simulation. To emulate real-world conditions, 5%, 10%, 15%, and 20% Gaussian noise are added. We consider scatterers characterized by 10 distinct dielectric constant profiles, positioned randomly across 50 predefined locations within the measurement area.

This means that we generate a dataset comprising 16,000 images for each configuration (calculated as 10 × 50 × 32). This dataset is subsequently partitioned into 80% for training and 20% for testing purposes. We use Stochastic Gradient Descent with Momentum (SGDM) to train the CNN. The training configuration includes a momentum factor of 0.99, a learning rate ranging from 10^{-4} to 10^{-5}, and a maximum of 40 epochs. The input of the U-Net is derived from both the BPS and DCS methods to estimate the preliminary dielectric distributions. The efficacy of the two approaches is assessed through a comparison of the U-Net reconstruction outcomes. We calculate the Root Mean Square Error (RMSE) and the Structural Similarity Index Measure (SSIM) using the formulations in Equations (59) and (60), respectively. These metrics are employed to assess the performance of each scenario comprehensively.

$$RMSE = \frac{1}{M_t}\sum_{i=1}^{M_t} \frac{\left\| \overline{\overline{\varepsilon}}_r - \overline{\overline{\varepsilon}}_r^{\,r} \right\|_F}{\left\| \overline{\overline{\varepsilon}}_r \right\|_F} \tag{64}$$

$$SSIM = \frac{\left(2\mu_{\tilde{y}}\mu_y + C_1\right)\left(2\sigma_{\tilde{y}y} + C_2\right)}{\left(\mu_{\tilde{y}}^2 + \mu_y^2 + C_1\right)\left(\sigma_{\tilde{y}}^2 + \sigma_y^2 + C_2\right)} \tag{65}$$

where $\overline{\overline{\varepsilon}}_r$ and $\overline{\overline{\varepsilon}}_r^{\,r}$ refer to the actual and reconstructed profiles of the relative permittivity, respectively. M_t represents the total number of experimental tests performed, while F signifies the Frobenius norm. $\tilde{y}$ and y represent the reconstructed and true relative permittivity profiles, respectively. μ_y denotes the mean of the reconstructed profile y, while $\sigma_{\tilde{y}}^2$ captures the variance, and $\sigma_{\tilde{y}y}$ signifies the covariance between $\tilde{y}$ and y. To address potential issues with zero denominators in the calculations, we introduce two small constraints, $C_1 = (K_1D)^2$ and $C_2 = (K_2D)^2$, where $K_1 = 0.01$ and $K_2 = 0.03$ serve as hyperparameters. Here, D represents the dynamic range of the pixel intensity values in the target image y.

4.1. Relative Permittivity Between 3.5 and 4

In this case, we define the permittivity distribution within a range of 3.5 to 4. We examine a scenario involving 10 scatterers, each with distinct permittivity profiles, positioned at 50 different locations throughout the designated measurement area. To simulate a realistic environment, we introduce 20% Gaussian noise into the scattered field data. The dielectric coefficient distribution is estimated using both the BPS and DCS techniques. These estimates are then input into a CNN to reconstruct electromagnetic images. Finally, we conduct a comparative analysis of the reconstruction outcomes from the two different sets of input data.

Figure 4a,d illustrate the ground truth for the ε_z and ε_x configurations, respectively. Figure 4b,e present the BPS reconstruction outcomes. Figure 4c,f showcase the DCS reconstruction results for the ε_z and ε_x cases, respectively. The findings indicate that when an object is buried within a half-space medium, the scattering of incident waves significantly impairs the reconstruction accuracy. This study demonstrates that despite employing two distinct input configurations and a well-trained CNN, the BPS method is limited to approximating the object's location. In contrast, the DCS approach achieves a more accurate determination of both the position and distribution of the buried materials. Additionally, Table 1 provides the Root Mean Square Error (RMSE) and Structural Similarity Index Measure (SSIM) metrics for this case.

Table 1. RMSE and SSIM for ε_z and ε_x for relative permittivity between 3.5 and 4.

Performance	ε_z		ε_x	
	BPS	**DCS**	**BPS**	**DCS**
RMSE	12.03%	1.43%	16.37%	1.25%
SSIM	57.32%	95.9%	29.19%	99.02%

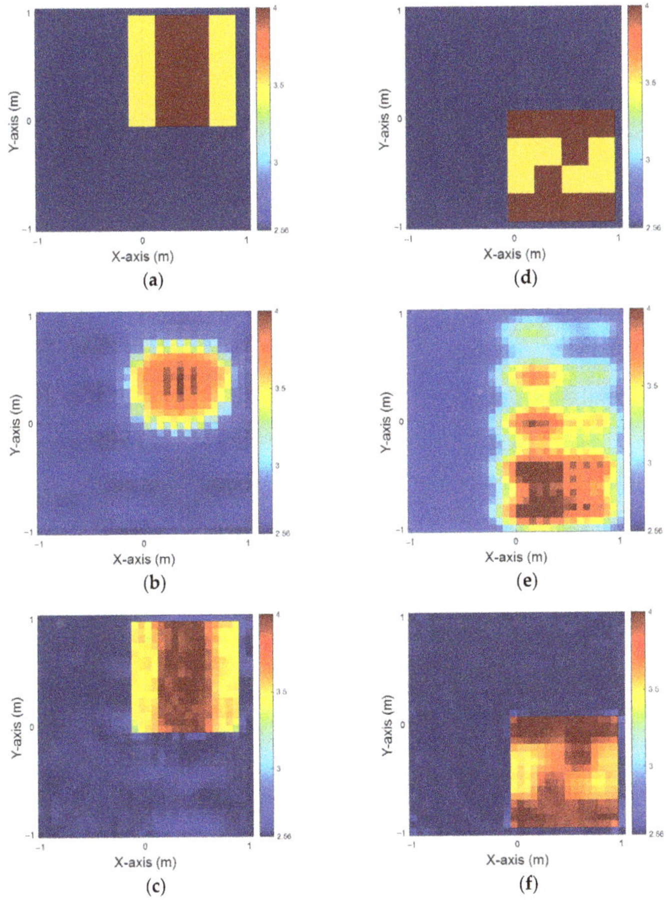

Figure 4. Relative permittivity between 3.5 and 4. (**a**) Ground truth for ε_z. (**b**) BPS reconstruction outcome for ε_z. (**c**) DCS reconstruction outcome for ε_z. (**d**) Ground truth for ε_x. (**e**) BPS reconstruction outcome for ε_x. (**f**) DCS reconstruction outcome for ε_x.

4.2. Relative Permittivity Between 4 and 4.5

In this case, the permittivity distribution is defined within the range of 4 to 4.5. We consider a scenario with 10 scatterers, each exhibiting unique permittivity profiles, positioned at 50 distinct locations across the designated measurement area. To mimic realistic conditions, 5% Gaussian noise is added to the scattered field data. The dielectric coefficient distribution is estimated using both the BPS and DCS methods. These estimates serve as the input for CNN to reconstruct electromagnetic images. Finally, we perform a comparative analysis of the reconstruction results obtained from the two different input datasets.

Figure 5a,d show the ground truth configurations for ε_z and ε_x, respectively. Figure 5b,e display the reconstruction results obtained using the BPS method, while Figure 5c,f present the DCS reconstruction outcomes for the ε_z and ε_x cases, respectively. The results reveal that the presence of an object buried in a half-space medium causes significant scattering of incident waves, adversely affecting reconstruction accuracy. In addition, the nonlinearity increases proportionally with the dielectric constant of the embedded object, posing significant challenges to reconstructing accurate electromagnetic images. Likewise in case A, despite utilizing two distinct input configurations and a well-trained CNN, the BPS method is limited to approximating the object's location. In contrast, the DCS method demonstrates superior performance, providing more accurate estimates of both the position and distribution of buried materials. Table 2 summarizes the RMSE and SSIM metrics for this scenario.

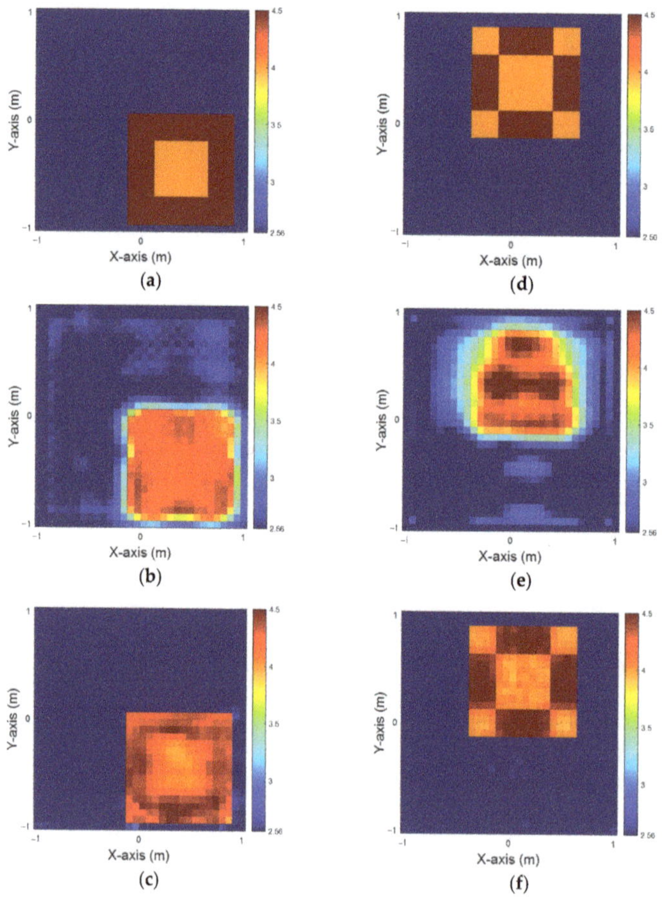

Figure 5. Relative permittivity between 4 and 4.5. (**a**) Ground truth for ε_z. (**b**) BPS reconstruction outcome for ε_z. (**c**) DCS reconstruction outcome for ε_z. (**d**) Ground truth for ε_x. (**e**) BPS reconstruction outcome for ε_x. (**f**) DCS reconstruction outcome for ε_x.

Table 2. RMSE and SSIM for ε_z and ε_x for relative permittivity between 4 and 4.5.

Performance	ε_z		ε_x	
	BPS	**DCS**	**BPS**	**DCS**
RMSE	7.8%	3.12%	9.72%	0.82%
SSIM	81.02%	97.46%	78.76%	99.62%

4.3. Relative Permittivity Between 4.5 and 5 Using the Model in Section 4.2

In this case, the permittivity distribution is defined within the range of 4.5 to 5. We consider a scenario with 10 scatterers, each exhibiting unique permittivity profiles, positioned at 50 distinct locations across the designated measurement area. To mimic realistic conditions, 5% Gaussian noise is added to the scattered field data. The dielectric coefficient distribution is estimated using the DCS method. The estimates serve as inputs to the CNN for image reconstruction. In practical scenarios, noise interference is prevalent and has a significant impact on reconstruction outcomes. We use the model in Section 4.2 to reconstruct the relative permittivity. Figure 6a,c are the ground truth for ε_z and ε_x, respectively. Figure 6b,d are the DCS reconstruction results of ε_z and ε_x using the Section 4.2 model, respectively. The numerical results show that our proposed method has good generalization ability. Table 3 summarizes the RMSE and SSIM metrics for this scenario.

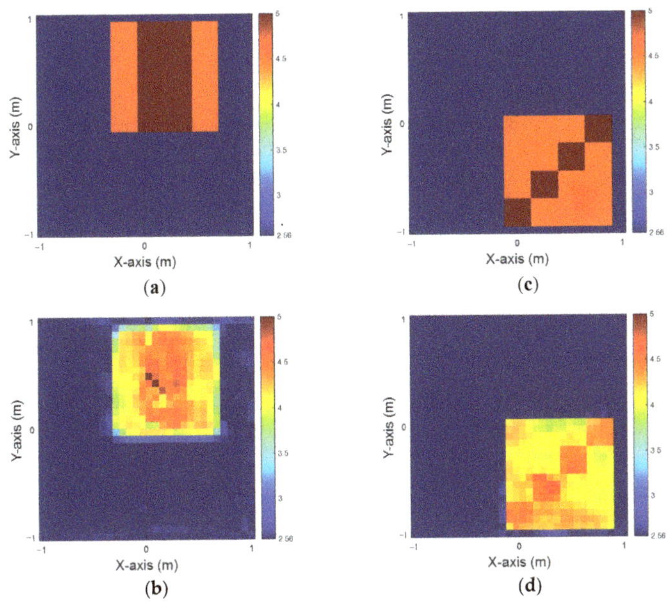

Figure 6. Relative permittivity between 4.5 and 5 using Section 4.2 model. (**a**) ground truth for ε_z (**b**) ground truth for ε_x (**c**) DCS reconstruction outcome for ε_z (**d**) DCS reconstruction outcome for ε_x.

Table 3. RMSE and SSIM for ε_z and ε_x for relative permittivity between 4.5 and 5.

Performance	ε_z	ε_x
RMSE	9.04%	8.02%
SSIM	91.28%	96.04%

5. Conclusions

We propose a novel method for the electromagnetic imaging of uniaxial objects buried in a half-space using CNNs. Measurement constraints make it difficult to gather sufficient information for improved reconstruction. We use the TM and TE waves to illuminate a uniaxial object, receive the scattered fields, and calculate the estimated permittivity distribution using the BPS and the DCS. The results show that the DCS, when processed through CNNs, provides a more accurate dielectric constant than the BPS. The TE case is more complex due to the vector nature of the TE electric field. Traditional algorithms are computationally expensive, whereas CNNs provide benefits like lower costs and faster processing in under one second, making it effective for electromagnetic inverse scattering problems. Compared to the BPS, a major drawback of the DCS is the significant computational resources and time required for estimating the initial dielectric constant distribution. However, when combined with a CNN, this approach produces impressive reconstruction results. It effectively tackles nonlinear phenomena and exhibits robust resistance to noise.

Future research will investigate the relationship between anisotropic parameters across various real-world objects and explore the potential of applying constraints to biaxial anisotropic parameters. There are promising directions in AI applications for electromagnetic imaging, such as using generative adversarial network and segmenting scatterers for high-resolution reconstruction.

Author Contributions: Conceptualization, H.J.; data curation, J.-S.C.; formal analysis, P.-H.C.; funding acquisition, J.-S.C.; investigation, H.J.; methodology, J.-S.C.; project administration, C.-C.C.; resources, H.J.; software, P.-H.C.; supervision, J.-S.C.; visualization, H.J.; writing—original draft, C.-C.C. and P.-H.C.; writing—review and editing, C.-C.C. All authors have read and agreed to the published version of the manuscript.

Funding: This research was funded by the National Science and Technology Council Taiwan, grant number NSTC 112-2221-E-032-014-MY2.

Institutional Review Board Statement: Not applicable.

Informed Consent Statement: Not applicable.

Data Availability Statement: The original contributions presented in this study are included in the article. Further inquiries can be directed to the corresponding author.

Conflicts of Interest: The authors declare no conflict of interest.

References

1. Uecker, M.; Hohage, T.; Block, K.T.; Frahm, J. Image reconstruction by regularized nonlinear inversion-joint estimation of coil sensitivities and image content. *Magn. Reson. Med.* **2008**, *60*, 674–682. [CrossRef]
2. Deisboeck, T.; Kresh, J.Y. *Complex Systems Science in Biomedicine*; Springer Science & Business Media: New York, NY, USA, 2007.
3. Song, L.P.; Yu, C.; Liu, Q.H. Through-wall imaging (TWI) by radar: 2-D tomographic results and analyses. *IEEE Trans. Geosci. Remote Sens.* **2005**, *43*, 2793–2798. [CrossRef]
4. Chen, X.; Wei, Z.; Li, M.; Rocca, P. A Review of Deep Learning Approaches for Inverse Scattering Problems. *Prog. Electromagn. Res.* **2020**, *167*, 67–81. [CrossRef]
5. Sanghvi, Y.; Kalepu, Y.; Khankhoje, U.K. Embedding Deep Learning in Inverse Scattering Problems. *IEEE Trans. Comput. Imaging* **2020**, *6*, 46–56. [CrossRef]
6. Habashy, T.M.; Groom, R.W.; Spies, B.R. Beyond the Born and Rytov approximations: A nonlinear approach to electromagnetic scattering. *J. Geophys. Res. Solid Earth* **1993**, *98*, 1759–1775. [CrossRef]
7. Belkebir, K.; Chaumet, P.C.; Sentenac, A. Super resolution in total internal reflection tomography. *J. Opt. Soc. Amer. A* **2005**, *22*, 1889–1897. [CrossRef]
8. Chen, X. *Computational Methods for Electromagnetic Inverse Scattering*; Wiley-IEEE Press: New York, NY, USA, 2018.
9. Chew, W.C.; Wang, Y.-M. Reconstruction of two-dimensional permittivity distribution using the distorted Born iterative method. *IEEE Trans. Med. Imaging* **1990**, *9*, 218–225. [CrossRef]

10. Van DenBerg, P.M.; Kleinman, R.E. A contrast source inversion method. *Inverse Probl.* **1997**, *13*, 1607.
11. Chen, X. Subspace-based optimization method for solving inverse scattering problems. *IEEE Trans. Geosci. Remote Sens.* **2010**, *48*, 42–49. [CrossRef]
12. Yao, H.M.; Sha, W.E.I.; Jiang, L. Two-Step Enhanced Deep Learning Approach for Electromagnetic Inverse Scattering Problems. *IEEE Antennas Wirel. Propag. Lett.* **2019**, *18*, 2254–2258. [CrossRef]
13. Zhang, H.H.; Yao, H.M.; Jiang, L.; Ng, M. Enhanced Two-Step Deep-Learning Approach for Electromagnetic-Inverse-Scattering Problems: Frequency Extrapolation and Scatterer Reconstruction. *IEEE Trans. Antennas Propag.* **2023**, *71*, 1662–1672. [CrossRef]
14. Yao, H.M.; Zhang, H.H.; Jiang, L.; Ng, M. Enhanced Deep Learning Approach for Electromagnetic Forward Modeling of Dielectric Target Within the Wide Frequency Band Using Deep Residual Convolutional Neural Network. *IEEE Antennas Wirel. Propag. Lett.* **2024**, *23*, 1884–1888. [CrossRef]
15. Wei, Z.; Chen, X. Deep-Learning Schemes for Full-Wave Nonlinear Inverse Scattering Problems. *IEEE Trans. Geosci. Remote Sens.* **2019**, *57*, 1849–1860. [CrossRef]
16. Zhang, L.; Xu, K.; Song, R.; Ye, X.; Wang, G.; Chen, X. Learning-Based Quantitative Microwave Imaging with a Hybrid Input Scheme. *IEEE Sens. J.* **2020**, *20*, 15007–15013. [CrossRef]
17. Zhou, Y.; Zhong, Y.; Wei, Z.; Yin, T.; Chen, X. An Improved Deep Learning Scheme for Solving 2-D and 3-D Inverse Scattering Problems. *IEEE Trans. Antennas Propag.* **2021**, *69*, 2853–2863. [CrossRef]
18. Song, R.; Huang, Y.; Ye, X.; Xu, K.; Li, C.; Chen, X. Learning-Based Inversion Method for Solving Electromagnetic Inverse Scattering with Mixed Boundary Conditions. *IEEE Trans. Antennas Propag.* **2022**, *70*, 6218–6228. [CrossRef]
19. Wang, Y.; Zhao, Y.; Wu, L.; Yin, X.; Zhou, H.; Hu, J.; Nie, Z. An Early Fusion Deep Learning Framework for Solving Electromagnetic Inverse Scattering Problems. *IEEE Trans. Geosci. Remote Sens.* **2023**, *61*, 2005914. [CrossRef]
20. Wu, Z.; Peng, Y.; Wang, P.; Wang, W.; Xiang, W. A Physics-Induced Deep Learning Scheme for Electromagnetic Inverse Scattering. *IEEE Trans. Microw. Theory Tech.* **2024**, *72*, 927–947. [CrossRef]
21. Chiu, C.C.; Lee, G.Z.; Jiang, H.; Hong, B.J. Microwave Imaging of a Periodic Homogeneous Dielectric Object Buried in Rough Surfaces. *J. Electromagn. Waves Appl.* **2019**, *33*, 1905–1919. [CrossRef]
22. Topbaş, T.O.; Alkumru, A. A hybrid method for an inverse scattering problem related to cylindrical bodies buried in a half-space. *Appl. Math. Sci. Eng.* **2023**, *31*, 2248355. [CrossRef]
23. Chiu, C.-C.; Chien, W.; Yu, K.-X.; Chen, P.-H.; Lim, E.H. Electromagnetic Imaging for Buried Conductors Using Deep Convolutional Neural Networks. *Appl. Sci.* **2023**, *13*, 6794. [CrossRef]
24. Chiu, C.C.; Lee, Y.H.; Chen, P.H.; Shih, Y.C.; Jiang, H. Application of Self-Attention Generative Adversarial Network for Electromagnetic Imaging in Half-Space. *Sensors* **2024**, *24*, 2322. [CrossRef] [PubMed]
25. Li, L.; Wang, L.G.; Teixeira, F.L.; Liu, C.; Nehorai, A.; Cui, T.J. DeepNIS: Deep Neural Network for Nonlinear Electromagnetic Inverse Scattering. *IEEE Trans. Antennas Propag.* **2019**, *67*, 1819–1825. [CrossRef]
26. Cheng, Y.; Xiao, L.-Y.; Zhao, L.-Y.; Hong, R.; Liu, Q.H. A 3-D Full Convolution Electromagnetic Reconstruction Neural Network (3-D FCERNN) for Fast Super-Resolution Electromagnetic Inversion of Human Brain. *Diagnostics* **2022**, *12*, 2786. [CrossRef]
27. Zhang, H.; Chen, Y.; Cui, T.J.; Teixeira, F.L.; Li, L. Probabilistic Deep Learning Solutions to Electromagnetic Inverse Scattering Problems Using Conditional Renormalization Group Flow. *IEEE Trans. Microw. Theory Tech.* **2022**, *70*, 4955–4965. [CrossRef]

sensors

Article

Leaky Wave Generation Through a Phased-Patch Array

Alessandro Calcaterra [1,†], Patrizio Simeoni [2,†], Marco Donald Migliore [3] and Fabrizio Frezza [1,*]

1 Department of Information Engineering, Electronics and Telecommunications, La Sapienza University of Rome, 00184 Rome, Italy
2 Department of Electronics Engineering and Communications, South East Technological University (SETU), Carlow Campus, R93 V960 Carlow, Ireland
3 Dipartimento di Ingegneria Elettrica e dell'Informazione (DIEI) "Maurizio Scarano", University of Cassino and Southern Lazio, 03043 Cassino, Italy
* Correspondence: fabrizio.frezza@uniroma1.it
† These authors contributed equally to this work.

Abstract: For this article, we approximated the field of a leaky-wave antenna (LWA) with the field produced by a uniform linear array (ULA). This article aims to provide an initial framework for applications where the generation of an inhomogeneous wave is wished, but, at the same time, a flexibility is required that is difficult to meet with the conventional LWA design. In particular, two different configurations were considered, one with a simple Menzel antenna operating at 12 GHz, and one, relevant for practical applications, with an antenna operating at 2.4 GHz. This study aimed, in both cases, to highlight the distance at which the field produced by the phased array with the chosen sampling method can approximate effectively the one produced by a leaky-wave antenna and to verify whether this could cause issues for the targeted application.

Keywords: leaky waves; leaky-wave antennas; inhomogeneous waves

1. Introduction

It has been demonstrated that inhomogeneous waves can enable deeper penetration in lossy media. Such waves can often be approximated by the field produced by a leaky-wave antenna. However, in specific applications—such as hyperthermia treatment, where the radiating environment (i.e., biological tissues) varies from patient to patient—generating a precise field configuration requires tuning that is challenging to achieve with classical leaky-wave antennas, particularly under deep penetration constraints [1]. In such cases, an alternative approach that maintains the desired field characteristics while offering greater flexibility may yield better results. This can be achieved through the use of a phased array. Leaky waves are the only inhomogeneous waves capable of delivering power in a lossless medium, as they are not bounded to a surface, as happens with surface waves, plasmonic polaritons, or others [2]. As with surface waves, on the other hand, leaky waves are waves that appear only in the presence of a discontinuity or an asymmetry, in general [3]. Leaky-wave antennas are well-known artificial structures used to generate near-field radiation that closely approximates a leaky wave, and they are traditionally employed to produce a leaky-wave field. In this article, we aim to replace the leaky-wave antenna with a phased array. As is well known in the literature, the field produced by a continuous aperture antenna can be approximated by that of an array by sampling the currents on the antenna aperture [4]. As stated in [4], the accuracy of this approximation increases as the distance between the array elements decreases. In the limit, the two fields become identical, though this is clearly impractical. When the spacing between elements is large, the fields between the two antennas may not match well and direct sampling may not provide good results. Solutions to this problem have been proposed in the literature, including root-matching methods and perturbation techniques [4,5], as well as more recent integrated sampling methods [4,6]. The need for these methods highlights the complexity of achieving a sufficiently accurate approximation of the field produced by an antenna array compared to that of a continuous antenna. For this reason, this paper investigated whether, in the case considered, simple sampling at a distance of $\lambda/2$ is sufficient for the application proposed in [7], by first comparing the far fields and then the near fields produced by the two antennas. This study represents an initial step in that direction. In the literature, many LWAs structures have been proposed, to address aspects such as miniaturisation, directivity, full scanning, and stop-band suppression [8], but they were not strictly relevant to the goal of this article. For this work, therefore, we chose a simple LWA—the Menzel antenna—due to its specific attenuation and phase vector properties [9]. A practical way to realise a leaky-wave antenna (LWA) is to consider a travelling-wave structure with several asymmetries properly designed to implement a given value for the attenuation and propagation constants [10–12].

In the literature, a well-known interpretation of the leaky-wave radiation phenomenon also explains the presence of a shadow cone where the field behaves improperly, as shown in Figure 1 [11]:

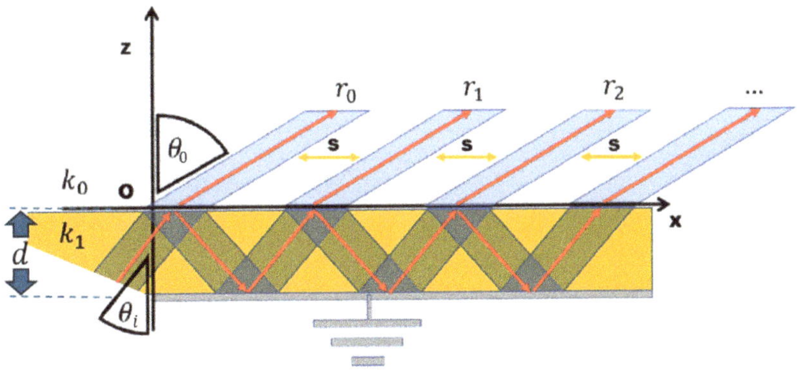

Figure 1. Pictorial representation of leaky radiation principle.

Inside the dielectric, as is well known from the electromagnetic theory [13–15], whenever the field reaches the dielectric–air interface, it is partially reflected and transmitted. The reflection and transmission coefficients, also known as Fresnel coefficients, depend on the angle of incidence, the electromagnetic characteristics of the two media involved, and on the polarization (TE or TM) of the considered field [13–15].

What is important is that at each incidence a leakage occurs, generating a propagating field in the air that moves away from the interface. The ratio of this field with respect to the impinging one is, by definition, the transmission coefficient [13–15]. The reflected field, on the other hand, keeps propagating inside the waveguide. Again, the remaining power will only be a portion of the initial impinging field, due to the leakage.

This leakage rate per meter is the attenuation constant that, due to the conservation of the tangential component of the electric field at the interface, is present in the field radiated in the air.

It is well known that the behaviour of the field inside the guide is frequency-dependent. In fact, starting from the cut-off frequency at which the field does not travel inside the dielectric (the stop-band region), by increasing the frequency the propagation vector points more and more towards the propagation direction shown in Figure 1 (i.e., x direction). Again, this is a good explanation of how a leaky-wave antenna operates by varying the frequency: the higher the frequency, the more the field is radiated towards endfire.

To analyse the numerous types of existing leaky-wave antennas was beyond the scope of this paper. We only considered a uniform monodimensional LWA, where the cut-off of the guided mode of interest coincides with the stop band.

The analysis and design of a given LWA requires the derivation of its dispersion diagram and the calculation of the attenuation and propagation constants related to the geometrical and electromagnetic characteristics of the waveguide and its asymmetries. Finally, the attenuation and propagation constant are related to the far-field properties.

The following approximate relationships are present in the literature, valid for $\alpha << \beta$, where α is the absolute value of the attenuation vector and β is the absolute value of the phase vector:

$$\theta_0 \cong sin^{-1}(\beta_z/k_0) \tag{1}$$

$$\theta_{3dB} \cong \frac{0.91}{(L/\lambda_0)cos\theta_0} \propto \frac{\alpha}{k_0} \tag{2}$$

In the equations illustrated above, θ_0 is the angle between the normal and the antenna aperture and the radiated beam, θ_{3dB} represents the beamwidth, λ is the wavelength, β_z is the z component of the phase vector, and k_0 is the wave number in vacuum. These relations imply that the pointing angle is decided by the value of the propagation constant β, while the beamwidth is decided by the attenuation constant α. The intuitive explanation of the frequency behaviour has been provided above. As far as the relationship between the attenuation constant and the beamwidth is concerned, it is sufficient to note that a larger attenuation results in an earlier decay of the electromagnetic field and so in a shorter electrical length and, as is known, a wider beam.

Phased Array

The properties of a phased array are well known and will not be analysed in detail here. The interested reader may refer to [4,16].

It is worth recalling here that the active electronically scanning array (AESA) can produce a beam that scans the space by simply modifying the phase of each element. In particular, the scanning is achieved by satisfying a condition on the phase shift to allow constructive interference between the fields produced by the radiating elements along the direction of interest.

For active linear uniform phased arrays, this condition is

$$\Delta\phi = kd[sin(\theta) - sin(\theta_0)] \tag{3}$$

where k is the wave number and d is the fixed distance between elements. The amplitude of the excitation phasor of each radiating element impacts the pattern shape. Typically, this property is used to reduce side lobes by applying proper tapering [16], i.e., reducing the amplitude of elements close to the edges. Intuitively, this happens because by reducing the currents on the edges the antenna results are electrically smaller, leading to a wider main beam and lower side lobes.

So, to achieve the desired pattern, i.e., to impose the desired set of coefficients for the radiating elements, it is necessary to modulate the amplitude and phase of each element. A typical solution that allows for achieving this employs a Digital Variable Attenuator to control the amplitudes and Digital Phase Shifters to modify the phases. These are components with a given number of bits (that ranges typically from 6 to 8) that introduce either an attenuation or a phase shift. Introducing a phase shift between radiating elements assures the beam pointing for a specific frequency. This means that for higher or lower frequencies, the relation (3) is no longer valid, resulting in a beam that squints around the direction of interest. This topic will be further developed in the Results section of this paper, where the behaviour of the phased array is compared to that of the LWA.

This article aims to establish an alternative and improved design of the antenna proposed in [7], which utilizes inhomogeneous waves for hyperthermia applications, as they may enable deeper tissue penetration. In this study, we focused on a numerically feasible scenario by considering an antenna operating at 12 GHz as a feasibility study. Then, we explored its application to hyperthermia by designing a prototype operating at 2.4 GHz, paving the way for more advanced antenna designs and experimental measurements.

2. Materials and Methods

All the simulations were performed with commercial electromagnetic software that implements finite integration techniques (FITs) in Time Domain [17,18]. The choice of the FIT Solver for all the structures considered here was due to its granting the same simulation conditions for the antennas. For the LWAs, the correct operating mode was excited through a waveguide port. For the array, coaxial connectors were designed to provide a 50 Ohm feeding line to the patch antennas. Finally, all the metallizations were perfect electric conductors (PECs), and the dielectrics (that were the same for both the LWAs and the patch arrays) were loss-free. These latter choices were made to improve the simulation time, and they do not represent a restriction of generality with respect to the paper's purposes. More specifically, the antenna has been designed and simulated by using the CST Microwave Studio Software [19] licensed to the DIET Department of "La Sapienza" University of Rome, and all figures shown in the paper have been obtained either by producing them with such a software, or re-designing them, eventually using MATLAB [20], with the addition of custom information, aimed at providing additional details which were not available in the original figures. Also, Microsoft Powerpoint [21] has been used to add additional information, e.g., geometrical dimensions, to the antenna designs produced.

2.1. The Menzel Antenna Operating at 12 GHz

The LWA chosen for this analysis is based on the Menzel antenna, already considered in [22,23]. Its geometrical characteristics are recalled in Figure 2. In this case, a waveguide port is placed on the antenna termination to absorb all the non-radiated field.

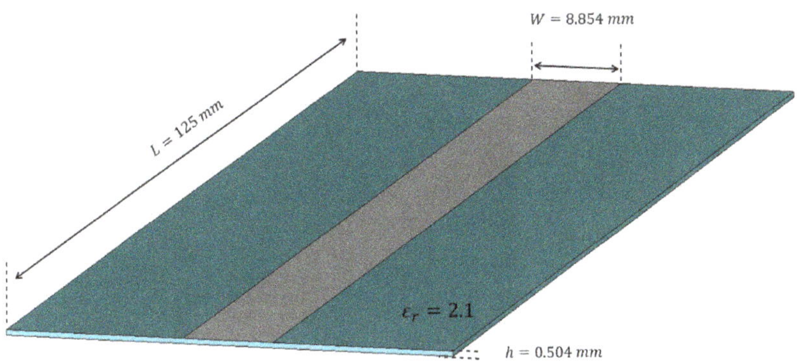

Figure 2. Design of the Menzel antenna.

This antenna radiates in its odd higher-order modes [22,23]; we chose here to excite the first higher-order odd mode, simulated and plotted in Figure 3.

In a simulation environment capable of separately analysing the modes of a waveguide it is possible to selectively excite the mode of interest.

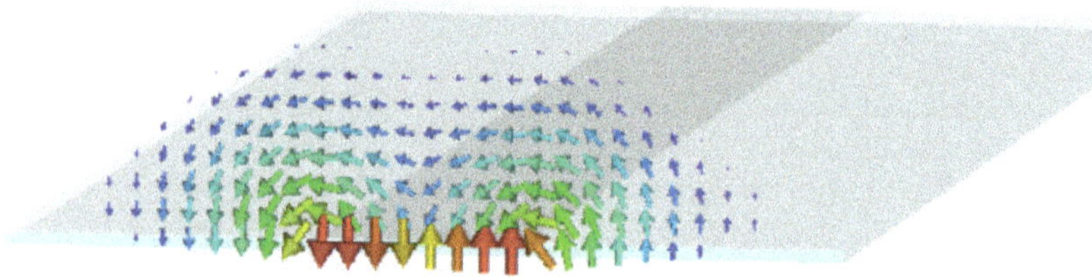

Figure 3. Radiating mode of the Menzel antenna.

At the prototyping stage, it is necessary to suppress the fundamental (Q-TEM) and the first even higher-order mode; however, as the scope of this paper was to highlight and demonstrate the similarities between a phased array and a leaky-wave antenna in a simulation environment we did not consider a more feasible model for the LWA, such as the one presented in [7]; we focused, instead, on the electromagnetic radiation of interest.

In Figures 4 and 5, respectively, the LWA near field and far field are shown:

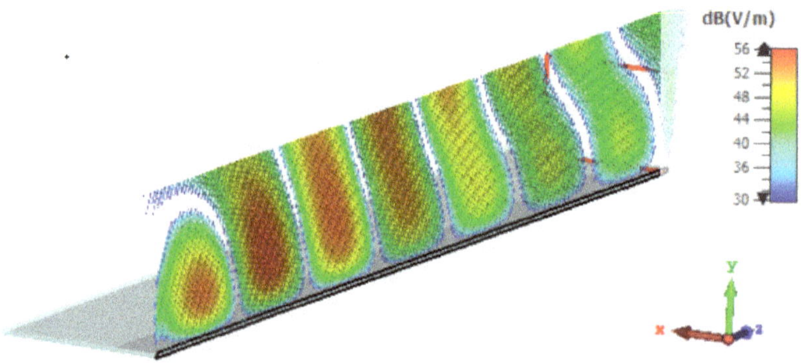

Figure 4. Near field of the Menzel antenna.

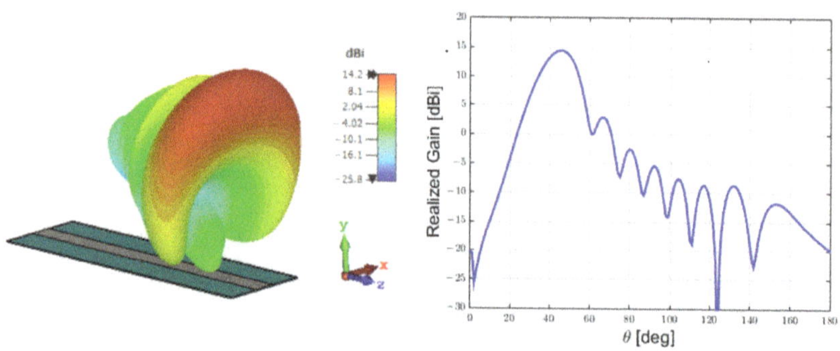

Figure 5. Far field of the Menzel antenna.

2.2. The Sampling Methodology

To sample the current on the aperture, we had to choose the elementary radiating elements and their locations, i.e., the spatial sampling. In regard to the radiating element, a simple patch antenna [24,25] was chosen, due to its similarity to the leaky-wave antenna considered.

Figure 6 shows the patch antenna that was designed as an element of the array: even if it was not necessary, the same substrate of the LWA was used:

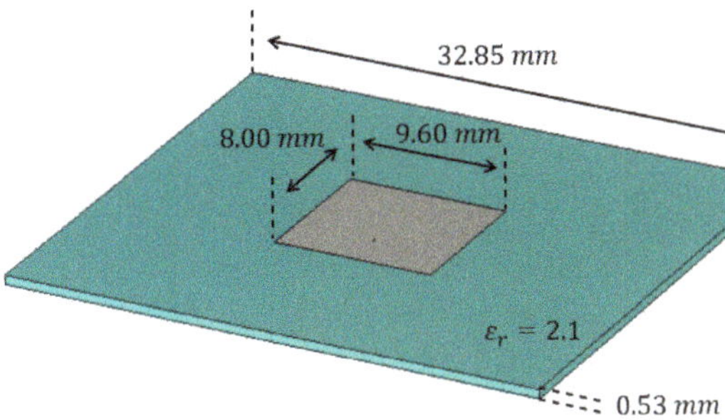

Figure 6. Patch antenna used as element of the array.

When in the array, the patch antenna had to be placed so that the field was polarized in the same direction as the one radiated by the leaky-wave antenna; see Figure 7. To determine the exact number of patches in the array, a straightforward observation could be made: in order to radiate the same field, it was necessary that the two structures had the same currents on the same electrical area. The simplest way to assure this for the two antennas considered was to make them the same geometrical length.

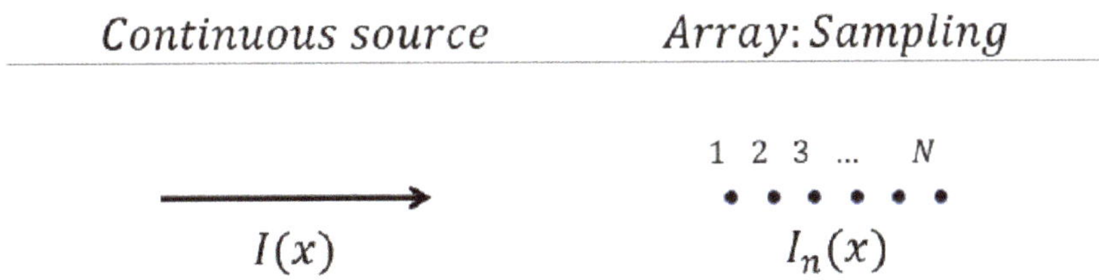

Figure 7. Schematic representation of a continuous current source (on the **left**) and of a discrete current source obtained by sampling the continuous source (on the **right**).

Placing the patches too close together would have resulted in a poor active reflection coefficient, due to the mutual coupling. On the other hand, it was not advisable to place them too far apart, since grating lobes might occur and, furthermore, a coarse sampling would have resulted in a poor representation of the field emitted by the LWA.

For these reasons, it was decided to keep $\lambda/2$ spacing between the patches: this was the maximum distance that allowed us to avoid grating the lobes in the visible region when scanning up to endfire. Given the geometry considered, the array that we designed consisted of 9 radiating elements (patches) placed at a constant distance.

To finalize the array design, it was essential to define the amplitudes and phases of the patch elements. Regarding the phases, we began by setting the LWA to radiate at a $\theta^{LWA} = 45°$ angle; then, the phases for the array were set to values equal to the ones required by a phased array to point the beam in the same direction as the LWA.

For the amplitudes, the normalised power in the dielectric of the LWA was simulated along the longitudinal direction. Then, it was sampled at each position in which a radiating element was present in the phased array along the symmetry axis. Figure 8 illustrates the resulting power distribution along the longitudinal direction of the patch antenna.

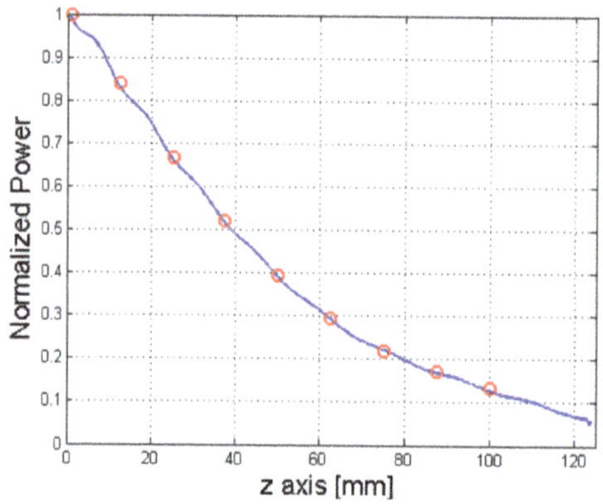

Figure 8. Sampling of the power inside the dielectric of the LWA.

The phasors used for the 9 elements are represented in Table 1:

Table 1. The table shows how the 9 radiating elements are excited in terms of power (in watts), amplitude (in volts), and phase (given as accumulated phase shift in degrees).

Element	Power [W]	Amplitude [V]	Phase [deg]
1	1	1.4142	0
2	0.8290	1.2876	127.37
3	0.6610	1.1498	254.74
4	0.5157	1.0155	382.11
5	0.3917	0.8851	509.48
6	0.2922	0.7645	636.85
7	0.2219	0.6661	764.22
8	0.1684	0.8504	891.59
9	0.1266	0.5032	1019.01

With these assumptions, the patch array of Figure 9 was realised:

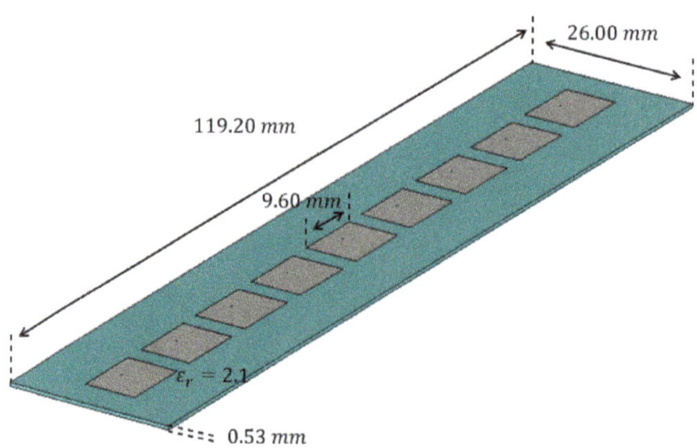

Figure 9. Layout of the phased array constituted by the patch antennas chosen.

The performances of the array obtained could be rapidly assessed by evaluating the active reflection coefficient of each element (Figure 10) and by the average embedded element pattern (Figure 11):

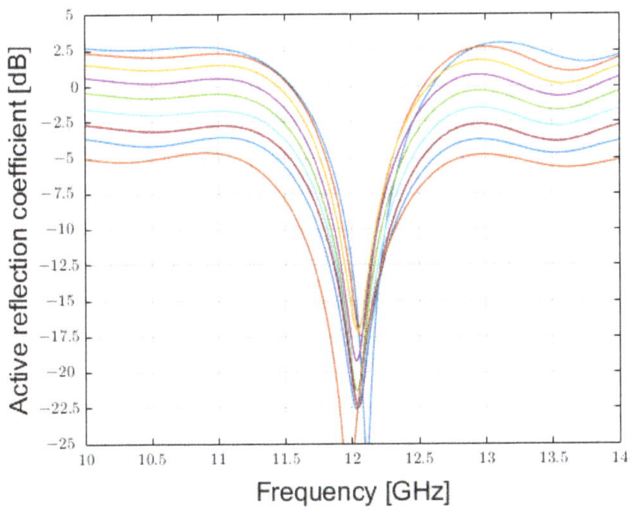

Figure 10. Active reflection coefficients for the 9 patch-antenna elements composing the phased array excited with the coefficients reported in Table 1.

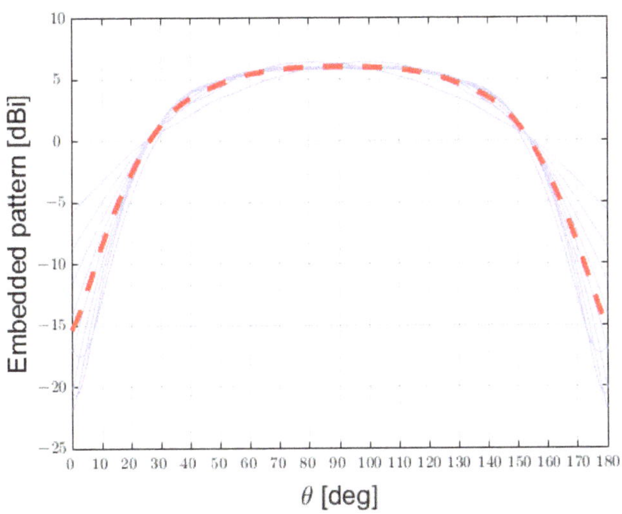

Figure 11. Embedded patterns of the 9 patch-antenna elements composing the phased array (in purple) and the average embedded pattern (in dashed red).

2.3. *Analysis of an Antenna Operating at 2.4 GHz*

We then focused on a practical application, i.e., an LWA for hyperthermia treatment operating at 2.4 GHz. It is well known that in LWAs there may be an excess of power at the end of the antenna; this problem becomes more relevant when trying to minimize antenna dimensions. In the literature, different methods have been employed, therefore, to try re-using the power that is available at the end of the LWA, in order to improve the efficiency [7,26–28]. Where frequency scanning is required, a design such as the one implemented in [27] could be considered, while in applications like the one in [7], where a specific incidence angle alongside a wave that well approximates a single inhomogeneous wave is desired, the alternative approach proposed here can also be explored. For a near-field application, it is pivotal to verify the distance from the antenna at which the phased array well approximates the leaky-wave antenna. To demonstrate whether an acceptable field can be achieved at a short distance from the array, we took into consideration the design proposed in [7], as a reference design, and we eliminated the reliance on the Wilkinson splitter, which has the downside of requiring an additional PCB, by designing the array shown in Figure 12:

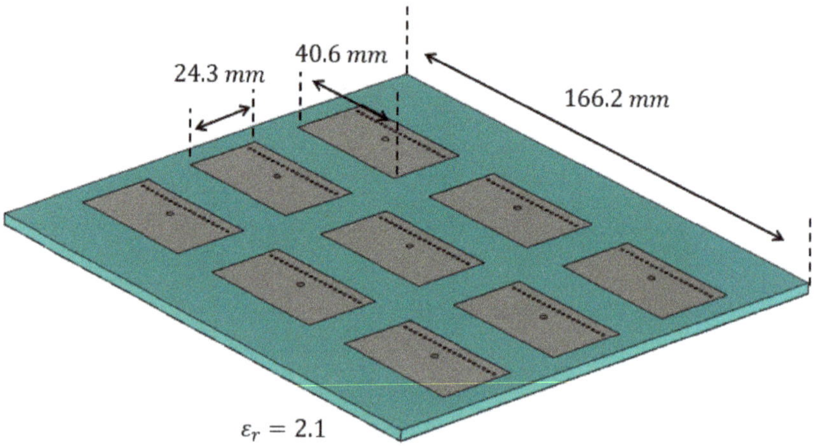

Figure 12. Design of a compact uniform LWA realised employing a phased array.

3. Results

3.1. Comparison Between the Antennas Operating at 12 GHz

We first compared the two antennas in terms of the generated far field, because the far field was obtained via an integral transformation of the near-field radiation, resulting in a preliminary estimator of the similarity among the two structures; see Figure 13. Also, this provided important information related to the radiating properties of the two antennas.

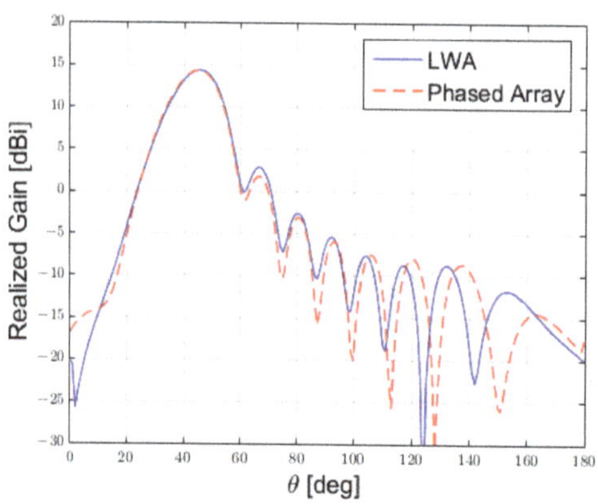

Figure 13. Comparison between the far field produced by the LWA versus the one produced by the phased array.

We then examined the near field, to identify the maximum vertical distance from the antenna where the array still produced a field that satisfactorily approximated the one generated by the Menzel antenna.

After having verified the similarity in the far field, before focusing on the near field, it is worth making some observations to extend the comparison between these two structures to the frequency-scan behaviour.

It is well known that the LWA radiation mechanism is given by the radiation losses provoked into a guiding structure. So, it is clear that the propagating mode inside the considered lossy waveguide is of chief importance as far as the radiation of the LWA is concerned. If this mode is in cut-off, there will be no radiation at all. By scanning the frequency, the mode will start propagating inside the structure (i.e., its wave number component along the direction of propagation passes from predominantly imaginary to predominantly real), and the LWA will start radiating near broadside. Again, increasing the frequency, the LWA will scan the beam up to endfire.

For the reasons just stated, for a one-dimensional LWA it is impossible to radiate at broadside, since it would correspond to the radiation when the mode inside the guide is in cut-off. This issue, known as stop-band [11], can be overcome in several ways that will not be considered here. The interested reader may refer to the relevant literature, e.g., Ref. [11].

For a phased array, the situation is quite different. In this case, a propagating mode inside the structure is not present, and, to enforce a radiation angle for the beam, a specific phase relation between the radiating elements has been super-imposed here. It is then interesting to analyse what happens when the frequency varies, in particular focusing on the resulting spatial geometrical scanning of the radiated beam. In this case, the direction of scanning can be computed manipulating Equation (3), which we write here for the sake of clarity, it is:

$$\Delta\phi = kd[sin(\theta) - sin(\theta_0)] = kdsin(\theta) - \beta \tag{4}$$

then it follows:

$$\theta_0 = arcsin\left(\frac{\beta}{kd}\right) = arcsin\left(\frac{\beta}{2\pi fd/c}\right) \tag{5}$$

where $\Delta\phi$ is the total phase between the adjacent antenna elements, while $\beta = kd\sin(\theta_0)$ represents the progressive phase shift between the elements of the array [4]. It follows that if the amplitude of β is chosen so that the antenna radiates at a $\theta_0 = 45^\circ$ angle for a frequency $f = 12$ GHz then as the frequency increases the radiated beam tends to point to broadside. If the frequency decreases, the beam is directed towards endfire. This behaviour is the opposite with respect to an LWA, but it is very interesting to note that in this case there is also a stop-band behaviour at broadside as, from the relation above, it is clear that to achieve a θ_0 equal to 0, the operating frequency needs to tend to infinity. To reach endfire, instead, it is sufficient to reach the frequency, as follows:

$$f = \frac{\beta}{2\pi \sin\theta_0 d/c} \tag{6}$$

To assess this behaviour, let us consider Figure 14, where the array factor of a linear equi-spaced array with a phased coefficient, so as to scan at 45° at 12 GHz, has been considered at different frequencies. Here, the squinting behaviour is exaggerated to show how, given a set of coefficients to scan at a given angle, a stop-band behaviour is experienced at broadside.

In Figure 14, only the array factor is shown, for three main reasons: firstly, it would not be feasible to realise a radiating element that could cover such a bandwidth. This is even more true for patches that typically cover fractional bandwidths of 1–5% when in standard configuration [29,30]. Secondly, this phenomenon is due to the relative amplitude and phase of the electromagnetic field for the elements in a given lattice. All these characteristics, according to the elementary theory of array, belong to the array factor and not to the element factor that is, or at least should be, common to all radiating elements [16].

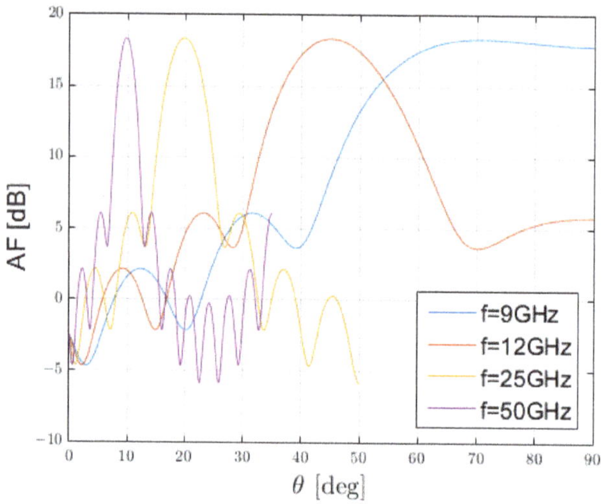

Figure 14. Array factor (AF) of a linear equi-spaced phased array with fixed phase-relation at different frequencies.

After having compared the far fields, we then analysed the differences in the near field, and, given the difference in behaviour in the case of the frequency scan, we concentrated on the situation in which both antennas pointed at a 45° angle, as in this case the two antennas operated at the same frequency, by design choice.

To make this evaluation, referring to the reference system shown in Figure 15, we firstly compared the amplitude of the electric field $|E_x(0, y_i, z)|$ as a function of z, sampling it at a different position, y_i, where the subscript i denoted a distinct sampling point along the y axis:

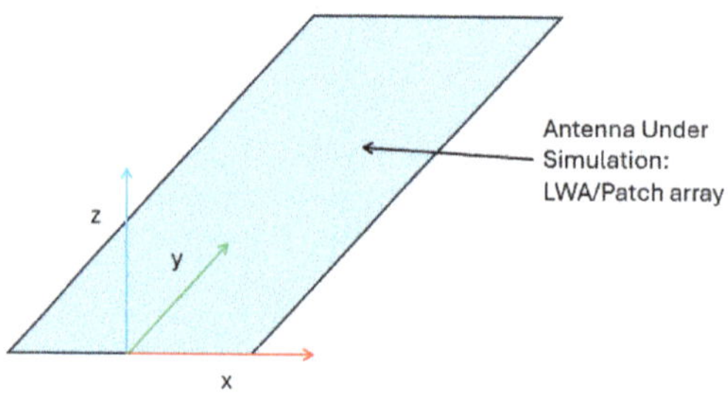

Figure 15. Reference system used for the comparison.

Our results are shown in Figure 16.

After this evaluation, we compared the field on planes for different constant values of the z coordinate, i.e., we plotted $|E_x(x, y, z_i)|$, choosing different values, z_i, along the z axis. Our results are visible in Figure 17.

The calculation of the amplitude of the field normalised by its maximum value was a useful estimator for evaluating the behaviour of the two structures without taking into account the different power radiated, i.e., without considering the efficiency of the antennas. Another interesting parameter that allowed us to obtain a deep understanding of the variation of the field was the derivative of such a field, as shown in Figure 18.

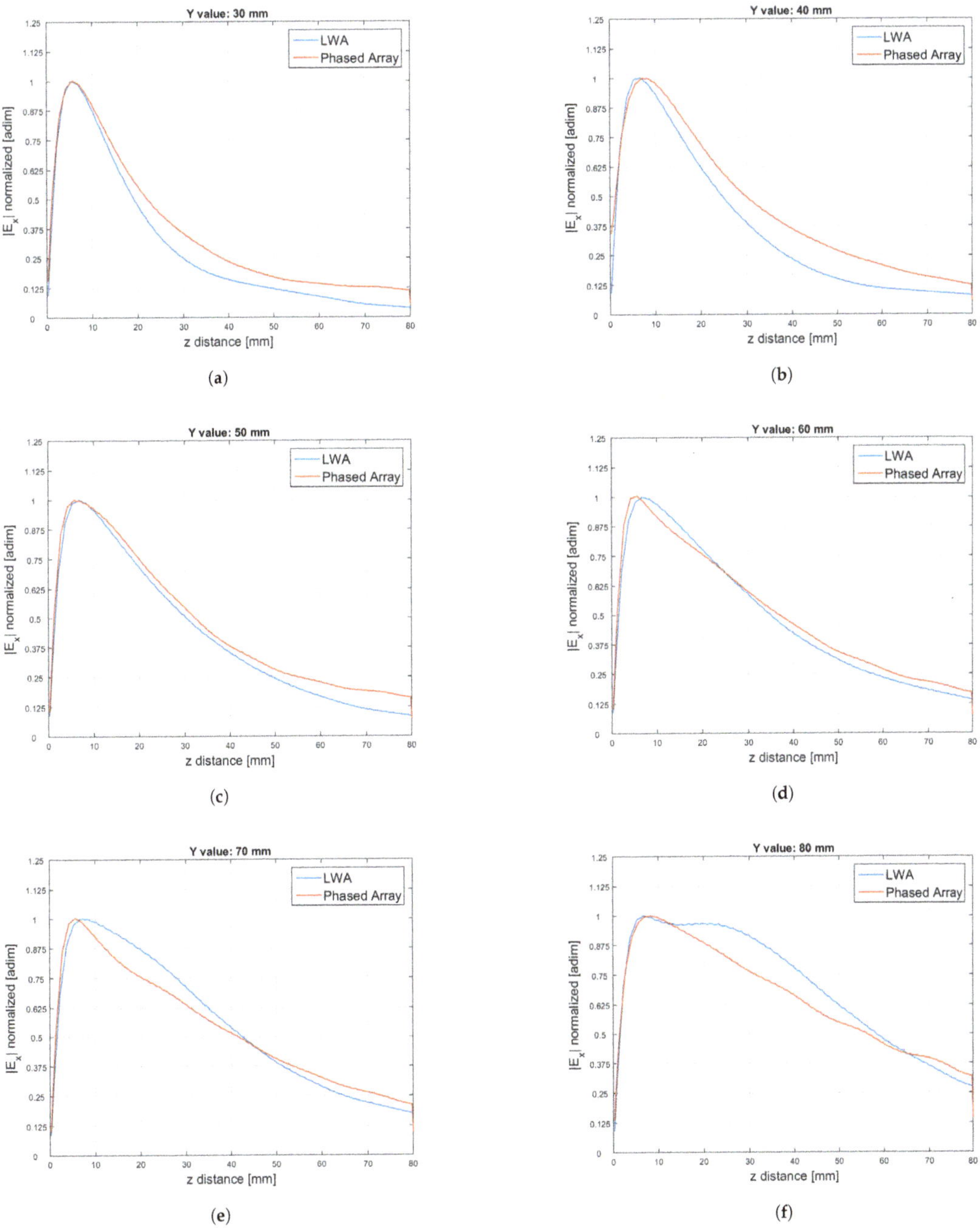

Figure 16. Comparison between the amplitude $|E_x|$ of the electric field normalised by its maximum for the LWA and the phased array along the z axis when $x = 0$ and for different constant values of the y coordinate: (**a**) $(x,y) = (0,30)$ mm; (**b**) $(x,y) = (0,40)$ mm; (**c**) $(x,y) = (0,50)$ mm; (**d**) $(x,y) = (0,60)$ mm; (**e**) $(x,y) = (0,70)$ mm; (**f**) $(x,y) = (0,80)$ mm.

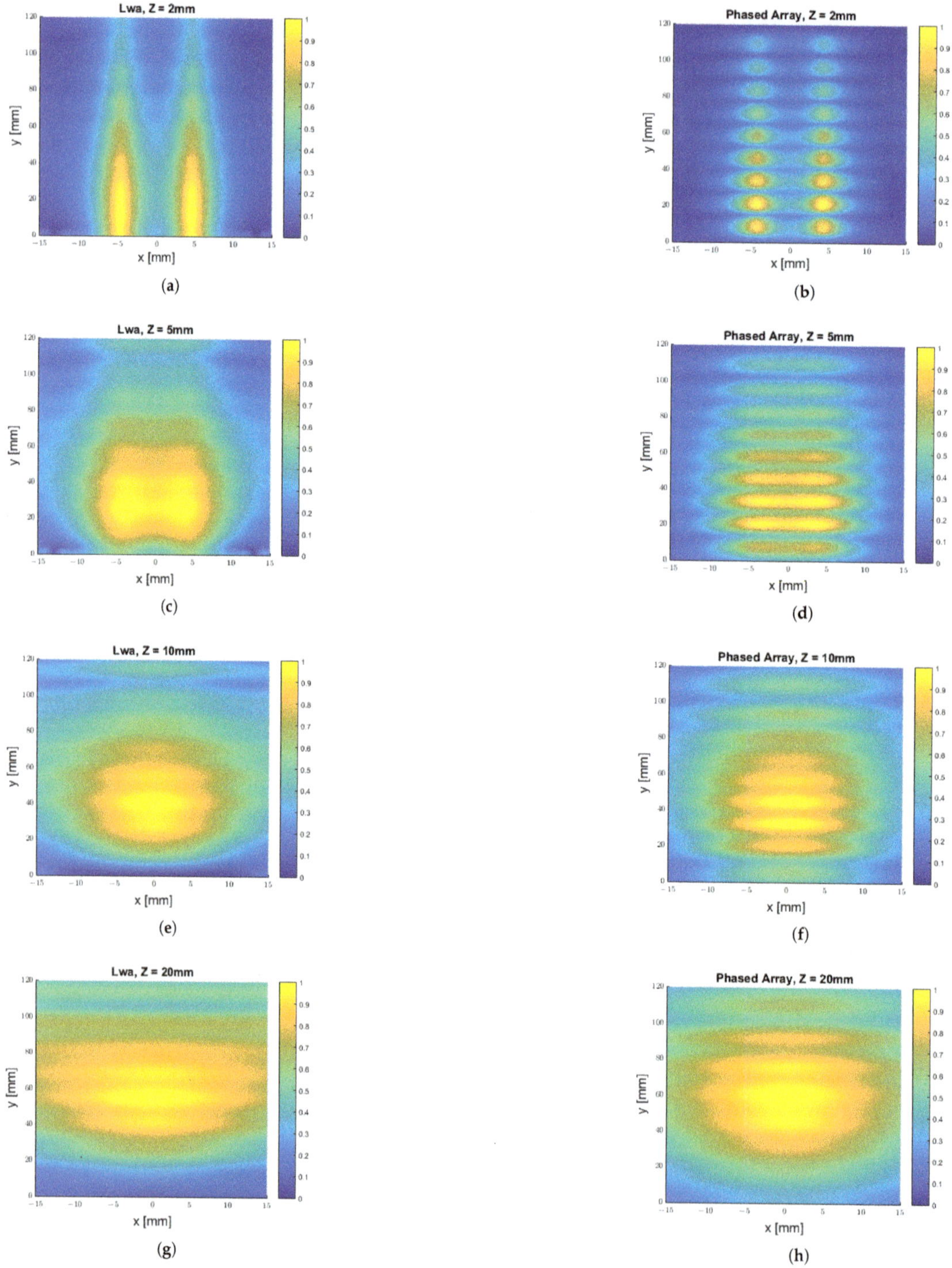

Figure 17. $|E_x|$ field component sampled at distinct planes $z = z_i$ for the LWA, on the left, and the phased array (PA), on the right: (**a**) $|E_x^{LWA}|$ for $z = 2$ mm; (**b**) $|E_x^{PA}|$ for $z = 2$ mm; (**c**) $|E_x^{LWA}|$ for $z = 5$ mm; (**d**) $|E_x^{PA}|$ for $z = 5$ mm; (**e**) $|E_x^{LWA}|$ for $z = 10$ mm; (**f**) $|E_x^{PA}|$ for $z = 10$ mm; (**g**) $|E_x^{LWA}|$ for $z = 20$ mm; (**h**) $|E_x^{PA}|$ for $z = 20$ mm.

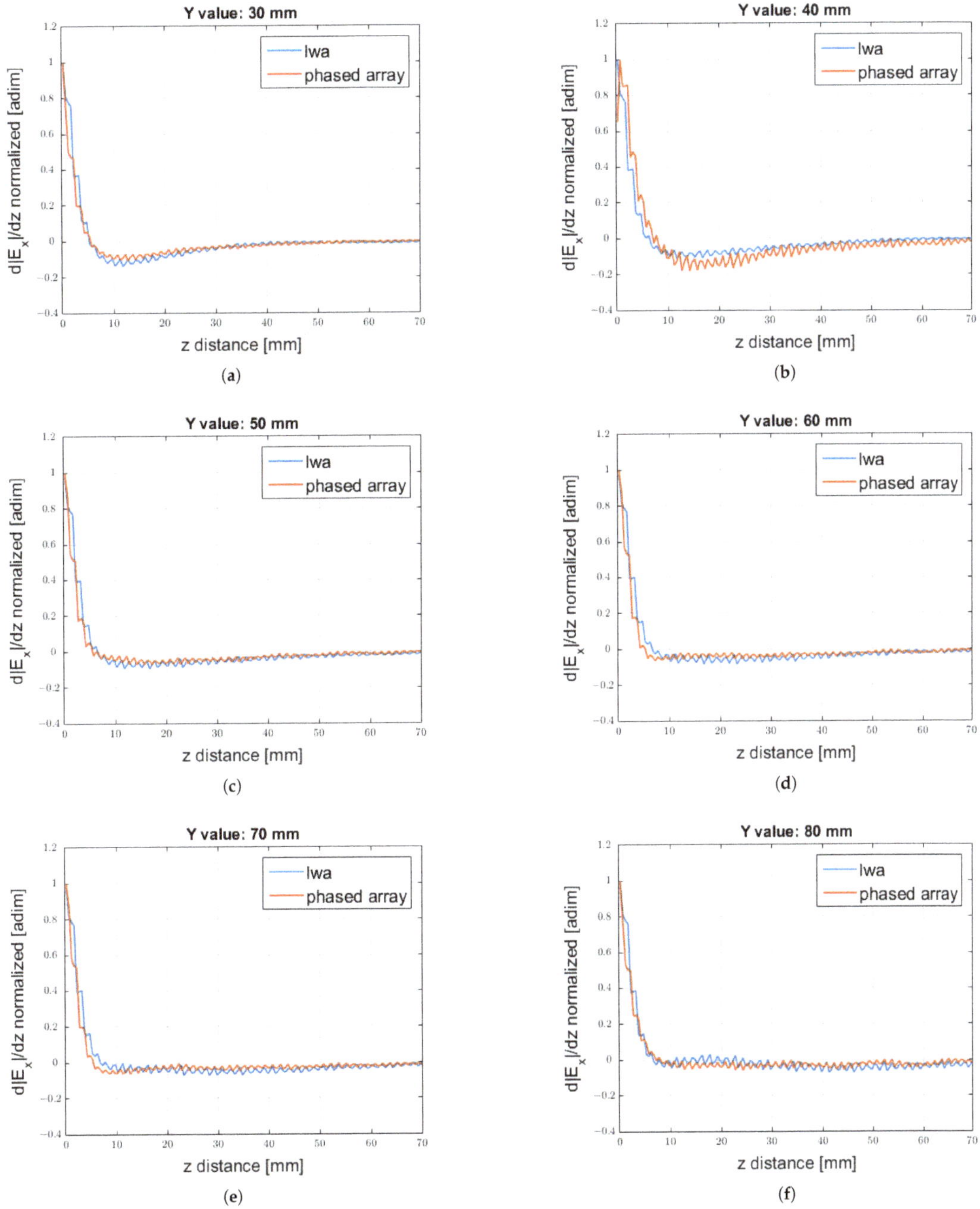

Figure 18. Comparison of the derivative of the $|E_x|$ component with respect to y along z for different constant values chosen along the y axis. Values: (**a**) $(x, y) = (0, 30)$ mm; (**b**) $(x, y) = (0, 40)$ mm; (**c**) $(x, y) = (0, 50)$ mm; (**d**) $(x, y) = (0, 60)$ mm; (**e**) $(x, y) = (0, 70)$ mm; (**f**) $(x, y) = (0, 80)$ mm.

To make a quantitative evaluation of the similarity, we computed the root mean square (RMS) value of the error, visible in Figure 19. For every point in the z direction (i.e., the longitudinal direction for the antenna), we evaluated the RMS error on the field values over the (x,y) plane.

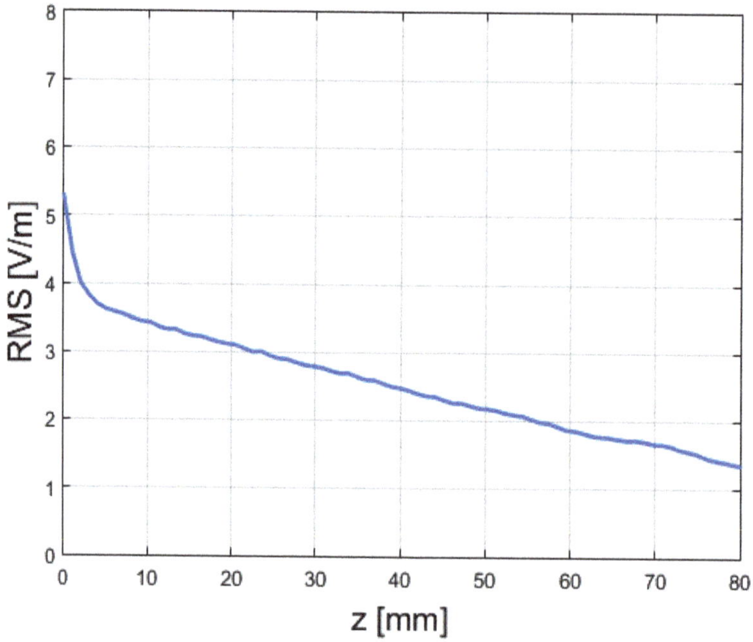

Figure 19. Root-mean square (RMS) error for the amplitude of the electric field between the LWA and the phased array, evaluated as specified by Equation (7).

The applied formula was

$$RMS_{err}(z) = \frac{1}{NM}\sqrt{\sum_{n=1}^{N}\sum_{m=1}^{M}\left(|E_x^{LWA}(x_i, y_j, z)| - |E_x^{pa}(x_i, y_j, z)|\right)^2} \quad (7)$$

where N, M were the numbers of samples on, respectively, the x and y axis.

It can be seen that the error rapidly increased as the distances along z decreased, where the different radiation mechanisms (continue vs. sampled) became more evident.

3.2. *Comparison Between the Antennas Operating at 2.4 GHz*

A procedure, analogous to the one just illustrated, was then applied to the LWA proposed in [7] for hyperthermia treatment, in comparison with the array of Figure 12. The comparison between the far field produced (an indication of the correct sampling), is demonstrated in Figure 20, from which one can easily verify that this early prototype suffered from a large backlobe, due to the limited number of patches employed to reduce the coupling between the array elements, while it produced a main lobe that well resembled the one produced by the antenna in [7].

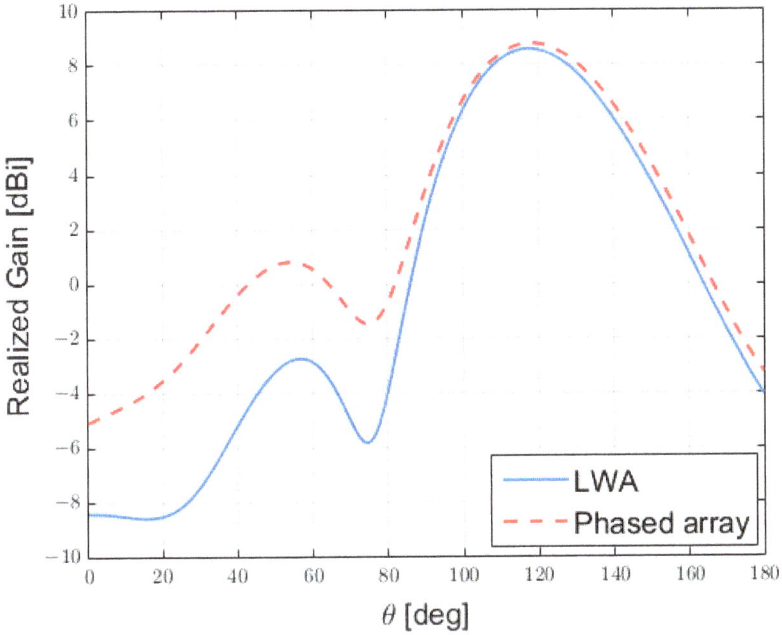

Figure 20. Gain produced by the phased array of Figure 12 compared to the one produced by the LWA of [7].

Moving to the near-field analysis, we considered the differences in the field produced by the two antennas at different distances from the aperture; see Figure 21. The antenna in [7] started approximating the behaviour of a uniform LWA (a full-size Menzel antenna having the same frequency) at a distance of $z = 21$ mm from the aperture: we observed a similar behaviour for the phased array, although the discretisation of the field caused by the presence of the patches remained noticeable. The approximation to the uniform LWA was instead very good for both antennas at a distance of $z = 50$ mm from the aperture.

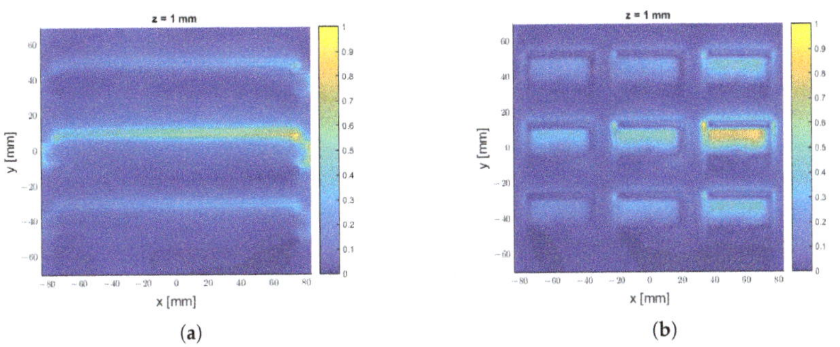

Figure 21. *Cont.*

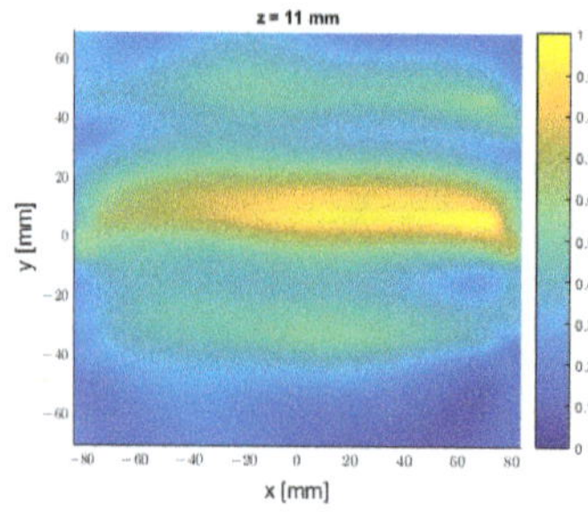

(**c**)

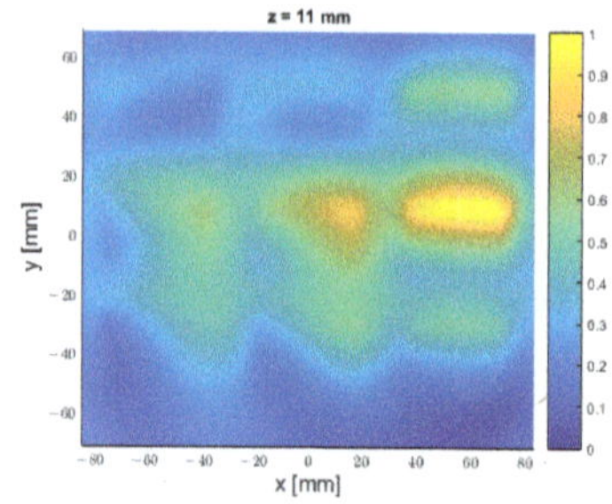

(**d**)

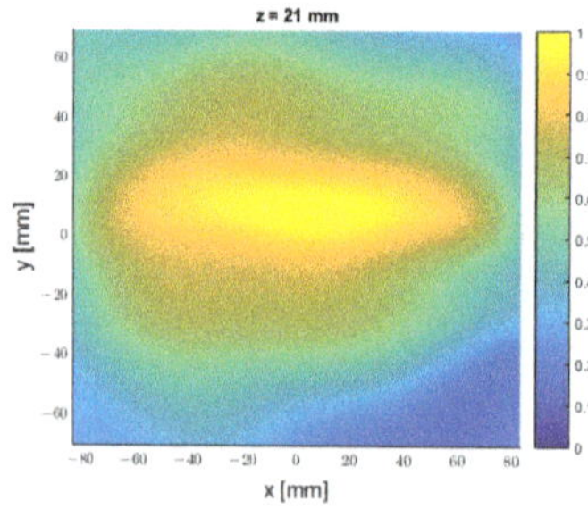

(**e**)

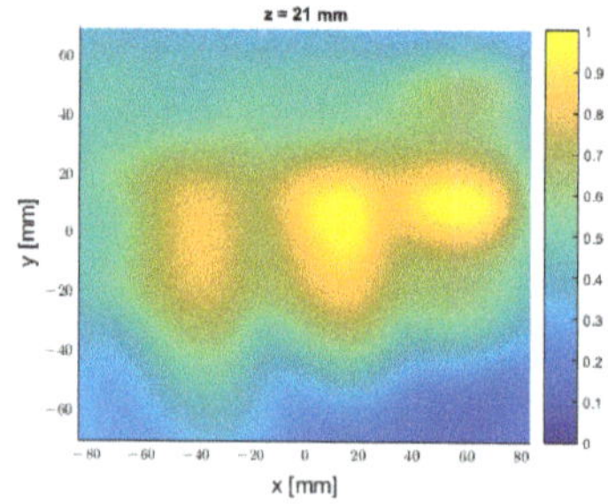

(**f**)

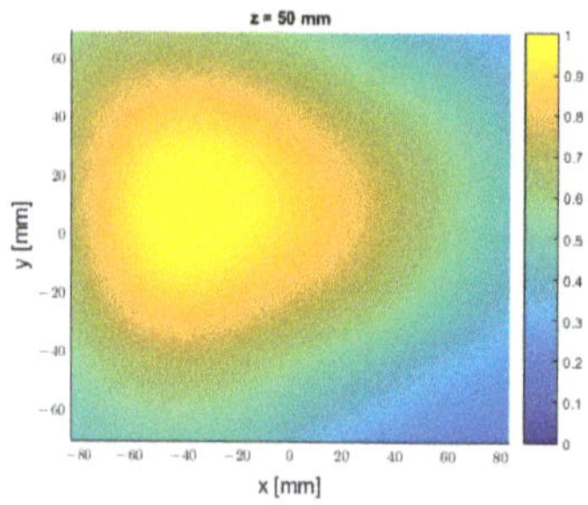

(**g**)

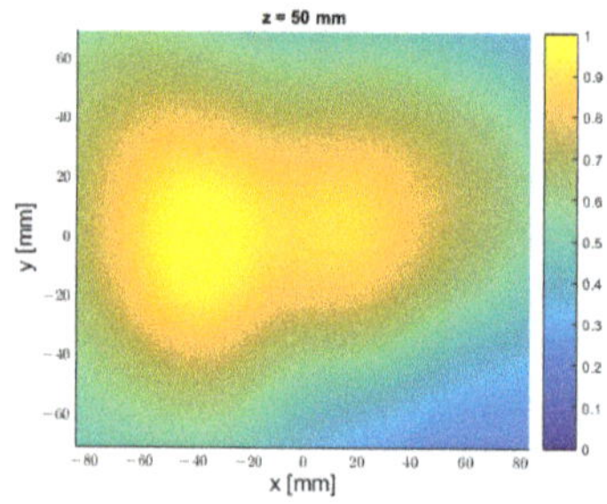

(**h**)

Figure 21. Comparison between the amplitude $|E_x|$ of the electric field normalised by its maximum for the LWA in [7] and the phased array designed here along the $z = k$ planes, where $z = 0$ represents the plane at the antenna aperture: (**a**) LWA field at $z = 1$ mm; (**b**) phased-array antenna field at $z = 1$ mm; (**c**) LWA field $z = 11$ mm; (**d**) phased-array antenna field at $z = 11$ mm; (**e**) LWA field $z = 21$ mm; (**f**) phased-array antenna field at $z = 21$ mm; (**g**) LWA field $z = 50$ mm; (**h**) phased-array antenna field at $z = 50$ mm.

4. Discussion

In this paper, we have considered the possibility of sampling a continuous distribution of current to achieve a discrete representation of the same field. We chose a leaky-wave antenna to be the first element of the comparison, due to the continuous current that flows on its metallization, and we compared its field with the one produced by a patch-antenna array. The coefficients that excited this array were chosen to be equal in amplitude and phase with the field samples taken from the LWA radiating at a 45-degree angle and at a frequency of 12 GHz, properly chosen to represent the total field. We used four different estimators, one performed in far field and three in near field, to evaluate the agreement of the two structures, also considering the RMS error as the "quality factor" to evaluate the goodness of the process. Depending on the degree of the acceptable agreement, this study has shown that it is possible, within certain limits, to consider an equivalence between the field radiated by a continuous current source and a discrete one.

After this preliminary analysis, we moved to a more practical scenario, where we tried to verify the suitability of the approach taken, by designing an array that mimicked the behaviour of an LWA design recently proposed for hyperthermia treatment.

In this paper, we made use of phase shifters, but other alternatives could be explored in future research. An interesting one would be the application of true time delays, which introduce a delay constant in the frequency. In this way, the squint phenomenon in the frequency does not occur.

Also, it should be noted that since phase shifters have a minimum phase step that can be introduced, there is a quantization error in the phase that can be applied. This is also true for variable attenuators, which cannot introduce any desired attenuation, but this is limited to a fixed step. These technical aspects were not developed in this paper, but a more comprehensive study taking into account these non-idealities will be conducted.

This study also demonstrated that with the approach taken it is possible to mimic the behaviour of a leaky-wave antenna, using only a patch array for a specific frequency. Additionally, the stop band was reproduced. However, while frequency-scanning behaviour can also be achieved with the patch array, it differs significantly from that of a leaky-wave antenna, both in terms of direction (which shifts toward broadside as frequency increases) and frequency range (which is much larger). Future research will need to explore alternative approaches to replicating this additional feature, if it is needed by the considered application.

5. Conclusions

This research demonstrates the feasibility of generating a field using a patch array that emulates the behaviour of a leaky-wave antenna (LWA). This finding allowed us to also propose an alternative approach to antenna design for hyperthermia treatment, with a prototype that produces a field that can be approximated with an inhomogeneous wave using a small phased-array antenna. The phased array obtained through the sampling method demonstrated a few challenges, especially in the application at 2.4 GHz, where we needed to control the coupling between the patches, which partially impacted the quality of the approximation to the LWA. This technique also demonstrated advantages, such as enhanced flexibility, that may allow easier adaptation for the amplitudes of the phase and attenuation vectors to specific tissues, and the elimination of both the need to re-use power at the end of the LWA to enhance antenna efficiency and the need to suppress unwanted modes. These advantages would come at the cost of a more challenging implementation, as an array is inherently more complex than an LWA structure. Additionally, our study highlighted significant differences in the scanning behaviour of the leaky-wave and phased-array antennas. Further research is required, to determine whether this aspect of an LWA can also be replicated with a patch array. Future research may focus on approaching the open challenges highlighted in this paper, to improve the design of this promising type of structure.

Author Contributions: Conceptualisation, A.C., P.S., M.D.M., and F.F.; methodology, A.C. and P.S.; software and simulation, A.C.; validation, M.D.M. and F.F.; formal analysis, P.S. and A.C.; data curation, A.C.; writing—original draft preparation, A.C. and P.S.; writing—review and editing, A.C., P.S., M.D.M., and F.F.; supervision, M.D.M., and F.F. All authors have read and agreed to the published version of the manuscript.

Funding: This research received no external funding.

Data Availability Statement: Data are contained within the article.

Acknowledgments: We would like to express our gratitude to the anonymous peer Reviewers who dedicated their expertise to this article. Their comments, constructive criticism, and suggestions have greatly contributed to the quality of this work. The authors also thank Giulia Buttazzoni and Roberto Vescovo († 2018, to whom the paper is dedicated) for the helpful discussions related to the topics of phased arrays and leaky-wave theory which inspired this work.

Conflicts of Interest: The authors declare no conflicts of interest.

Abbreviations

The following abbreviations are used in this manuscript:

LWA	leaky-wave antenna
EM	electromagnetic
RMS	root mean square

References

1. Baccarelli, P.; Frezza, F.; Simeoni, P.; Tedeschi, N. An Analytical Study of Electromagnetic Deep Penetration Conditions and Implications in Lossy Media through Inhomogeneous Waves. *Materials* **2018**, *11*, 1595. [CrossRef] [PubMed]
2. Zenneck, J. Über die Fortpflanzung ebener elektromagnetischer Wellen längs einer ebenen Leiterfläche und ihre Beziehung zur drahtlosen Telegraphie. *Ann. der Phys.* **1907**, *328*, 846–866. [CrossRef]
3. Frezza, F.; Tedeschi, N. Electromagnetic inhomogeneous waves at planar boundaries: Tutorial. *J. Opt. Soc. Am. A* **2015**, *32*, 1485–1501. [CrossRef] [PubMed]
4. Balanis, C.A. *Antenna Theory: Analysis and Design*, 4th ed.; John Wiley & Sons: Hoboken, NJ, USA, 2015.
5. Elliott, R. On discretizing continuous aperture distributions. *IEEE Trans. Antennas Propag.* **1977**, *25*, 617–621. [CrossRef]
6. Hodges, R.E.; Rahmat-Samii, Y. On sampling continuous aperture distributions for discrete planar arrays. *IEEE Trans. Antennas Propag.* **1996**, *44*, 1499–1508. [CrossRef]
7. Calcaterra, A.; Simeoni, P.; Migliore, M.D.; Mangini, F.; Frezza, F. Optimized Leaky-Wave Antenna for Hyperthermia in Biological Tissue Theoretical Model. *Sensors* **2023**, *23*, 8923. [CrossRef] [PubMed]
8. Monticone, F.; Alu, A. Leaky-wave theory, techniques, and applications: From microwaves to visible frequencies. *Proc. IEEE* **2015**, *103*, 793–821. [CrossRef]
9. Oliner, A.A. Leakage from higher modes on microstrip line with application to antennas. *Radio Sci.* **1987**, *22*, 907–912. [CrossRef]
10. Jackson, D.R.; Caloz, C.; Itoh, T. Leaky-Wave Antennas. *Proc. IEEE* **2012**, *100*, 2194–2206. [CrossRef]
11. Oliner, A.A.; Jackson, D.R.; Volakis, J. Leaky-wave antennas. *Antenna Eng. Handb.* **2007**, *4*, 12.
12. Arya, V.; Garg, T. Leaky Wave Antenna: A Historical Development. *Microw. Rev.* **2021**, *27*, 3–16.
13. Balanis, C.A. *Balanis' Advanced Engineering Electromagnetics*; John Wiley & Sons: Hoboken, NJ, USA, 2024.
14. Franceschetti, G. *Electromagnetics: Theory, Techniques, and Engineering Paradigms*; Springer Science & Business Media: Berlin/Heidelberg, Germany, 2013.
15. Fano, R.M.; Adler, R.B.; Chu, L.J. *Electromagnetic Energy Transmission and Radiation*; The M.I.T. Press: Cambridge, MA, USA, 1969.
16. Mailloux, R.J. *Phased Array Antenna Handbook*; Artech House: Norwood, MA, USA, 2017.
17. Weiland, T. Time domain electromagnetic field computation with finite difference methods. *Int. J. Numer. Model. Electron. Networks, Devices Fields* **1996**, *9*, 295–319. [CrossRef]
18. Bondeson, A.; Rylander, T.; Ingelström, P. *Computational Electromagnetics*; Springer: Berlin/Heidelberg, Germany, 2012.
19. CST Studio Suite. *CST Microwave Studio (2023)*; Simulia: Providence, RI, USA, 2023.
20. *MATLAB (R2024b)*; The MathWorks Inc.: Natick, MA, USA, 2024.
21. *Microsoft PowerPoint (Version 365)*; Microsoft Corporation: Redmond, WA, USA, 2024.
22. Menzel, W. A new travelling wave antenna in microstrip. In Proceedings of the 8th European Microwave Conference, Paris, France, 4–8 September 1978; pp. 302–306.
23. Oliner, A.; Lee, K. Microstrip leaky wave strip antennas. In Proceedings of the 1986 Antennas and Propagation Society International Symposium, Philadelphia, PA, USA, 8–13 June 1986; Volume 24, pp. 443–446.
24. James, J. *Handbook of microstrip antennas*; Peter Peregrinus Ltd.: London, UK, 1989.
25. James, J.R.; Hall, P.S.; Wood, C. *Microstrip Antenna: Theory and Design*; IET: Stevenage, UK, 1981.

26. Geng, Y.; Wang, J.; Li, Y.; Li, Z.; Chen, M.; Zhang, Z. Leaky-wave antenna array with a power-recycling feeding network for radiation efficiency improvement. *IEEE Trans. Antennas Propag.* **2017**, *65*, 2689–2694. [CrossRef]
27. Poveda-García, M.; Algaba-Brazalez, A.; Cañete-Rebenaque, D.; Gómez-Tornero, J. Series Arrangement Technique for Highly-Directive PCB Leaky-Wave Antennas With Application to RFID UHF Frequency Scanning Systems. *IEEE Open J. Antennas Propag.* **2024**, *5*, 1193–1208. [CrossRef]
28. Nguyen, H.V.; Parsa, A.; Caloz, C. Power-recycling feedback system for maximization of leaky-wave antennas' radiation efficiency. *IEEE Trans. Microw. Theory Tech.* **2010**, *58*, 1641–1650. [CrossRef]
29. Barthia, P.; Bahl, I. *Microstrip Antenna Design Handbook*; Artech House: London, UK, 2003.
30. Gupta, K.C. *Microstrip Antenna Design*; Artech House: London, UK, 1988.

sensors

MDPI

Article

Performance Evaluation and Calibration of Electromagnetic Field (EMF) Area Monitors Using a Multi-Wire Transverse Electromagnetic (MWTEM) Transmission Line

Renzo Azaro [1,*], Roberto Franchelli [2] and Alessandro Gandolfo [2]

1 EMC S.r.l., Erzelli Science and Technology Park, Via Sant'Elia 242, 16152 Genova, Italy
2 NARDA Safety Test Solutions S.r.l., Via Benessea 29/B, Cisano sul Neva, 17035 Savona, Italy; roberto.franchelli@narda-sts.it (R.F.); alessandro.gandolfo@narda-sts.it (A.G.)
* Correspondence: azaro@emclab.it

Abstract: The exposure levels generated by environmental electromagnetic field (EMF) sources can be measured and monitored by employing EMF area monitors. The operating spectrum of environmental EMF sources is not limited to high frequencies ($f > 30$ MHz) but also extends to low frequencies ($f < 30$ MHz), where sources associated, for example, with radio transmitters typically generate non-negligible field contributions. For this reason, professional EMF area monitors can be equipped with different field sensors, properly calibrated according to standardized procedures. Because low-frequency electric fields are very sensitive to environmental boundary conditions, equipping an EMF area monitor with electric field sensors, previously calibrated as stand-alone devices, can lead to measurement errors due to field perturbations introduced by the physical structure of the area monitor itself. This paper describes the activities carried out to assess the performance of an EMF area monitor in simulated realistic conditions and calibrate it in the 300 kHz–20 MHz frequency band. The activities were conducted using a multi-wire transverse electromagnetic (MWTEM) transmission line as a controlled electric field source, with dimensions suitable for exposure of the entire structure of the EMF area monitor. In view of using this approach to calibrate the area monitors as a whole instead of the individual sensors, the uniformity of the electric field generated by the available MWTEM transmission line was analyzed in detail both numerically and experimentally. Finally, the results of the evaluation and calibration of an area monitor are reported and discussed.

Keywords: electromagnetic measurements; EM field probe calibration; exposure to electromagnetic fields; multi-wire TEM transmission line

1. Introduction

The rapid and continuous growth of wireless communications and data transmissions requires a great number of stations that generate non-negligible environmental electromagnetic fields. This significantly increases the density of radiating sources while at the same time increasing public concerns about exposure levels to the electromagnetic fields generated by the necessary systems. Similarly, the massive use of wireless technologies in a great variety of equipment and the pervasive distribution of electrical power networks implies a large number of intentional and non-intentional sources of electromagnetic energy being distributed in the environment and operating in a very wide frequency range, ranging from tens of Hz to tens of GHz. In this highly complex electromagnetic scenario, exposure to electromagnetic fields is a topic of great importance, especially for people living near infrastructures recognized as EMF sources. Local authorities and government agencies have an interest in the real-time, continuous monitoring of the exposure levels generated by different sources to ensure that nearby areas comply with applicable national or international standards [1–4]. Typical areas of concern are schools, hospitals, and residential and public areas, not to mention exposure in the workplace. Classical electromagnetic field measurement instruments, despite their optimized performance and accuracy, are not suitable for this kind of task as they generally require the presence of an operator; in addition, due to limited autonomous power supplies and a lack of weatherproofing, they are not suitable for long-term or continuous unattended outdoor monitoring. These limitations have led to the development of EMF area monitors, i.e., electromagnetic field measurement systems that can autonomously measure field levels and record and transmit data. EMF area monitors are typically battery-operated and equipped with solar panels for an independent power supply. In general, an EMF area monitor can be fitted with a set of field probes covering all parts of the spectrum of interest for a particular site to be monitored. An example of an EMF area monitor in a real application scenario is shown in Figure 1.

Figure 1. An example of a battery-operated EMF area monitor in a real application scenario (courtesy of NARDA Safety Test Solutions S.r.l.).

By using an EMF area monitor, the measurements taken with a certain time step can be read remotely at any time or automatically sent at preset intervals to a remote PC or a data server via the mobile network. Alarms can be automatically sent to selected recipients if preset limits are exceeded. When measurements are required from several positions, an effective solution is to implement a geographically distributed network of EMF area monitors that can continuously detect exposure levels, present the results to the public in an easily accessible format, and compare the results with the applicable standards [1–4]. Several researchers, from both academia and industry, have investigated ways of satisfying the requirements set by the International Telecommunication Union (ITU) [5] for the monitoring of environmental electromagnetic fields. Various studies based on the integration of field probes with wireless sensor networks (WSNs) [6–8] or with radio interfaces based on the GSM/GPRS standard [9–14] can be found in the literature. Recently, as described in [15], extensive geographical monitoring has been experimented by integrating an area monitor inside a car. The constant and current interest in the topic is demonstrated by the publication of scientific and technical papers, even in recent years [16–20]. In this framework, considering that an EMF area monitor is a more complex measuring system than a classical EMF probe and considering that it can operate in scenarios with different electrical boundary conditions (e.g., with or without cabling and a power supply unit for connection to the mains), researchers have studied the device's behavior in different simulated environments [21].

In this paper, a description of the architecture of a typical EMF area monitor is followed by a presentation of the activities carried out in order to calibrate it as a whole system and validate its behavior. To this end, the electric field generated by a multi-wire TEM (MWTEM) transmission line was analyzed in detail both numerically and experimentally in order to evaluate its capability to generate controlled electric field distributions that are sufficiently uniform for the exposure of the entire structure of the EMF area monitor for frequencies up to 30 MHz.

In most cases, sources capable of generating non-negligible field contributions are associated with mobile phone systems or radio transmitters operating at frequencies higher than 30 MHz. However, in certain cases—often involving high emission levels—high-power RF sources operating at frequencies below 30 MHz are used for long-distance communication transmitters (e.g., for naval or military communications) or for industrial, scientific, and medical (ISM) applications, authorized by the ITU, such as dielectric heating, plastic welding, food processing, and short-wave diathermy.

2. The Structure of an EMF Area Monitor

Figure 2 presents the general structure of the AMB-8059 area monitor. Three orthogonal broadband probes in the field probe module measure the field amplitude, and the resulting signal is converted by the analog to digital converter (ADC) and sent to the micro-controller unit (MCU). The area monitor, managed by the control module, is able to autonomously perform, acquire, memorize, and transmit measurements. The area monitor is equipped with a set of radio interfaces (Wi-Fi, GSM/GPRS, LTE) necessary for remote control and data exchange. The power supply module optimizes power consumption for long-time autonomous operation by managing battery use and charging via solar panel or, if available, via a power supply unit connected to the mains. Due to its non-negligible structure and the use of cables, the presence of an area monitor can modify the field under measurement, causing errors in the measurement results. This issue is magnified for frequencies below 30 MHz, as analyzed in [22] for electric field measurement with a rod antenna. To solve this kind of problem, the AMB-8059 is equipped with a relay that automatically disconnects the electrical connection when not needed. For the same reason, to improve measurement accuracy, the calibration of the area monitor has to be reconsidered as a whole system, as typically the field probes are calibrated as stand-alone devices and the effects of integrating them within the area monitor have to be evaluated and, if necessary, taken into account.

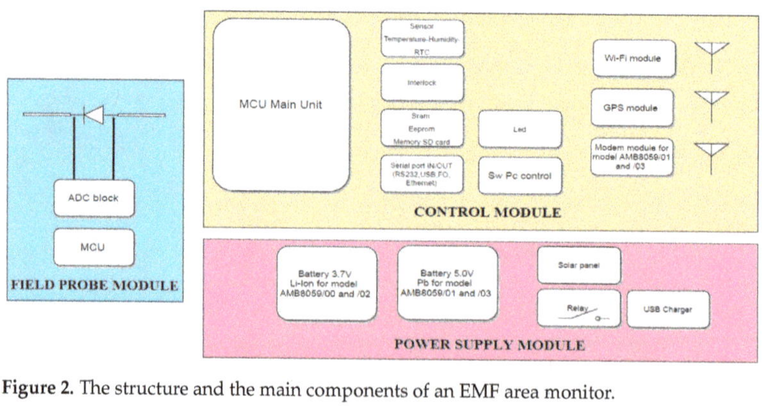

Figure 2. The structure and the main components of an EMF area monitor.

3. Experimental Setup

To calibrate and evaluate the performance of an area monitor as a whole system, we need to generate a controlled and sufficiently uniform electric field in a volume suitable for hosting the entire structure of the device under calibration. At higher frequencies ($f > 30$ MHz), this is easily accomplished by means of standard radiating devices like simple dipoles, biconical, log-periodic, and horn antennas. The same approach cannot be used at lower frequencies ($f < 30$ MHz), as in this frequency range, the physical size of dipolar antennas becomes impractical, and in order to operate under realistic electromagnetic field conditions, it is necessary to stay far from the antenna itself. This implies an overly challenging test setup while requiring the generation of intense electromagnetic fields over a large area.

A possible approach for the generation of high field levels in a confined region while avoiding conventional antennas is based on the use of transverse electromagnetic (TEM) transmission lines that in the enclosed volume support a uniform and linearly polarized plane wave [23–25].

Each transmission line composed of two separate conductors for $f < f_{C1}$ can support the propagation of the TEM field configuration, while for $f > f_{C1}$, other field configurations associated with high-order modes (f_{C1} being the cut-off frequency of the first high-order mode) can occur and heavily modify the field configuration. The main properties of a TEM wave are that the electric and magnetic fields are perpendicular to each other and both are transverse to the wave direction of propagation. This type of wave, known as a plane wave, is also present in the far field of an antenna radiating in free space. Coaxial lines, parallel plate lines, TEM cells, and similar structures (all characterized by the presence of two separate conductors or groups of conductors) can support this field configuration. For this reason, they are widely used to generate uniform fields for electromagnetic compatibility testing [26,27] and calibration. Wire array cells belong to this category, and thus, the multi-wire TEM line has been evaluated and validated for calibration activities.

Because classical TEM cells generally have small dimensions and a closed structure, it is very difficult to insert an entire area monitor inside their useful volume. They are therefore a good solution for calibrating smaller instruments like classical field probes, but impractical for equipment as large as an area monitor.

An adequately sized alternative structure, composed of wire arrays and capable of supporting a TEM mode, is the multi-wire TEM (MWTEM) transmission line [28]. This device is less popular than TEM cells and, to the best of the authors' knowledge, has never been used for the calibration and validation of field probes. The structure of the MWTEM line used for the activities described in this paper is shown in Figure 3, where three different sections can be observed: a central uniform multi-conductor transmission line enclosing the volume used for testing, and two lateral tapered transmission lines. At one end (the input port), the MWTEM line is equipped with a balun/matching network (50 Ω/200 Ω) and at the other end with a 200 Ω dummy load. The balun is necessary as no metallic element of the line supporting the TEM wave is connected to a constant potential (the device operates in balanced mode). The MWTEM line is equipped with a nominal 200 Ω load as this matches its characteristic impedance; according to transmission line theory, this is the optimal choice for minimizing reflections from the load and reducing standing wave phenomena along the line in the higher frequency band.

As shown in Figure 3, instead of metallic plates (as in classical TEM structures, e.g., the parallel plate line [26]), the two halves of the line are composed of an array of seven aluminum pipes (2 cm in diameter). These pipes are parallel in the central section and taper between the central section and the two ends. The MWTEM line used for experiments is pictured in Figure 4. In particular, Figure 4b,c show the two ends of the line with the balun and the load, respectively.

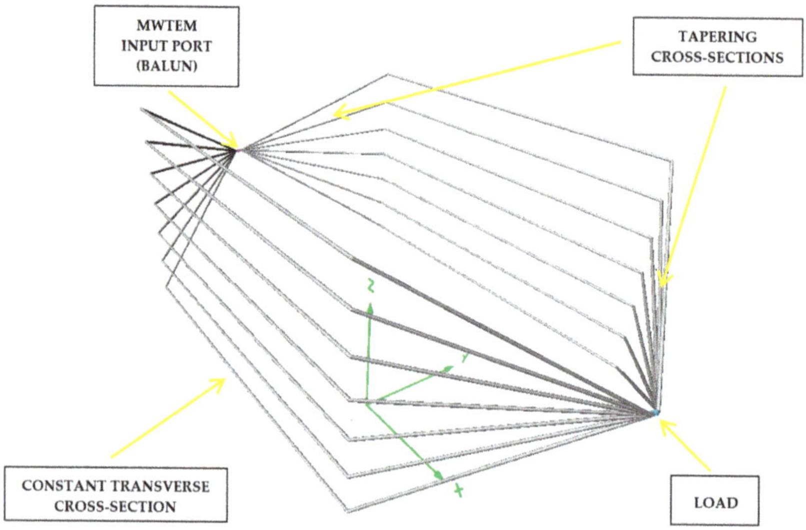

Figure 3. The geometry of the multi-wire TEM (MWTEM) transmission line.

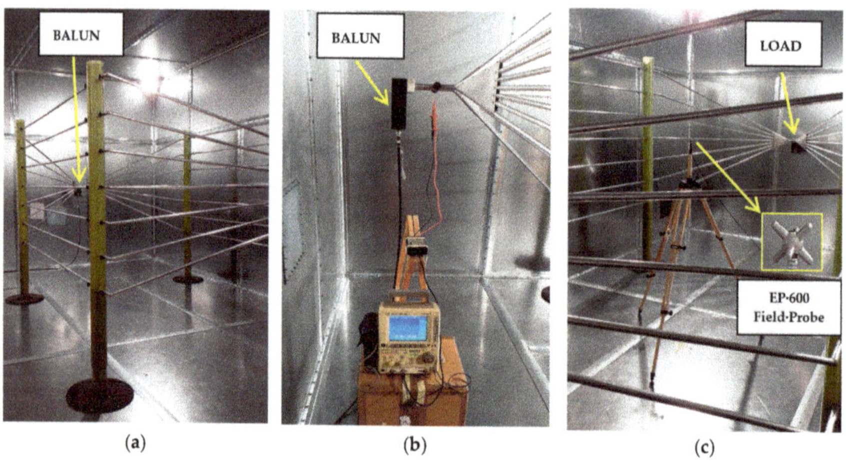

Figure 4. Experimental assessment of the electric field distribution generated by the MWTEM transmission line: (**a**) the line inside a shielded room; (**b**) measurement of the voltage amplitudes at the balanced input port; (**c**) electric field amplitude measurements with the EP600 probe in the central position of the line.

The use of a TEM cell composed of wire arrays instead of planar elements greatly complexifies the electromagnetic problem to be solved in order to find the electrical parameters of the device and accurately analyze the uniformity of the generated electric field. Because no analytical formula is available for the field values and characteristic impedance, a simulation software has been used to obtain a detailed and accurate characterization of the device. The results of the characterization by numerical simulation are reported and discussed in Section 4.

An outline of the theory of MWTEM cells is reported in Appendix A. Starting from the description of the TEM mode field components existing in the basic structure of the parallel plate line, an approximated theory of the TEM mode valid for an MWTEM in a quasi-static condition is reported. The relationships obtained according to the quasi-static model can be profitably employed for the design of a generic N-wire MWTEM line in terms of transversal electric field components and line impedance.

The MWTEM transmission line is an open structure, so when driven by radiofrequency power, it can radiate non-negligible fields in the surrounding area. During the experimental activities, it was driven using a 75 W power amplifier, then placed inside a shielded chamber to minimize interference with sensitive equipment in the nearby area. The technical characteristics of all equipment used during the experiments are summarized in Table 1.

Table 1. Characteristics of equipment used for the experimental activities.

Equipment	Manufacturer	Model	Characteristics
Shielded Chamber	Beltech	-	Dimensions: 6.1 × 4.9 × 3.6 m
MWTEM	Comtest	-	7 + 7 wires balanced transm. line, 4.5 × 2.3 × 1.5 m, Z_0 = 200 Ω, f_{max} = 20 MHz
MWTEM Balun	Comtest	-	Impedances levels: 50 Ω /200 Ω
MWTEM Load	Comtest	-	Impedance: 200 Ω (nominal value)
RF Signal Generator	Hewlett Packard	E4400B	Freq. band: 250 kHz–1 GHz
Power Amplifier	Amplifier Research	75A250	Freq. band: 10 kHz–250 MHz, P_{max} = 75 W
Digital Oscilloscope	Yokogawa	DL1620	f_{max} = 200 MHz
High Voltage Diff. Probe	Sapphire Instruments	SI-9010	V_{max} = ±7000 V, f_{max} = 70 MHz
Electric Field Probe	Narda STS	EP600	Freq. band: 100 kHz–9.25 GHz, 0.14–140 V/m
Network Analyzer	Agilent	E5061B	Freq. band: 5 Hz–3 GHz

4. Numerical and Experimental Assessment of the Electric Field Generated by the MWTEM Transmission Line

Due to the absence of detailed technical documentation about the spatial distribution of the field generated by the MWTEM transmission line, before its use for area monitor calibration and evaluation, the distribution and uniformity of the electric field generated inside its useful volume was analyzed by means of both numerical simulations and measurements. For simulations, at the input port of the line (between the balun and the tapered section), the same set of applied peak-to-peak voltage amplitudes $V_{p\text{-}p}$ used during the experimental activities was considered. As shown in Figure 4b, the voltages were detected by means of a suitable differential probe connected to the balanced input port of the MWTEM transmission line and with the help of a digital oscilloscope. The numerical simulations were performed with a method of moment (MoM) algorithm by using a model of the MWTEM transmission line developed according to the guidelines of the employed software, i.e., the Numerical Electromagnetics Code (NEC). As described in detail in [29–31], NEC uses both an electric field integral equation (EFIE) and a magnetic field integral equation (MFIE) to model the electromagnetic response of systems composed of conductive wires and surfaces. Each formulation has advantages for specific cases: EFIE is well suited for small-volume thin-wire structures, while MFIE works well for large surfaces. Assuming a time-harmonic dependance $e^{j\omega t}$, under the assumption of thin wires, the electric field integral equation (EFIE) can be written as [29–31]:

$$-\hat{n}\left(\vec{r}\right) \times \vec{E}^{I}\left(\vec{r}\right) = -\frac{j\eta}{4\pi k}\hat{n}\left(\vec{r}\right) \times \int_L I(s')\left(k^2\hat{s}' - \nabla\frac{\partial}{\partial s'}\right)g\left(\vec{r},\vec{r}\prime\right)ds\prime \quad (1)$$

where:

$$g(\vec{r},\vec{r}') = e^{-jk|\vec{r}-\vec{r}'|}\Big/\left|\vec{r}-\vec{r}'\right| \quad (1)$$

$\vec{E}^{I}$ is an incident electric field;
$k = \omega\sqrt{\mu_0\varepsilon_0}$, $\eta = \sqrt{\mu_0/\varepsilon_0}$;
$\vec{r}$ is the observation point;
$\hat{n}\left(\vec{r}\right)$ is the unit normal vector of the surface at $\vec{r}$;
I is the current flowing in the wire having length L;
s is the distance parameter along a wire axis;
$\hat{s}$ is the unit vector tangent to wire axis.

The magnetic field integral equation (MFIE) can be written as:

$$-\hat{n}\left(\vec{r}_0\right) \times \vec{H}^{I}\left(\vec{r}_0\right) = -\frac{1}{2}\vec{J}_s\left(\vec{r}_0\right) + \frac{1}{4\pi}\int_S \hat{n}\left(\vec{r}_0\right) \times \left[\vec{J}_s\left(\vec{r}'\right) \times \nabla' g\left(\vec{r}_0, \vec{r}'\right)\right] dA' \quad (2)$$

where:
$\vec{H}^{I}$ is an incident magnetic field;
$\vec{J}_S$ is a surface current density induced on a surface S;
$\vec{r}_0$ is a surface point;
$\hat{n}\left(\vec{r}_0\right)$ is the unit vector normal to S at $\vec{r}_0$.

The integral Equations (1) and (2) are solved numerically in NEC by the method of moments (MoM) [32]. A brief review of this method is provided in Appendix B.

The $\vec{E}_{tot}$ amplitudes were calculated for a discrete set of frequencies using the NEC implementation available in the 4NEC2 (V. 5.9.3) software package [31], considering a plane transversal to the line axis (x-axis) being placed in the center of the length of line (x = 0 m) and a plane parallel to the line axis (x-axis) and orthogonal to the y-axis (i.e., a longitudinal plane). The geometry of the MWTEM transmission line and the axes of the coordinate system are shown in Figure 3. The main parameters of the simulations and the results obtained in the central position of the transversal and longitudinal sections (x = 0, y = 0, z = 1.55 m) are summarized in Tables 2 and 3. In Table 2, for each frequency value considered, the following data are reported: the peak-to-peak voltage values $V_{p\text{-}p}$ applied at the balanced input port of the MWTEM line, the electric field amplitudes measured with an EP600 field probe, the field value calculated with the formula $\left|\vec{E}\right| = \frac{V_{rms}}{d}$ (where d = 2.32 m is the distance between the two halves of the MWTEM line in the central section), and the simulated field values obtained assuming a constant 200 Ω load.

Table 2. MWTEM transmission line—simulation parameters and results.

Freq. [MHz]	Applied Voltage [$V_{p\text{-}p}$]	EP600 Meas. [V/m]	$\left\|\vec{E}_{tot}\right\|$ Calc. [1] [V/m]	$\left\|\vec{E}_{tot}\right\|$ Simul. (Z_{nom} = 200 Ω) [V/m]
0.3	162.5	17.93	24.52	18.86
0.5	162.5	18.02	24.74	19.01
1	164.5	18.57	24.91	19.17
2	158.3	17.95	23.83	18.40
3	153.1	17.48	23.18	17.97
5	133.3	15.75	20.17	15.85
7	113.3	13.30	17.37	13.82
9	139.5	13.42	21.25	16.93
10	145.8	12.75	22.37	17.71
12	160.4	11.81	24.26	18.57
15	176.1	12.99	26.81	18.58
18	150.0	14.62	22.50	13.71
20	114.5	15.31	17.37	9.80

[1] Value calculated with the formula $\left|\vec{E}\right| = \frac{V_{rms}}{d}$, where d = 2.32 m.

Table 3. MWTEM transmission line—effects of measured load impedance values and of a shielded chamber.

Freq. [MHz]	$Z_{meas.}$ [Ω]	$\left\|\vec{E}_{tot}\right\|$ Simul. [V/m]	$\left\|\vec{E}_{tot}\right\|$ Simul. (with Shield) [V/m]
0.3	201-j5	18.86	17.67
0.5	201-j8	19.04	17.93
1	200-j15	19.25	18.20
2	196-j30	18.71	17.63
3	189-j43	18.64	17.67
5	171-j58	16.71	16.59
7	153-j72	13.53	14.70
9	133-j79	12.87	14.74
10	124-j80	11.69	13.62
12	107-j80	10.23	11.07
15	86-j74	11.32	11.10
18	68-j64	12.04	12.25
20	58-j51	12.14	14.11

For a more accurate characterization of the device, the complex values of the load impedance (composed of non-inductive high-power resistances) were measured with a network analyzer, because, as expected, they differed markedly from the nominal value (200 Ω) because of the non-ideal behavior of the lumped components making up the load. Table 3 reports, for each frequency, the measured complex load impedance values and the simulated electric field amplitudes (obtained with the measured load impedances). To evaluate the effects and take them into account, simulations were also performed with the MWTEM line being placed inside a shielded chamber having the characteristics reported in Table 1 and pictured in Figure 4.

An initial observation from Table 2 is that the formula $\left|\vec{E}\right| = \frac{V_{rms}}{d}$ always overestimates the actual field values. In Figure 5, the results obtained with the MoM algorithm are graphed and compared with the measured values. By analyzing the plots in Figure 5, we can observe that the simulations assuming a constant resistive load (200 Ω) provide a good representation of the field amplitude up to about 7 MHz, while for higher frequencies, it is necessary to use the actual load impedance.

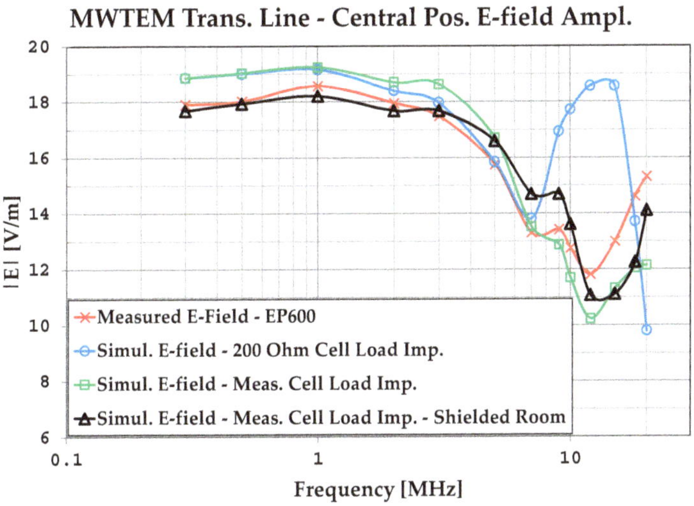

Figure 5. Simulated and measured E-field amplitude in the central position of the MWTEM transmission line.

We can also observe good accordance between measurements and simulations, in particular between the values measured with the EP600 probe and those obtained by simulations considering the measured load impedances and the effects of the shielded chamber. These results confirm the effectiveness of the model used for the simulations, in particular for the field uniformity analysis. The measurements taken with the EP600 probe can be assumed to be reliable as this device is very small, battery-operated, and connected to a PC with control software through a fiber optic cable, allowing for high measurement accuracy. Taking the field amplitude measured with the EP600 probe as a reference, Figure 6 shows the deviations observed with three different boundary conditions assumed during the simulations: the MWTEM line with the nominal load impedance (200 Ω), MWTEM line with the measured load impedance, and MWTEM line with the measured load impedance inside the shielded chamber. The plots in Figure 6 confirm the important role of the load and indicate that the use of the effective impedance values of the load greatly increases the accuracy of the calculated field values; in addition, the shielded chamber appears to have had an appreciable but limited effect. Excluding the configuration with nominal load impedance, the effects of the boundary conditions are evident for frequencies higher than about 8 MHz, where deviations of about 2 dB can be observed. For lower frequencies, the deviations are very small (about ± 0.5 dB); the limited effect of the shielding chamber confirms that the electric field is mainly confined within the useful volume of the MWTEM line, as occurs in static conditions. To evaluate the uniformity of the electric field generated by the MWTEM line, the results obtained in the two orthogonal planes were analyzed in detail. Figure 7b–d presents the plots of simulated $\left|\vec{E}_{tot}\right|$ values in the transversal plane, while the corresponding plots in the longitudinal plane can be observed in Figure 8b–d. The simulations were performed for the same set of frequency values considered during the experimental activities, listed in Table 2, with a step of 1.5 cm (along x, y and z) for the electric field calculation in both the transversal and the longitudinal planes. As can be observed in Figures 7 and 8, in the central volume of the MWTEM transmission line, the electric field amplitude is sufficiently uniform. In more detail, the plots in Figure 9 show the difference (in dB) between the field amplitude in the central position (x = 0, y = 0, z = 1.55 m) and the values calculated moving along the y-direction (−1.1 < y < 1.1 m, Figure 9a) and along the x-direction (−2.5 < x < 2.5 m, Figure 9b).

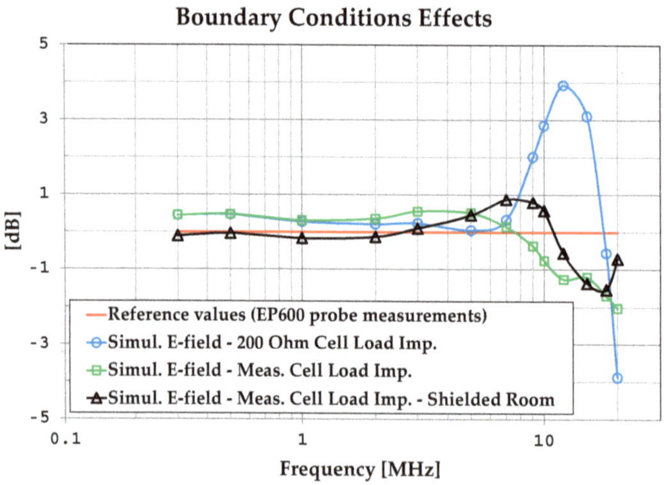

Figure 6. Effects of boundary conditions: MWTEM line load value (nominal 200 Ω, measured) and interaction with the shielded chamber.

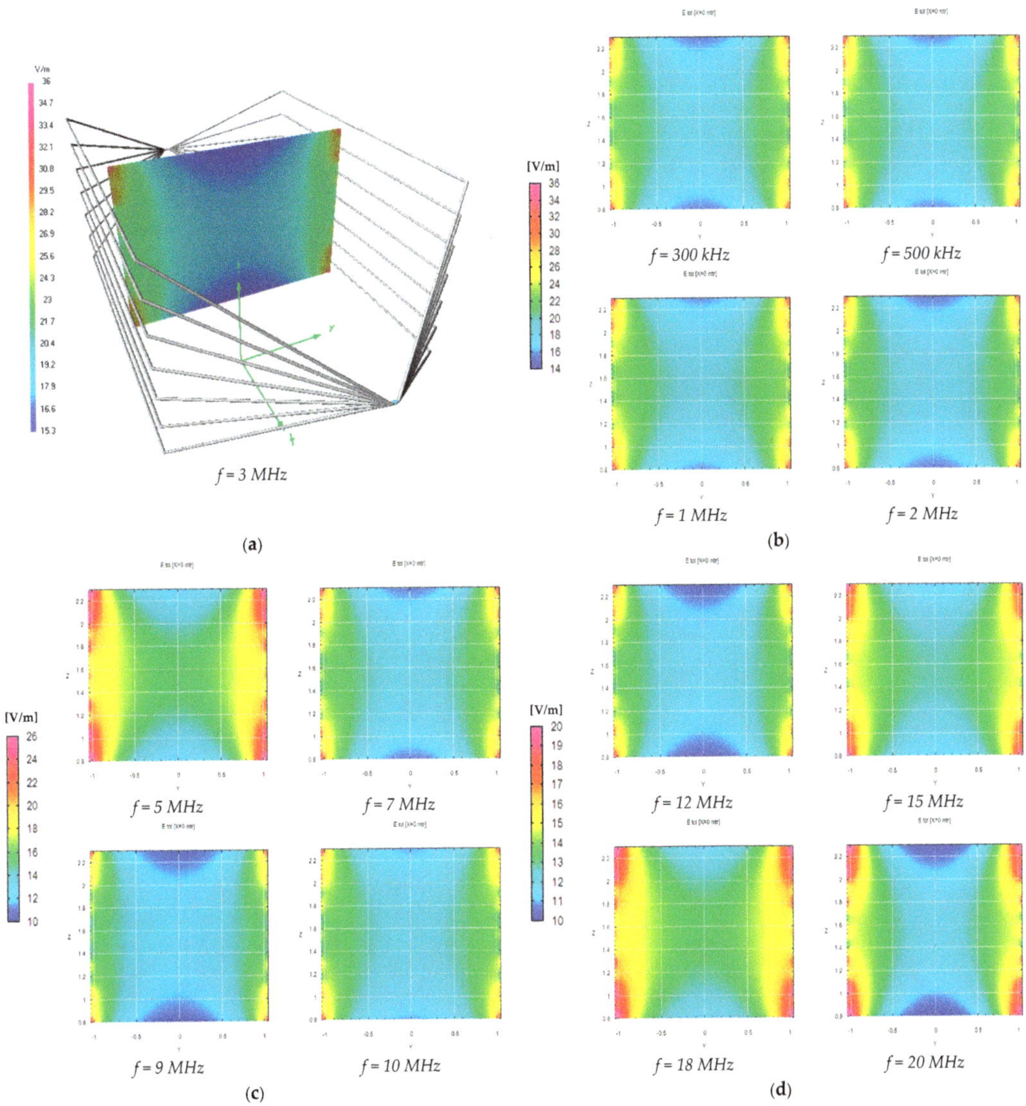

Figure 7. Numerical evaluation of the uniformity of the electric field generated inside the volume formed by the MWTEM transmission line: (**a**) geometry of the line and the transversal section considered for the simulations (results shown for f = 3 MHz); (**b**) results for f = 0.3 MHz, f = 0.5 MHz, f = 1 MHz, and f = 2 MHz; (**c**) results for f = 5 MHz, f = 7 MHz, f = 9 MHz, and f = 10 MHz; (**d**) results for f = 12 MHz, f = 15 MHz, f = 18 MHz, and f = 20 MHz.

By analyzing these data, it can be observed that the field in the central zone shows small and regular variations: about 1.5 dB for a position variation of ±0.9 m in the transversal plane and about 2 dB for a position variation of ±0.5 m in the longitudinal plane. Taking into account that in a TEM structure the electric field is almost transversal, its proper use is to host the area monitor with the axis lying in the transversal plane, where the electric field amplitude is sufficiently uniform in a spatial interval adequate to host the device under testing. The electric field generated for the area monitor calibration is in practice in the near-field region with respect to the field sources; then, to verify the existence in the volume formed by the MWTEM transmission line of a TEM field configuration, a detailed numerical analysis of the field components was also carried out. The analysis was performed on a transversal and a longitudinal cross-section, having the dimensions 1×1 m. The results, reported in detail in Appendix C, allowed us to verify that, according to the TEM configuration definition, the longitudinal components of both the electric field vector and the magnetic field vector are negligible if compared with the transversal components. In addition, it was also possible to confirm that the electric field vector is mainly directed along the transversal y-direction and the magnetic field along the transversal z-direction (i.e. the electric and the magnetic fields are mutually orthogonal).

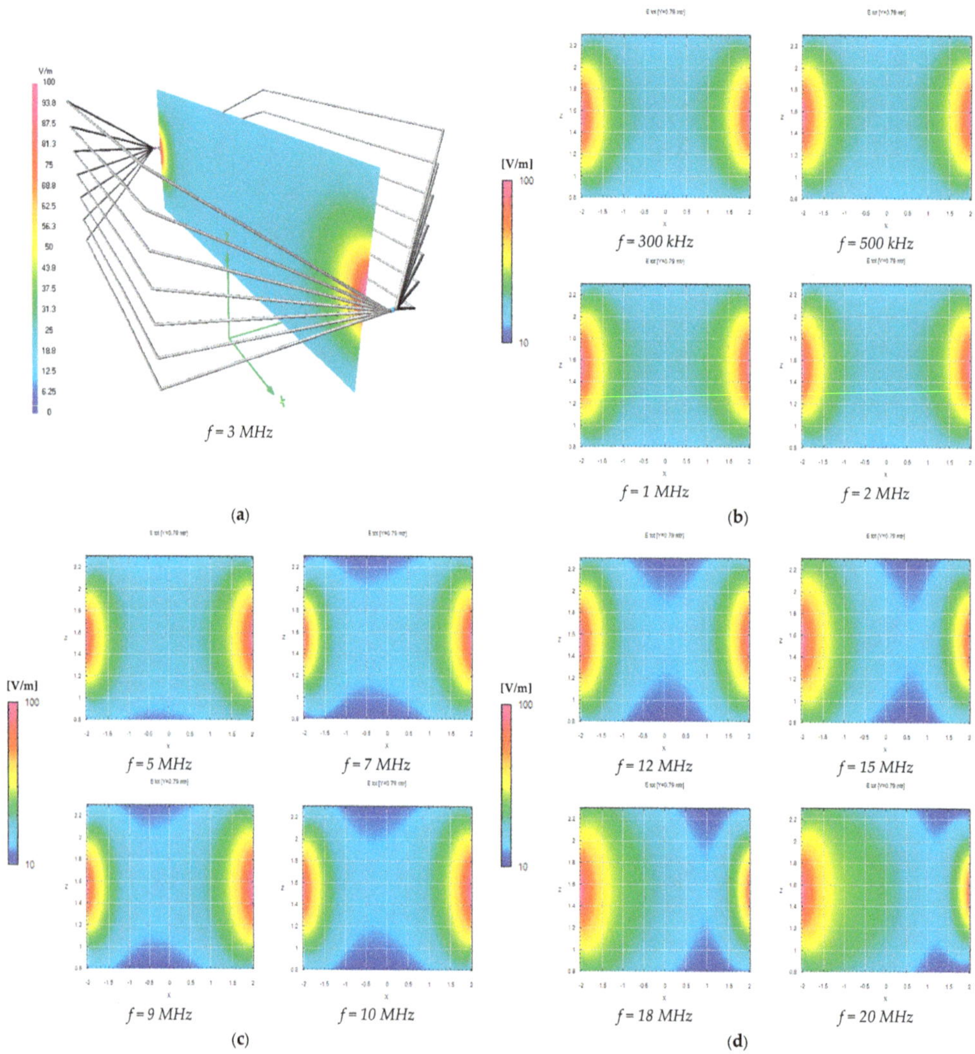

Figure 8. Numerical evaluation of the uniformity of the electric field generated inside the volume formed by the MWTEM transmission line: (**a**) geometry of the line and the longitudinal section considered for the simulations (results shown for f = 3 MHz); (**b**) results for f = 0.3 MHz, f = 0.5 MHz, f = 1 MHz, and f = 2 MHz; (**c**) results for f = 5 MHz, f = 7 MHz, f = 9 MHz, and f = 10 MHz; (**d**) results for f = 12 MHz, f = 15 MHz, f = 18 MHz, and f = 20 MHz.

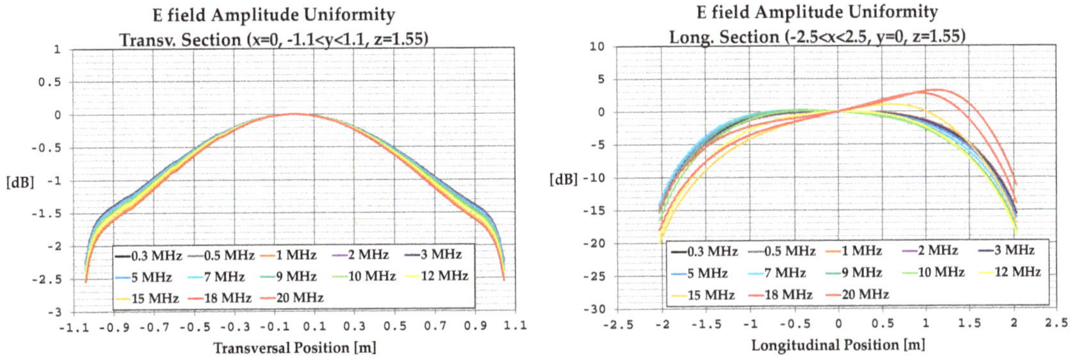

(a) (b)

Figure 9. Numerical evaluation of the uniformity of the electric field generated inside the volume formed by the MWTEM transmission line: (**a**) field behavior along the transversal direction ($x = 0$, $-1.1 < y < 1.1$, $z = 1.55$ m); (**b**) field behavior along the longitudinal direction ($-2.5 < x < 2.5$, $y = 0$, $z = 1.55$ m).

The models of the experimental setup, in free space and inside the shielded chamber, followed the modeling guidelines of the NEC software [31]. The metallic enclosure of the shielded chamber was formed with wire grids according to the modelling concepts given in [33–35] for conductive surfaces. The use of the discretization criteria given in [31] (expressed in terms of the length of the segments composing the wires of the MWTEM structure, normalized to the wavelength) guarantees the high accuracy of the results. Some details about the models and the computational effort are summarized in Table 4. All of the simulations were carried out on a personal computer equipped with an Intel Core i7 CPU 2.10 GHz and with 8 GB of RAM.

Table 4. NEC model parameters and simulation effort details.

NEC Model	Number of Segments	Simulation Time [1] [s]
MWTEM in free space	2100	30
MWTEM inside shielded chamber	9062	1743

[1] Simulation time necessary for each frequency value.

5. Experimental Evaluation and Calibration of the EMF Area Monitor

After completing the assessment of the uniformity of the electric field generated in the central volume of the MWTEM transmission line using an EP600 field probe as a reference, the field generated by the MWTEM line was employed to evaluate and calibrate the AMB-8059 area monitor (provided by NARDA STS S.r.l., Cisano sul Neva, Savona, Italy) equipped with a solar panel and an EP-1B-02 wideband probe (frequency band 0.1–3000 MHz, amplitude range 0.2–200 V/m). This made it possible to evaluate the frequency behavior of the area monitor (equipped with the EP-1B-02 probe as a unique field sensor) and determine its correction factor (C.F.), defined, as in (3), at a given frequency and with a fixed electric field amplitude, as the ratio of the simulated electric field amplitude in the central position of the MWTEM line (assumed as the reference field level $E_{REF.}$) to the measurement ($E_{MEAS.}$) obtained with the area monitor placed with the EP-1B-02 probe in the same position:

$$C.F.\ [dB] = 20Log_{10}\left(\frac{E_{REF.}}{E_{MEAS.}}\right) \quad (3)$$

In practice, during real on-site measurement activities, the correction factor has to be applied to the measured field to obtain the effective value of the field under measurement. This is performed manually, or automatically if the C.F. is stored in the area monitor's control module. In Figure 10, the C.F. obtained for the area monitor as a whole instrument is compared with the correction factor of the EP-1B-02 probe calibrated as a stand-alone device. As expected, the frequency response of the field probe integrated into the area monitor is not the same as that of the probe alone; when integrated within the area monitor, the probe shows a less uniform frequency behavior and also proves to be a bit more sensitive (differences vary between 0.5 dB and 2.7 dB).

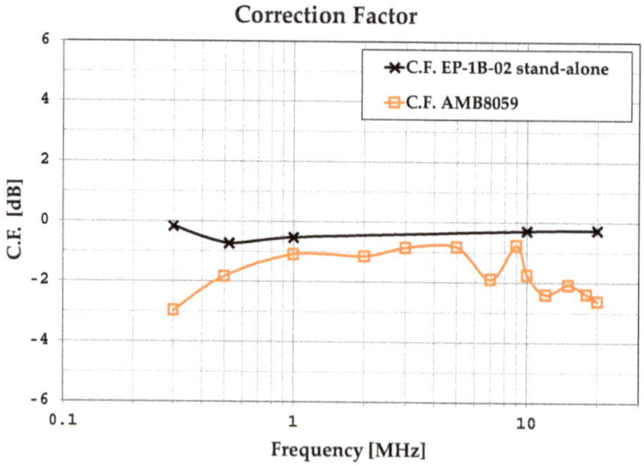

Figure 10. Comparison between the C.F. of the EP-1B-02 probe integrated into the area monitor and the C.F. of the EP-1B-02 probe as a stand-alone device.

Knowing the C.F. makes it possible to improve measurement accuracy when a source with a known spectrum position is under measurement; the availability of a correction factor is of paramount importance in several practical cases where the area monitor is placed in the proximity of a known source deemed to be continuously monitored. When the spectral position of the source is known, the use of the precise C.F.—valid for the emission frequency of the source—allows for the optimum improvement of measurement accuracy, especially if the C.F. is available with a sufficiently small frequency step. In addition, when electromagnetic pollution is generated by unknown sources, the frequency behavior of the correction factor can be used to define a mean value to be applied to measurements in order to reduce uncertainty. Regarding different types of sources, we assume a source to be known if its emission frequency band is identified. For example, a source is considered to be known if it generates an exposure level at a site and is associated with a broadcasting system whose emission frequency is known either from the system owner or from preliminary spectral narrow-band measurements. Conversely, a source is considered unknown at a site where no information on its emission spectrum is available from the owner or from narrow-band measurements. In addition, knowing the emission frequency bands of multiple sources is ineffective if they all contribute to a non-negligible exposure level at the same site. In such cases, it is impossible to apply different C.F.s to measurements taken with wideband probes (such as those integrated into the area monitor) as they provide only a single measurement value, representing the combined effects of all sources. Therefore, when dealing with unknown sources, using a single C.F. value—such as the mean C.F.—is likely the best approach.

To verify the effectiveness of using the mean C.F. value, an electric field was generated in the MWTEM line and measured in its central position with the area monitor at the same frequencies considered for the calibration. The deviations of the measurements taken with the area monitor after the application of the mean C.F. value from the measurements taken under the same conditions with the EP600 probe (assumed as references) are shown in Figure 11. We can observe that the application of a mean constant C.F. for frequencies higher than 1 MHz makes the sensor's frequency response uniform and implies errors of about 1 dB; for frequencies lower than 1 MHz, it implies higher yet still limited errors of around 2 dB.

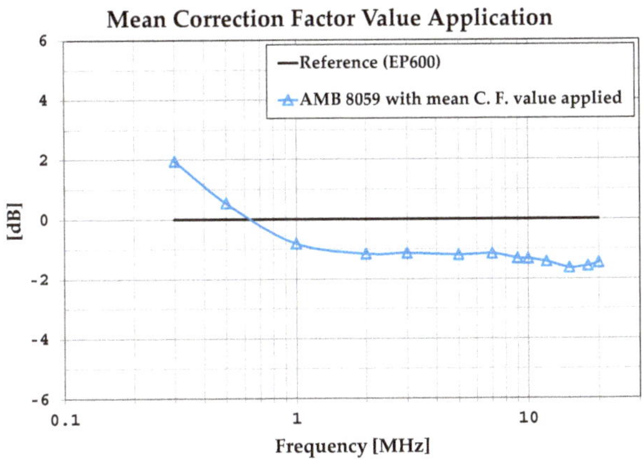

Figure 11. Effect of applying the mean C.F. value.

Future activities will investigate the feasibility of specific correction factors associated with different ambient area configurations, e.g., those equipped with devices and cables for specific applications. Further analysis will also be performed to evaluate the uncertainty associated with the correction factor measurement procedure.

6. Conclusions

The electric field generated by an MWTEM transmission line was analyzed in detail, both numerically and experimentally, to evaluate its capability to generate for $f < 30$ MHz a controlled electric field sufficiently uniform for the exposure of the entire structure of an EMF area monitor. Typically, the field probe integrated in an area monitor is calibrated as a stand-alone device without taking into account possible effects on the probe response due to electromagnetic interaction with the area monitor structure and materials. To avoid such problems and improve measurement accuracy, the ambient area monitor was calibrated by exposing the whole structure of the device to the electric field generated by the previously characterized MWTEM line. A frequency-varying correction factor (C.F.) to be applied in the case of known field sources was determined, and the effectiveness of using a mean (constant) C.F. value in the case of unknown sources was evaluated.

Author Contributions: Conceptualization, R.A., R.F. and A.G.; Investigation, R.A., R.F. and A.G.; Writing—review & editing, R.A., R.F. and A.G. The authors contributed equally to this work. All authors have read and agreed to the published version of the manuscript.

Funding: This research received no external funding.

Institutional Review Board Statement: Not applicable.

Informed Consent Statement: Not applicable.

Data Availability Statement: Data are contained within the article.

Acknowledgments: The authors wish to thank NARDA STS S.r.l. for providing an AMB-8059 area monitor and an EP600 field probe; and EMC S.r.l. for providing a shielded chamber, the MWTEM transmission line, and other devices necessary for the experimental activities.

Conflicts of Interest: Author Renzo Azaro was employed by the company EMC S.r.l., Roberto Franchelli and Alessandro Gandolfo were employed by the company NARDA Safety Test Solutions S.r.l. All authors' declares that the research was conducted in the absence of any commercial or financial relationships that could be construed as a potential conflict of interest.

Abbreviations

The following abbreviations are used in this manuscript:

EMF	Electromagnetic field
HF	High frequency
LF	Low frequency
MWTEM	Multi-wire transverse electromagnetic
MoM	Method of moments

Appendix A

The basic structure that can be employed to explain the operation of the MWTEM line from a theoretical point of view is the parallel plates transmission line. Because in practice it is relatively inexpensive and easy to construct, even for large products, the parallel plane line has long been used in electromagnetic compatibility activities for the susceptibility testing of electrical and electronic products and, in general, for the generation of a reference electromagnetic field. The geometry of a parallel plate transmission line is shown in Figure A1, where it is assumed that the plates' width w is larger than their spacing h, so for a simplified analysis of the structure the fields, it can be assumed to be the same as if the plates were of infinite width, which means that any edge effects can be neglected.

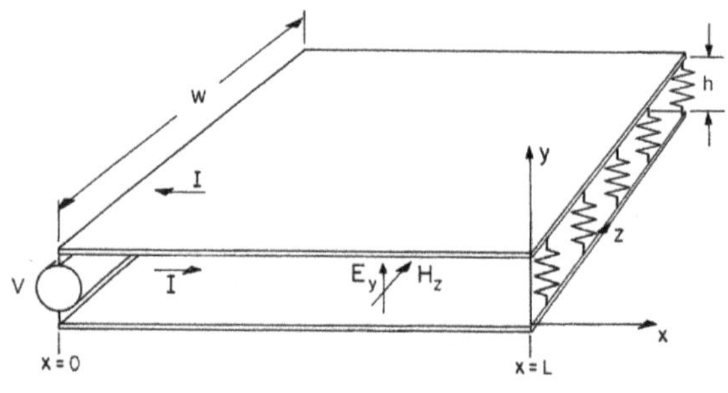

Figure A1. The parallel plate transmission line.

It can be shown [36,37] that the solution of Maxwell equations allows for three different classes of waves for the parallel plate transmission line: transverse electromagnetic (TEM, with $E_x = 0$, and $H_x = 0$), transverse electric (TE, with $E_x = 0$) and transverse magnetic (TM, with $H_x = 0$). Below the cut-off frequencies of the TE and TM modes, it can be assumed that between the parallel plates there is only the TEM mode; thus, in the sinusoidal steady state, the field components are:

$$E_y(x) = E_0 e^{\mp jkx} \quad \text{(A1)}$$

$$\eta H_z(x) = \pm E_0 e^{\mp jkx} \quad \text{(A2)}$$

where $\eta = \sqrt{\mu/\epsilon}$ and $k = \omega\sqrt{\mu\varepsilon}$ and (A1) and (A2) represent both progressive and regressive waves, whose ratio depends on the load connected at the end of the line ($x = L$).

Because $H_x = 0$, the voltage between the two plates can be calculated by integrating E_y along any path drawn between them:

$$V(x) = -\int_0^h E_y(x)dy = -hE_0 e^{\mp jkx} \tag{A3}$$

By applying the relationship $\vec{J} = \hat{n} \times \vec{H}$, where $\hat{n}$ is a unit vector normal to plates, the current density on one plate can be calculated from the magnetic field as:

$$J_x = -H_z \tag{A4}$$

Then, if the plate width is w, the current on one plate is:

$$I(x) = -wH_z = \mp wE_0/_{\eta}\, e^{\mp jkx} \tag{A5}$$

In terms of transmission line voltage and current waves, the following relationships hold:

$$V(x) = V_0 e^{\mp j\beta x} \tag{A6}$$

$$I(x) = \pm V_0/_{Z_0}\, e^{\mp j\beta x} \tag{A7}$$

$$Z_0 = \sqrt{L/_C}\quad \beta = \omega\sqrt{LC} \tag{A8}$$

The electrical parameters for a parallel plane transmission line are:

$$C = \varepsilon^{w}/_{h}\ [\mathrm{F/m}],\ L = \mu^{h}/_{w}\ [\mathrm{H/m}] \tag{A9}$$

Then, from (A8):

$$Z_0 = \eta^{h}/_{w}\quad \beta = k \tag{A10}$$

If the driving frequency is so low that a wavelength is much longer than the length of the line L, i.e.:

$$\beta L = {}^{2\pi L}/_{\lambda} \ll 1 \tag{A11}$$

a quasi-static approach can be adopted and the following approximated relationships hold:

$$|E_y(x)| = {}^{|V(x)|}/_{h} \tag{A12}$$

$$|H_z(x)| = {}^{|I(x)|}/_{w} \tag{A13}$$

Both costs and mechanical considerations have led to the development of MWTEM lines, i.e., parallel plate lines where the conducting planes are replaced by grids of parallel wires aligned along the direction of current flow.

The theoretical analysis of this kind of structure is not simple; however, under the assumption of validity of a quasi-static approximation, an approach based on potential function calculation can be profitably employed. According to [38,39], the calculation (on the transversal section) of the electric field components of the quasi-static TEM field distribution and the line impedance of an MWTEM line can be carried out according to the procedure described below.

Figure A2 shows the transversal section of an N-wire MWTEM line above an infinite ground plane, composed of a number of parallel wires lying along two planes and where:

$$\frac{r_0}{a} \ll 1, \frac{r_0}{b} \ll 1, \frac{r_0}{d} \ll 1, \tag{A14}$$

Each plate consists of N wires of radius r_0, equally spaced by d, and the ground is at a distance b below the bottom wires. The wires in the plane $y = -a$ are at the potential $-V/2$, and the wires in the plane $y = +a$ are at the potential $+V/2$.

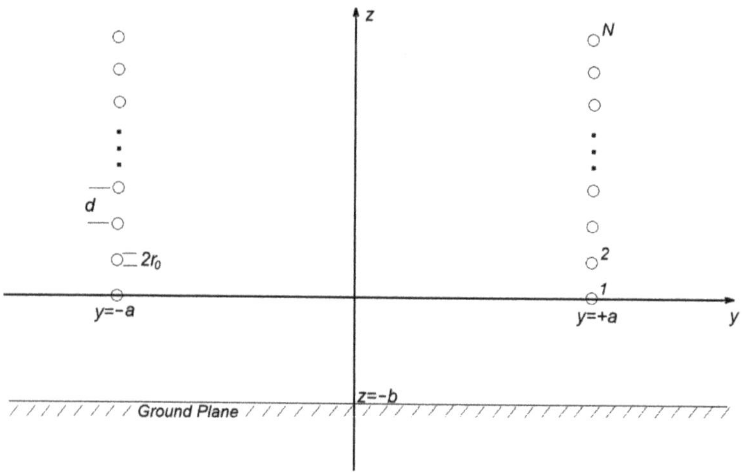

Figure A2. The transversal section of an N-wire MWTEM transmission line.

Under a quasi-static condition and thin wires hypothesis, the solution can be based on the superposition of the static fields generated by the parallel wires with a given charge per unit length.

The potential on the nth wire u_n is:

$$u_n = p_{n1}q_1 + p_{n2}q_2 + \cdots + p_{nN}q_N \tag{A15}$$

where q_i is the charge on the ith wire and each p_{ij}, named Maxwell's potential coefficient, is a function of the wire locations, $p_{ij} = p_{ji}$ and $p_{ij} \geq 0$ for all i,j.

By inverting the p_{ij} matrix (N × N), one obtains the k_{pq} matrix (N × N), whose elements are defined as capacitance coefficients:

$$q_n = k_{n1}u_1 + k_{n2}u_2 + \cdots + k_{nN}u_N \tag{A16}$$

where k_{rs} is the capacitance coefficient and

$$k_{rs} = k_{sr},$$

$$k_{rs} \geq 0 \text{ for } r = s$$

$$k_{rs} \leq 0 \text{ for } r \neq s.$$

By assuming the same potential on all the N wires:

$$u_1 = u_2 = \cdots = u_N = u_0 \tag{A17}$$

According to (A16), the charge on the nth wire is:

$$q_n = u_0 \sum_{m=1}^{N} k_{nm} \tag{A18}$$

If, according to Figure A2, the transmission line is composed of N pairs of thin wires above a perfectly conducting ground, it is possible to show that [38,39]:

$$p_{nm} = \frac{1}{2}\ln\left[\frac{\left[(n-m)^2 d^2 + 4a^2\right]^2}{(n-m)^4 d^4 + 4(n-m)^2 a^2 d^2}\right] - \frac{1}{2}\ln\left[1 + \frac{a^2}{\left[b + \left(\frac{n+m-2}{2}\right)d\right]^2}\right] \text{ for } m \neq n \tag{A19}$$

$$p_{nn} = \ln\left[\frac{2a}{r_0}\right] - \frac{1}{2}\ln\left[1 + \frac{a^2}{\left[b + \left(\frac{n-1}{2}\right)d\right]^2}\right] \text{ for } m = n \tag{A20}$$

By applying a numerical inversion technique to the p_{ij} matrix (N × N), it is possible to find the k_{pq} matrix (N × N) and then define and calculate the effective plate separation a_{eff} as:

$$\frac{1}{a_{eff}} = \sum_{n=1}^{N}\left(\sum_{m=1}^{N} k_{nm}\right)\left\{\frac{\frac{2}{a}}{1 + \left[\frac{(N-1)d}{2a} - (n-1)\frac{d}{a}\right]^2} - \frac{\frac{2}{a}}{1 + \left[\frac{(N-1)d}{2} - (n-1)d + 2b\right]^2}\right\} \tag{A21}$$

If one defines an electric field efficiency factor

$$f_E = \frac{a}{a_{eff}} \tag{A22}$$

it is then possible to calculate the y component of the electric field in the center of the transversal section of the line according to the relationship:

$$E_y\Big|_{\substack{y=0 \\ z=\frac{(N-1)}{2}d}} = \frac{\Delta V}{2a_{eff}} = \frac{\Delta V}{2a} f_E \tag{A23}$$

where ΔV is the difference of potential across the two plates of the line.

The transversal electric field components $E_{y\,rel}$, $E_{z\,rel}$ (normalized to the field strength in the y-direction at the center of the line) can be calculated according to the following relationships:

$$E_{y_{rel}}(y,z) = \frac{E_y}{E_y\Big|_{\substack{y=0 \\ z=\frac{(N-1)}{2}d}}} = \frac{a}{f_E}\sum_{n=1}^{N}\left\{\left(\sum_{m=1}^{N} k_{nm}\right), \left[\frac{y+a}{(y+a)^2+[z-(n-1)d]^2} - \frac{y-a}{(y-a)^2+[z+(n-1)d]^2}\right], + \left(\sum_{m=1}^{N} k_{nm}\right)\left|\left[\frac{y-a}{(y-a)^2+[z+(n-1)d+2b]^2} - \frac{y+a}{(y+a)^2+[z+(n-1)d+2b]^2}\right]\right\} \tag{A24}$$

$$E_{z_{rel}}(y,z) = \frac{E_z}{E_y\Big|_{\substack{y=0 \\ z=\frac{(N-1)}{2}d}}} = \frac{a}{f_E}\sum_{n=1}^{N}\left\{\left(\sum_{m=1}^{N} k_{nm}\right)\left[\frac{z-(n-1)d}{(y+a)^2+[z-(n-1)d]^2} - \frac{z-(n-1)d}{(y-a)^2+[z-(n-1)d]^2}\right] - \left(\sum_{m=1}^{N} k_{nm}\right)\left|\left[\frac{z+(n-1)d+2b}{(y+a)^2+[z+(n-1)d+2b]^2} - \frac{z+(n-1)d+2b}{(y-a)^2+[z+(n-1)d+2b]^2}\right]\right\} \tag{A25}$$

Under the assumption of the existence of only a TEM mode, the impedance of the transmission line is related to the free-space characteristic impedance by the following relationship [39]:

$$Z = f_g \sqrt{\frac{\mu_0}{\varepsilon_0}} \cong 120\pi f_g \tag{A26}$$

where f_g, defined as geometrical factor, is:

$$f_g = \frac{u_1}{\pi \sum_{n=1}^{N} \left(\sum_{m=1}^{N} k_{nm} \right)} \tag{A27}$$

and where:

$$\begin{aligned} u_1 = \left(\sum_{m=1}^{N} k_{1m}\right) ln\left(\frac{2a}{r_0}\right) + \frac{1}{2}\sum_{n=2}^{N} \left\{ ln\left[\frac{\left[(n-1)^2 d^2 + 4a^2\right]^2}{(n-1)^4 d^4 + 4(n-1)^2 a^2 d^2}\right] \left(\sum_{m=1}^{N} k_{nm}\right) \right\} \\ - \frac{1}{2}\sum_{n=1}^{N} \left\{ ln\left[1 + \frac{a^2}{\left[b + \left(\frac{n-1}{2}\right)d\right]^2}\right] \left(\sum_{m=1}^{N} k_{nm}\right) \right\} \end{aligned} \tag{A28}$$

By substituting in (A19), (A20), (A27) and (A28) the geometrical parameters of the MWTEM line used for the experimental activities described in the paper, i.e., $N = 7$ (number of wires per plate), $d = 0.25$ m (wire spacing), $r_0 = 0.01$ m (wire radius), $a = 1.16$ m (half the distance between the wire plates), and $b = 0.8$ m (distance between the bottom wires and the ground), it is possible to calculate the geometrical factor ($f_g = 0.5511$) and then $Z \cong 120\,\pi f_g = 207.75\ \Omega$, which is in good accordance with the value of the MWTEM line load (200 Ω, value equal to the line impedance, according to the common criterion used for a transmission line load selection).

Appendix B

Electric and magnetic field integral equations (EFIE and MFIE) have the following general forms:

$$\vec{E} = f_e\left(\vec{J}\right), \text{ (EFIE)} \tag{A29}$$

$$\vec{H} = f_m\left(\vec{J}\right), \text{ (MFIE)} \tag{A30}$$

where $\vec{E}$ and $\vec{H}$ are the electric and magnetic field vectors and the currents $\vec{J}$ are the field sources, which in the method of moments (MoM) [32] formulation are expressed as a sum of basis function:

$$\vec{J} = \sum_{i=1}^{M} J_i \vec{b}_i \tag{A31}$$

where $\vec{b}_i$ are the basis functions and J_i their coefficients.

For a general problem of determining the electromagnetic field generated by currents flowing on a conductive structure, some current contributions are known a priori (i.e., forced by sources), while others (on both the same structure and different ones) are the unknowns of the problem. Once the currents have been calculated, the electric and magnetic vector fields can be determined, both in the near field and in the far field.

The electric and magnetic field integral equations that have to be solved take the following form [29]:

$$F\left(g\left(\vec{r}\prime\right)\right) = h \tag{A32}$$

where F is a linear operator, h is a known excitation function, and g is the system response to be found. According to the MoM formulation, the unknown function g is expressed as a linear combination of N terms:

$$g\left(\vec{r\prime}\right) \approx a_1 g_1\left(\vec{r\prime}\right) + a_2 g_2\left(\vec{r\prime}\right) + \cdots + a_N g_N\left(\vec{r\prime}\right) = \sum_{n=1}^{N} a_n g_n\left(\vec{r\prime}\right) \quad \text{(A33)}$$

where a_n are unknown constants and $g_n\left(\vec{r\prime}\right)$ are known functions, namely the basis functions, belonging to the same domain of $g\left(\vec{r\prime}\right)$. Then, (A32) can be written as:

$$\sum_{n=1}^{N} a_n F\left(g\left(\vec{r\prime}\right)\right) = h \quad \text{(A34)}$$

The solution of (A34) is based on the linearity of the operator; by expanding the expression in N linearly independent equations, imposing the boundary conditions in N different points by means of a point-matching technique and employing an inner product is defined as:

$$< w, g >= \int_S w^* g ds \quad \text{(A35)}$$

where S is the surface of the structure under analysis and w is the test function, defined in the same domain of the F function (usually a Dirac function):

$$[w_m] = \left[\delta\left(\vec{r} - \vec{r}_m\right)\right] \quad m = 1, 2, \ldots, N \quad \text{(A36)}$$

In (A36), $\vec{r}$ is a given position in the reference coordinate system and $\vec{r}_m$ is the position of application of the boundary conditions.

By applying (A35) and (A36) in (A34), the following relationship is obtained:

$$\sum_{n=1}^{N} a_n < w_m, F(g_n) >=< w_m, h > \quad m = 1, 2, \ldots, N \quad \text{(A37)}$$

which is written in the matrix form and solved for the unknown coefficients by means of a matrix inversion operation, which leads to:

$$[a_n] = [G_{mn}]^{-1}[h_m] \quad \text{(A38)}$$

where:

$$[G_{mn}] = \begin{bmatrix} < w_1, F(g_1) > & < w_1, F(g_2) > & \cdots \\ < w_2, F(g_1) > & < w_2, F(g_2) > & \\ \vdots & & \ddots \end{bmatrix}, \ [a_n] = \begin{bmatrix} a_1 \\ a_2 \\ \vdots \\ a_N \end{bmatrix} \text{ e } [h_m] = \begin{bmatrix} < w_1, h > \\ < w_2, h > \\ \vdots \\ < w_N, h > \end{bmatrix} \quad \text{(A39)}$$

In the case of structures composed of conductive filamentary elements (wires), for the Numerical Electromagnetics Code (NEC), the unknown function $g\left(\vec{r\prime}\right)$ is the linear current $I(s)$ that flows along the elements. According to the MoM, each conductive element is assumed to be subdivided in segments, and on each of them, the current $I(s)$ is expanded in the following form [29]:

$$I_i(s) = A_i + B_i sink(s - s_i) + C_i cosk(s - s_i), \text{ dove } |s - s_i| < \Delta_i/2 \quad \text{(A40)}$$

where s is the position coordinate along the segment of length Δ and s_i is the value of s in the center of the segment. As discussed in the literature [40,41], using this kind of form guarantees a rapid convergence of the solution.

Appendix C

A detailed numerical analysis of the electric and magnetic field components was carried out for a thorough understanding of the characteristics of the field generated inside the volume formed by the MWTEM transmission line. The field generated inside the MWTEM volume is at an electrical short distance from the sources, and the properties of a far-field TEM mode cannot be a priori assumed.

According to the international standard IEC 61000-4-20 [42], which specifies the characteristics of TEM waveguides both for the generation and measurement of electromagnetic fields during their use for electromagnetic compatibility activities, a TEM is a mode in which the components of the electric and magnetic fields in the propagation direction (E_x and H_x in the case of interest) are much less than the primary field components (E_y and H_z) across any transverse cross-section. The TEM mode is considered to be dominant if statistically at least 75% of the secondary (unintended) electric field components are at least 6 dB less than the primary component of the electric field. Given the good accordance of simulations and measurements observed during the MWTEM validation activity, the TEM mode verification was carried out by simulation by using the numerical model described in the paper, using the input voltages and the load impedance values of Table 2.

With reference to the geometry of the line shown in Figures 3, 7 and 8, a transversal and a longitudinal section of 1 × 1 m were defined, and on them, the x, y and z components of the electric and magnetic field vectors were calculated and analyzed. The results for the transversal section for four frequency values (0.3 MHz, 5 MHz, 10 MHz, and 20 MHz) are shown in Figures A3 and A4 for the electric field and magnetic vector components, respectively, while the same kind of results for the longitudinal section are shown in Figures A5 and A6. Even if the primary requirement was on the electric field (to which the field probes of interest are sensitive), in view of using the MWTEM line also for magnetic field probe calibration, the analysis was also carried out on the magnetic field vector components.

As can be observed in Figures A3–A6, the primary field components E_y and H_z are always greater than the secondary ones (E_x, E_z, H_x, H_y). All of the simulated results obtained on the transversal and longitudinal planes were analysed in detail, and the differences (in dB) between the primary and the secondary components are reported in Table A1. All of the secondary components are 6 dB less than the primary components, both for the electric and the magnetic field vectors. Moreover, the transversal components E_z and H_y, orthogonal to the principal ones (E_y, H_z), are at least 15 dB below, thus confirming a uniform linear polarization. Similar results can be hypothesized in all of the 0.3–20 MHz frequency band.

Table A1. MWTEM transmission line—primary and secondary field vector components analysis.

Freq. [MHz]	Transversal Section (x = 0, −0.5 < y < 0.5, 1.05 < z < 2.05)				Longitudinal Section (−0.5 < x < 0.5, y = 0, 1.05 < z < 2.05)			
	$\frac{E_y}{E_x}$ [dB]	$\frac{E_y}{E_z}$ [dB]	$\frac{H_z}{H_x}$ [dB]	$\frac{H_z}{H_y}$ [dB]	$\frac{E_y}{E_x}$ [dB]	$\frac{E_y}{E_z}$ [dB]	$\frac{H_z}{H_x}$ [dB]	$\frac{H_z}{H_y}$ [dB]
0.3	>40	15.5	>40	18.7	>40	>40	23.9	>40
5	>40	15.4	35.3	18.8	>40	>40	23.1	>40
10	38.8	15.2	32.9	18.9	>40	>40	23.7	>40
20	>40	14.6	17.9	18.4	>40	>40	15.1	>40

For the sake of completeness, the wave impedance has also been calculated along two transversal paths (x = 0, −0.5 < y < 0.5, z = 1.55 m and x = 0, y = 0, 1.05 < z < 2.05) and on a longitudinal path (−0.5 < x < 0.5, y = 0, z = 1.55 m). The results are shown in

Figures A7 and A8, where impedance values varying around the free space impedance $Z_0 = 120\ \pi\ \Omega$ can be observed. It is worth noting that the free space impedance could be observed only in a line having an electrical length sufficient to evidence progressive and regressive waves; because of the dimensions of the transversal section of the MWTEM line, this condition would occur for frequency values at which non-negligible field perturbations due to high-order modes would arise.

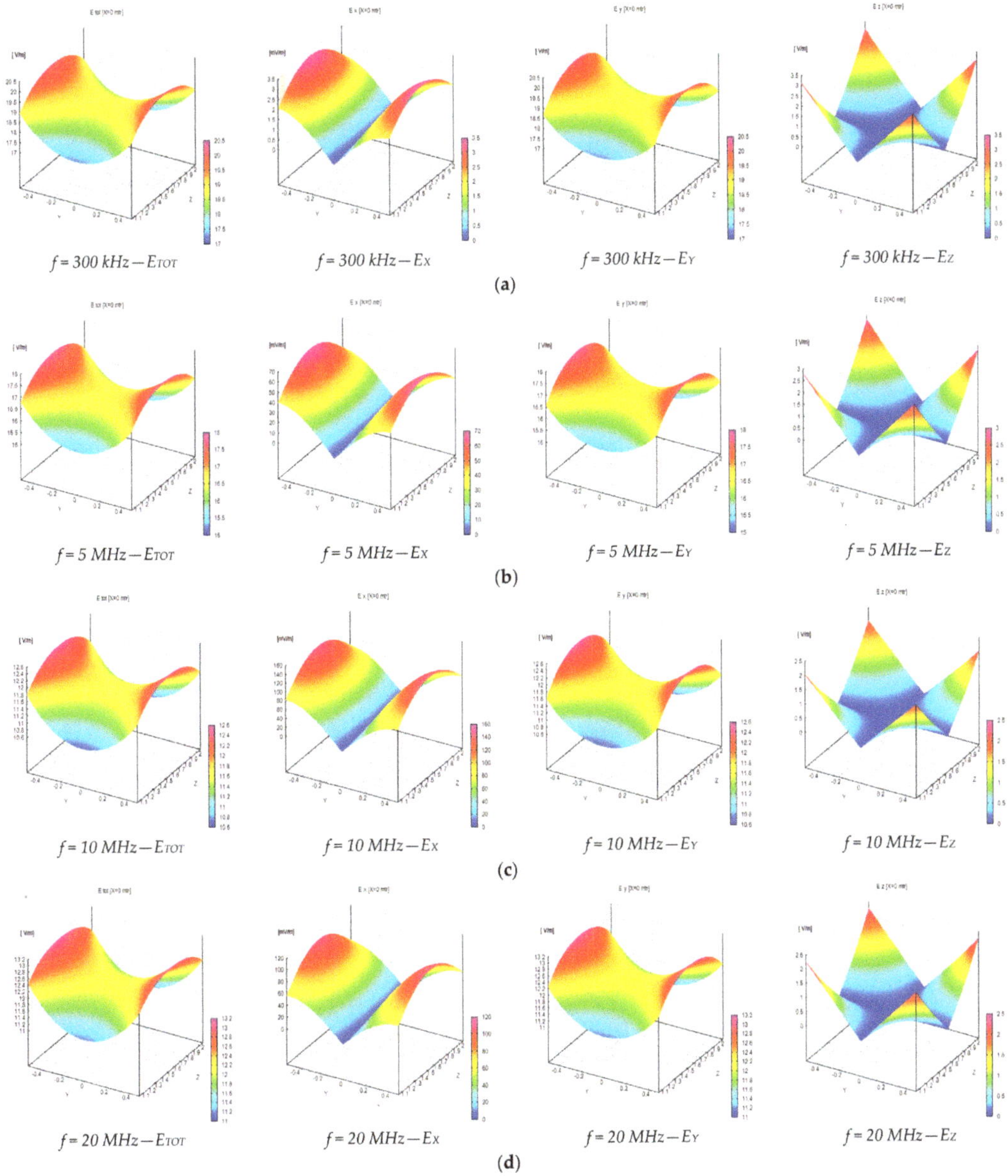

Figure A3. Numerical evaluation of the components (E_X, E_Y, E_Z) of the electric field generated on the central transversal section inside the volume formed by the MWTEM transmission line: (**a**) f = 0.3 MHz, (**b**) f = 5 MHz, (**c**) f = 10 MHz, (**d**) f = 20 MHz.

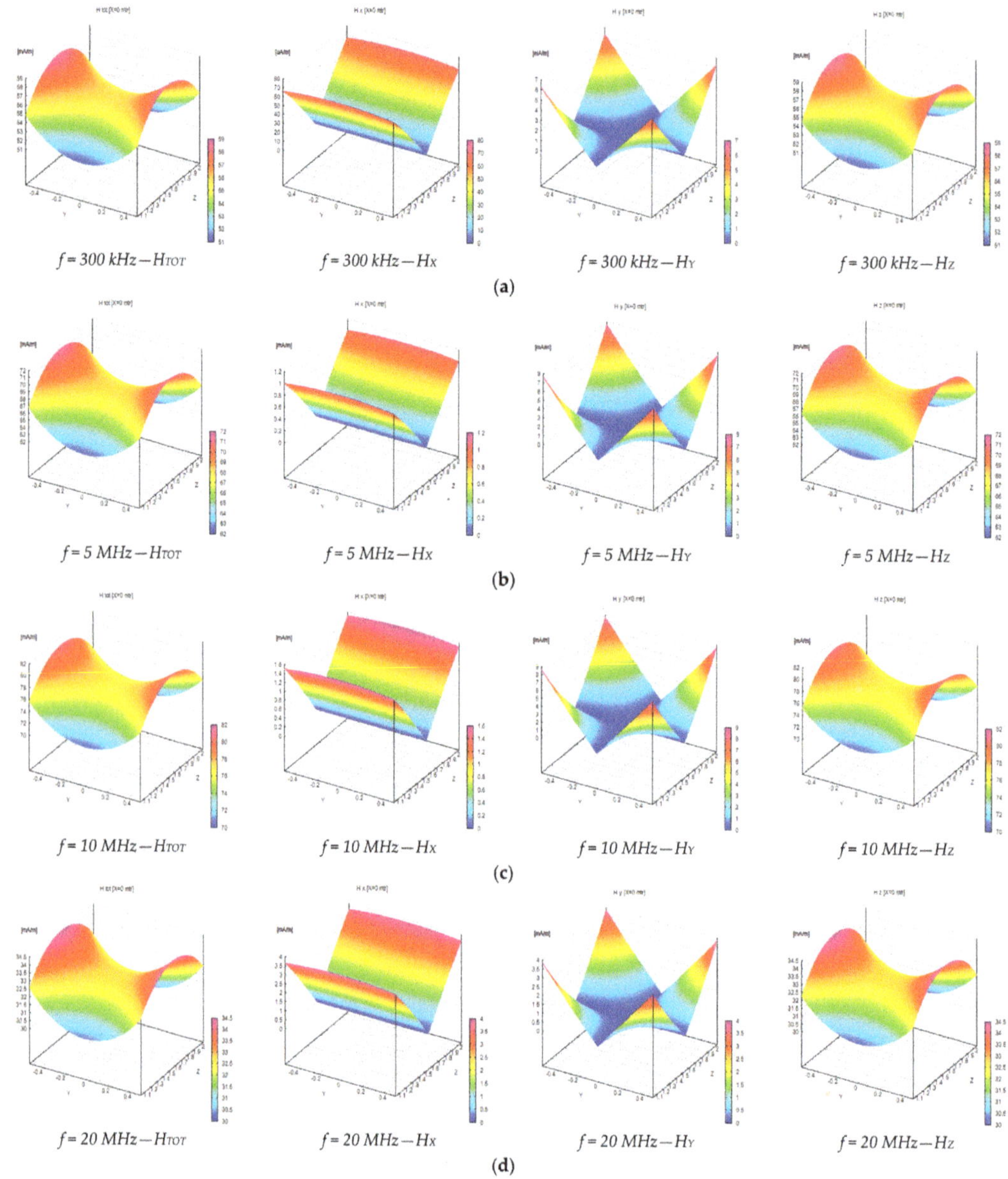

Figure A4. Numerical evaluation of the components (H_X, H_Y, H_Z) of the magnetic field generated on the central transversal section inside the volume formed by the MWTEM transmission line: (**a**) f = 0.3 MHz, (**b**) f = 5 MHz, (**c**) f = 10 MHz, (**d**) f = 20 MHz.

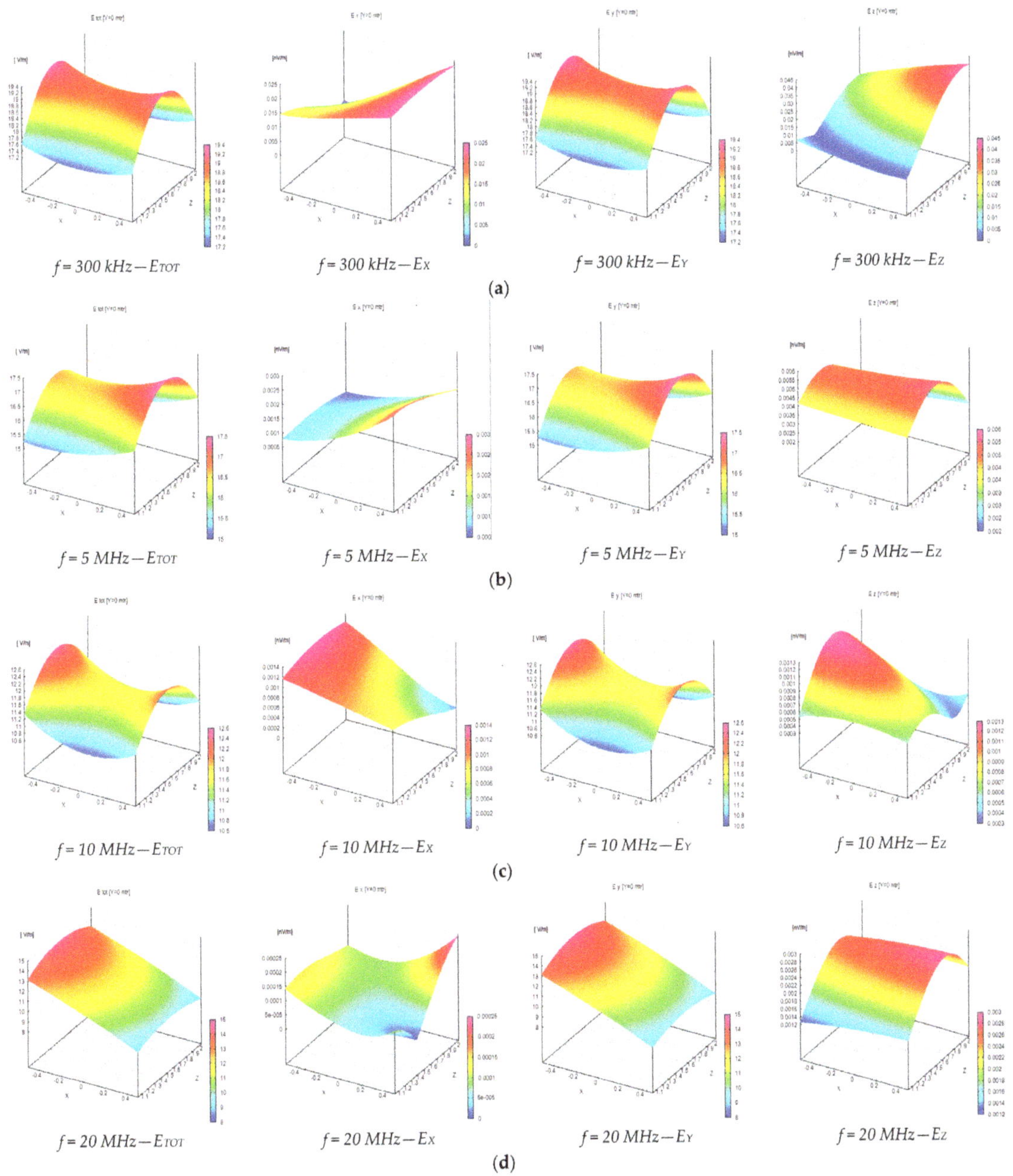

Figure A5. Numerical evaluation of the components (E_X, E_Y, E_Z) of the electric field generated on the central longitudinal section inside the volume formed by the MWTEM transmission line: (**a**) f = 0.3 MHz, (**b**) f = 5 MHz, (**c**) f = 10 MHz, (**d**) f = 20 MHz.

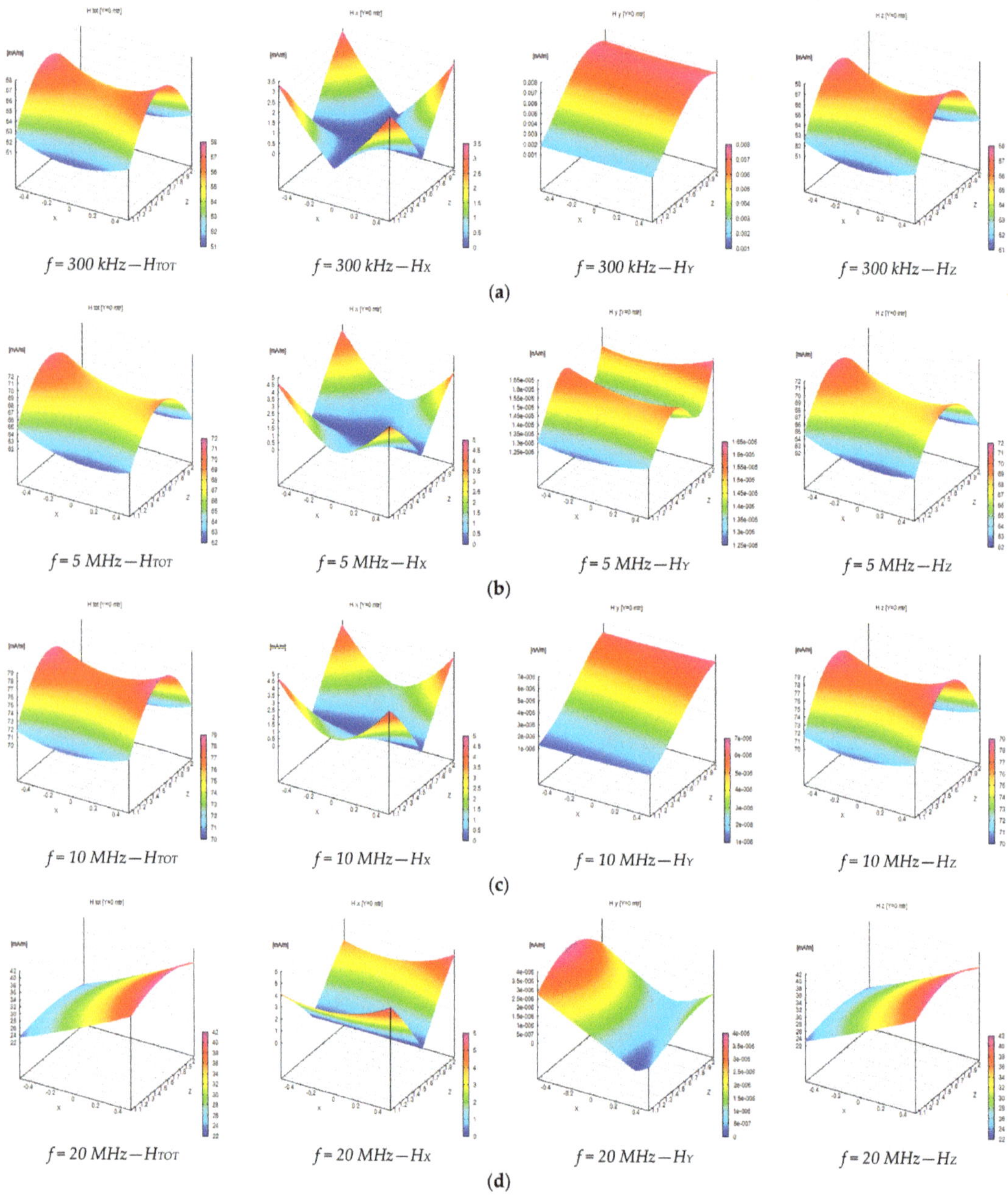

Figure A6. Numerical evaluation of the components (H_X, H_Y, H_Z) of the magnetic field generated on the central longitudinal section inside the volume formed by the MWTEM transmission line: (**a**) f = 0.3 MHz, (**b**) f = 5 MHz, (**c**) f = 10 MHz, (**d**) f = 20 MHz.

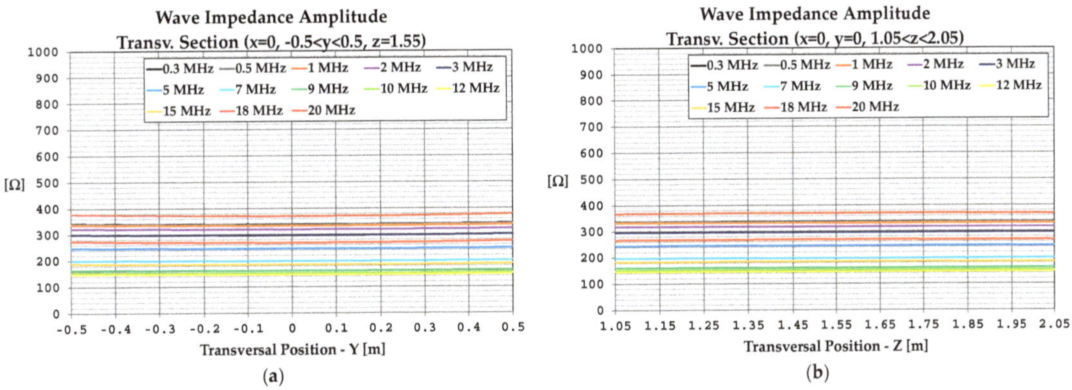

Figure A7. Numerical evaluation of the near field inside the volume formed by the MWTEM transmission line: (**a**) wave impedance along the transversal direction (x = 0, −0.5 < y < 0.5, z = 1.55 m), (**b**) wave impedance along the transversal direction (x = 0, y = 0, 1.05 < z < 2.05 m).

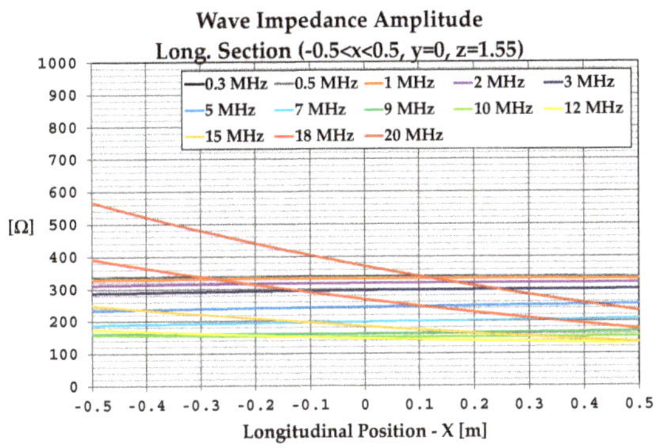

Figure A8. Numerical evaluation of the near field inside the volume formed by the MWTEM transmission line: wave impedance along the longitudinal direction (−0.5 < x < 0.5, y = 0, z = 1.55 m).

References

1. International Commission on Non-Ionizing Radiation Protection. Guidelines for limiting exposure to time-varying electric and magnetic fields (1 Hz–100 kHz). *Health Phys.* **2010**, *99*, 818–836. [CrossRef] [PubMed]
2. International Commission on Non-Ionizing Radiation Protection. Guidelines for limiting exposure to electromagnetic fields (100 kHz to 300 GHz). *Health Phys.* **2020**, *118*, 483–524. [CrossRef] [PubMed]
3. European Union. *Directive 2013/35/EU on the Minimum Health and Safety Requirements Regarding the Exposure of Workers to the Risks Arising from Physical Agents (Electromagnetic Fields) and Repealing Directive 2004/40/EC*; Council of the European Union: Brussels, Belgium, 2013.
4. European Union (UE) Council Recommendation. *Council Recommendation of 12 July 1999 on the Limitation of Exposure of the General Public to Electromagnetic Fields (0 Hz to 300 GHz)—(1999/519/EC)*; Council of the European Union: Brussels, Belgium, 1999.
5. International Telecommunication Union (ITU). *Recommendation ITU-T K.83—Monitoring Electromagnetic Field*; International Telecommunication Union (ITU): Geneva, Switzerland, 2011.
6. Ioriatti, L.; Martinelli, M.; Viani, F.; Benedetti, M.; Massa, A. Real-time distributed monitoring of electromagnetic pollution in urban environments. In Proceedings of the IEEE Geoscience and Remote Sensing Symposium (IGARSS 2009), Cape Town, South Africa, 12–17 July 2009.
7. Viani, F.; Donelli, M.; Oliveri, G.; Massa, A.; Trinchero, D. A WSN-based system for real-time electromagnetic monitoring. In Proceedings of the IEEE International Symposium on Antennas and Propagation (APSURSI), Spokane, DC, USA, 3–8 July 2011.
8. Viani, F.; Polo, A.; Donelli, M.; Giarola, E. A Relocable and Resilient Distributed Measurement System for Electromagnetic Exposure Assessment. *IEEE Sens. J.* **2016**, *16*, 4595–4604. [CrossRef]

9. Antic, D.; Djuric, N.; Kljajic, D. Environmental EMF monitoring in the SEMONT system using quad-band AMB 8057/03 sensor. In Proceedings of the IEEE 10th International Conference on Wireless and Mobile Computing, Networking and Communications (WiMob), Larnaca, Cyprus, 8–10 October 2014.
10. Antic, D.; Djuric, N.; Kljajic, D. The AMB 8057-03 sensor node implementation in the SEMONT EMF monitoring system. In Proceedings of the IEEE 12th International Symposium on Intelligent Systems and Informatics (SISY), Subotica, Serbia, 11–13 September 2014.
11. Kundacina, O.; Vincan, V.; Antic, D.; Kljajic, D.; Kasas-Lazetic, K. The AMB 8057 utilization for SEMONT's EMF monitoring in a closed room. In Proceedings of the 24th Telecommunications Forum (TELFOR), Belgrade, Serbia, 22–23 November 2016.
12. Djuric, N.; Kavecan, N.; Kljajic, D.; Mijatovic, G.; Djuric, S. Data Acquisition in Narda's Wireless Stations based EMF RATEL Monitoring Network. In Proceedings of the 2019 International Conference on Sensing and Instrumentation in IoT Era (ISSI), Lisbon, Portugal, 29–30 August 2019.
13. Djuric, N.; Kavecan, N.; Radosavljevic, N.; Djuric, S. The Wideband Approach of 5G EMF Monitoring. In Proceedings of the 12th EAI International Conference (AFRICOMM 2020), Ebène City, Mauritius, 2–4 December 2020.
14. Jadhav, A.V.; Patil, A.P.; Ajwilkar, P.G.; Nadkarni, R.N.; Bhise, S. Continuous RF monitoring. In Proceedings of the 2nd International Conference on Applied and Theoretical Computing and Communication Technology (iCATccT), Bangalore, India, 21–23 July 2016.
15. Carciofi, C.; Garzia, A.; Valbonesi, S.; Gandolfo, A.; Franchelli, R. RF electromagnetic field levels extensive geographical monitoring in 5G scenarios: Dynamic and standard measurements comparison. In Proceedings of the 2020 International Conference on Technology and Entrepreneurship (ICTE), Bologna, Italy, 21–23 September 2020.
16. Wang, S.; Chikha, W.B.; Zhang, Y.; Liu, J.; Conil, E.; Jawad, O.; Ourak, L.; Wiart, J. RF Electromagnetic Fields Exposure Monitoring using Drive Test and Sensors in a French City. In Proceedings of the XXXVth General Assembly and Scientific Symposium of the International Union of Radio Science (URSI GASS), Sapporo, Japan, 19–26 August 2023.
17. Lee, A.K.; Jeon, S.; Wang, S.; Wiart, J.; Choi, H.D.; Moon, J.I. Population Density and DL EMF Exposure Levels by Region in Korea. In Proceedings of the 2024 IEEE Asia-Pacific Microwave Conference (APMC), Bali, Indonesia, 17–20 November 2024.
18. Liu, S.; Tobita, K.; Onishi, T.; Taki, M.; Watanabe, S. Electromagnetic field exposure monitoring of commercial 28-GHz band 5G base stations in Tokyo. *Bioelectromagnetics* **2024**, *45*, 281–292. [CrossRef] [PubMed]
19. Dikmen, I.C. Continuous Monitoring and Modeling of High-Frequency EM Pollution Using Cutting-Edge Regression Techniques. *IEEE Access* **2025**, *13*, 46627–46637. [CrossRef]
20. Kiouvrekis, Y.; Psomadakis, I.; Vavouranakis, K.; Zikas, S.; Katis, I.; Tsilikas, I.; Panagiotakopoulos, T.; Filippopoulos, I. Explainable Machine Learning-Based Electric Field Strength Mapping for Urban Environmental Monitoring: A Case Study in Paris Integrating Geographical Features and Explainable AI. *Electronics* **2025**, *14*, 254. [CrossRef]
21. Azaro, R.; Franchelli, R.; Gandolfo, A. Networks of EMF Area Monitor for Distributed Human Exposure Monitoring: Assessment of Performances in Simulated Realistic Scenarios. In Proceedings of the IEEE International Symposium on Measurements & Networking (M&N), Padua, Italy, 18–20 July 2022.
22. Azaro, R.; Gandolfo, A. On the Electric Field Distribution in Radiated Emission Rod Antenna Test Setups: Numerical Analysis and Experimental Validation. *IEEE Trans. Electromagn. Compat.* **2020**, *62*, 1611–1618. [CrossRef]
23. Crawford, M.L. Generation of Standard EM Fields Using TEM Transmission Cells. *IEEE Trans. Electromagn. Compat.* **1974**, *16*, 189–195. [CrossRef]
24. Groh, C.; Karst, J.P.; Koch, M.; Garbe, H. TEM waveguides for EMC measurements. *IEEE Trans. Electromagn. Compat.* **1999**, *41*, 440–445. [CrossRef]
25. Karst, J.P.; Groh, C.; Garbe, H. Calculable field generation using TEM cells applied to the calibration of a novel E-field probe. *IEEE Trans. Electromagn. Compat.* **2002**, *44*, 59–71. [CrossRef]
26. Macnamara, T.M. *Handbook of Antennas for EMC*, 2nd ed.; Artech House: Norwood, MA, USA, 2018; pp. 273–334.
27. Caorsi, S.; Azaro, R. Field Uniformity Measurements in a TEM Cell using the Modulated Scattering Technique. In Proceedings of the International Symposium on Electromagnetic Compatibility EMC 98 ROMA, Rome, Italy, 14–18 September 1998.
28. Carbonini, L. Comparison of analysis of a WTEM cell with standard TEM cells for generating EM fields. *IEEE Trans. Electromagn. Compat.* **1993**, *35*, 255–263. [CrossRef]
29. Burke, G.J.; Poggio, A.J. Numerical Electromagnetics Code—Method of Moments. Available online: https://apps.dtic.mil/sti/tr/pdf/ADA956129.pdf (accessed on 18 January 2025).
30. Poggio, A.J.; Miller, E.K. Integral Equation Solutions of Three-Dimensional Scattering Problems. In *Computer Techniques for Electromagnetics*; Mittra, R., Ed.; Pergamon Press: New York, NY, USA, 2007; pp. 159–264.
31. NEC Based Antenna Modeler and Optimizer. Available online: https://www.qsl.net/4nec2/ (accessed on 18 January 2025).
32. Harrington, R.F. *Field Computation by Moment Methods*; MacMillan: New York, NY, USA, 1968.
33. Ludwig, A. Wire grid modeling of surfaces. *IEEE Trans. Antennas Propagat.* **1987**, *35*, 1045–1048. [CrossRef]

34. Rubinstein, A.; Rachidi, F.; Rubinstein, M. On wire-grid representation of solid metallic surfaces. *IEEE Trans. Electromagn. Compat.* **2005**, *47*, 192–195. [CrossRef]
35. Rubinstein, A.; Rostamzadeh, C.; Rubinstein, M.; Rachidi, F. On the use of the equal area rule for the wire-grid representation of metallic surfaces. In Proceedings of the 7th International Zurich Symposium on Electromagnetic Compatibility, Singapore, 28 February–3 March 2006.
36. Ramo, S.; Whinnery, J.R.; Van Duzer, T. *Fields and Waves in Communication Electronics*; Wiley: New York, NY, USA, 1994.
37. Haus, H.A.; Melcher, J.R. *Electromagnetic Fields and Energy*; Prentice-Hall: Englewood Cliffs, NJ, USA, 1989.
38. Higgins, D.F. A Method of Calculating Impedance and Field Distribution of a Multi-Wire Parallel Plate Transmission Line Above a Perfectly Conducting Ground. Air Force Weapons Laboratory, Sensor and Simulation Notes, Note 150, February 1972. Available online: https://summa.unm.edu/notes/SSN/note150.pdf (accessed on 10 April 2025).
39. Baum, C.E.; Higgins, D.; Cummings, D.B.; Faulkner, J.E.; Heinberg, M.; Moore, P.; Nelson, E.B.; Price, M.L.; Crewson, W.F.; Naff, J.T. Electromagnetic Design Calculation for ATLAS I, Design 1. Air Force Weapons Laboratory, Sensor and Simulation Notes, Note 153, June 1972. Available online: https://summa.unm.edu/notes/SSN/note153.pdf (accessed on 10 April 2025).
40. Neureuther, A.R.; Fuller, B.D.; Hakke, G.D.; Hohmann, G. A Comparison of Numerical Methods for Thin Wire Antennas. In Proceedings of the URSI Meeting, Boston, MA, USA, 10–12 September 1968.
41. Miller, E.K.; Bevensee, R.M.; Poggio, A.J.; Adams, R.; Deadrick, F.J.; Landt, J.A. An Evaluation of Computer Programs Using Integral Equation for the Electromagnetics Analysis of Thin Wire Structures. Lawrence Livermore Laboratory, Interaction Notes, Note 177, March 1974. Available online: http://ece-research.unm.edu/summa/notes/In/0177.pdf (accessed on 25 February 2025).
42. *IEC 61000-4-20:2022-11*; Electromagnetic Compatibility (EMC)—Part 4–20: Testing and Measurement Techniques—Emission and Immunity Testing in Transverse Electromagnetic (TEM) Waveguides. IEC: Geneva, Switzerland, 2022.

sensors

MDPI

Article

Sub-MHz EMAR for Non-Contact Thickness Measurement: How Ultrasonic Wave Directivity Affects Accuracy

Alexander Siegl [1,*], David Auer [1], Bernhard Schweighofer [1], Andre Hochfellner [2], Gerald Klösch [2] and Hannes Wegleiter [1]

1 Christian Doppler Laboratory for Measurement Systems for Harsh Operating Conditions, Graz University of Technology, Inffeldgasse 23/2, 8010 Graz, Austria; bernhard.schweighofer@tugraz.at (B.S.); wegleiter@tugraz.at (H.W.)

2 voestalpine Stahl Donawitz GmbH, Kerpelystrasse 199, 8700 Leoben, Austria

* Correspondence: alexander.siegl@tugraz.at

Abstract: Electromagnetic acoustic resonance (EMAR) is a well-established non-contact method for ultrasonic thickness measurement, typically operated at frequencies above 1 MHz using an electromagnetic acoustic transducer (EMAT). This study successfully extends EMAR into the sub-MHz range, allowing supply voltages below 60 V and thus offering safer and more cost-effective operation. Experiments were conducted on copper blocks approximately 20 mm thick, where a relative thickness accuracy of better than 0.2% is obtained. Regarding this result, the research identifies a critical design principle: Stable thickness resonances and subsequently accurate thickness measurement are achieved when the ratio of ultrasonic wavelength to EMAT track width (λ/w) falls below 1. This minimizes the excitation and interactions with structural eigenmodes, ensuring consistent measurement reliability. To support this, the study introduces a system-based model to simulate the EMAR method. The model provides detailed insights into how wave propagation affects the accuracy of EMAR measurements. Experimental results align well with the simulation outcome and confirm the feasibility of EMAR in the sub-MHz regime without compromising precision. These findings highlight the potential of low-voltage EMAR as a safer, cost-effective, and highly accurate approach for industrial ultrasonic thickness measurements.

Keywords: electromagnetic acoustic resonance; thickness gauging; electromagnetic acoustic transducer; EMAR limitations; EMAR simulation

1. Introduction

In the field of non-destructive testing, thickness gauging of solid materials is used to determine wall thickness or to assess material wear over time [1–3]. One approach of measuring the thickness is to generate and measure ultrasonic waves inside the specimen with ultrasonic transducers [4,5]. A common way to induce ultrasonic waves is by a piezoelectric transducer [6–8]. It is attached to the specimen with a couplant to achieve proper mechanical coupling between the transducer and the specimen. However, there are industrial applications where the surface of the specimen is rough or has a temperature higher than 100 °C [9]. Those harsh environments limit the use of a piezoelectric transducer. One possible solution is to use a laser as a transducer. However, lasers are costly and bulky, and therefore, challenging to install in actual measurement applications [10,11].

An alternative is provided by the so called electromagnetic acoustic transducer (EMAT) [12–14]. It is used to generate and measure ultrasonic bulk waves as well as surface or guided waves without contact to the sample [15–18]. An EMAT can be built in a compact and cost efficient way and is suitable to conduct ultrasonic thickness gauging for applications in harsh environments [19–22].

An EMAT can be operated in two ways. The first method, which is illustrated in Figure 1a is the pulse-echo method.

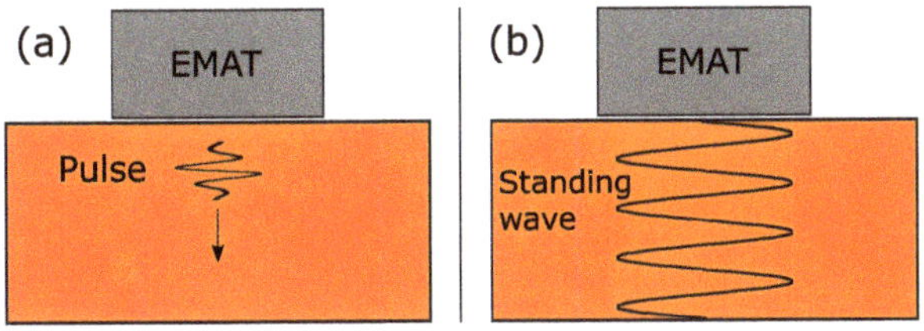

Figure 1. Illustration of EMAT operating methods. (**a**) Pulse-echo method: A short pulse is emitted and caught again by the EMAT. By measuring the ToF of the wave packet, the thickness is estimated. (**b**) EMAR method: A standing wave is implied beneath the EMAT, where the thickness can be estimated from the resonance frequency.

Here, the EMAT emits a short mechanical wave pulse into the specimen and catches it again after a certain time of flight (ToF), which is proportional to the thickness of the specimen [23–25]. Regarding the pulse-echo method, the poor coupling efficiency of the EMAT often makes it a challenging task to measure a signal with sufficient quality for the thickness estimation [26,27]. However, there is a further approach to the pulse-echo method to measure a specimen thickness with the EMAT, which is illustrated in Figure 1b. It is referred to as "electromagnetic acoustic resonance" (EMAR) [12,28]. The EMAR method implies that a standing wave along the thickness of the block is generated. This standing wave is the result of creating a mechanical resonance along the thickness, which leads to amplified mechanical movement inside the specimen. In this respect, EMAR produces a higher quality measurement signal compared to the pulse-echo method [12]. In order to excite the mechanical resonance, the excitation frequency has to match the resonance frequency

$$f_{res,n} = n\frac{c}{2d}. \tag{1}$$

Here, c is the propagation velocity of the ultrasonic wave in the specimen, d is the thickness of the specimen and n is any positive integer.

Table 1 shows a brief review of related work regarding EMAR for thickness gauging, given the operating frequency, sample thickness, reported accuracy, and the highlights of each work.

As can be seen, EMAR is typically operated above 1 MHz. Thereby, it is not only used on thin plates or foils in the sub-millimeter to millimeter range where detection of wall thinning in the µm-region is achieved, but also on thicker walls or plates up to a thickness of 10 mm. Regarding the measurement time, EMAR measurements are also fast, e.g., Dixon et al. state real-time online monitoring is possible with their approach by only one single shot measurement; however, measurement accuracy decreases as plate thickness increases [29]. While Table 1 is focused on non-contact EMAR studies, it is acknowledged that other approaches also utilize thickness resonances. Notably, Heinlein et al. [3] demonstrated thickness gauging using guided waves with sub-MHz cut-off frequencies, which coincide with the thickness resonance frequencies exploited in the EMAR approach. Although their configuration involves spatially separated piezoelectric transducers, the underlying resonance condition remains the same. For completeness of the literature review, EMAR is not only used for thickness gauging, but also in the area of material characterization [28,30] and thin layer detection [31,32].

Table 1. Literature review on non-contact EMAR-based thickness gauging.

Author	Operating Frequency in MHz	Sample Thickness	Accuracy	Highlights
Kawashima et al. [33]	Up to 150	20 µm to 1 mm	N/A	High frequency
Hobbis et al. [34]	>5	0.28 mm to 2.8 mm	0.08 µm (max. std)	Moving specimen
Dixon et al. [29]	>5	100 µm to 500 µm	0.2 % (relative)	Rapid 'single shot' measurement
Chen et al. [35]	>1	0.5 mm to 3 mm	<4 % (relative)	Laser/EMAT configuration
Li et al. [36]	>2	up to 8 mm	<2.5 % (relative)	Sloped specimen
Yusa et al. [37]	1 to 4	up to 10 mm	±0.2 mm (absolute)	Probabilistic evaluation
Cai et al. [38]	2 to 6	up to 10 mm	0.9 % (relative)	Specimen thickness step change
This work	0.04 to 1	20 mm	<0.2 % (relative)	Wave directivity impact on accuracy

As seen in Table 1, the published results for thickness gauging are obtained by operating the EMAR method at frequencies in the MHz range, typically higher than 1 MHz. The selected operating frequency depends, among other things, on the thickness of the specimen according to Equation (1). A corresponding brief numerical example to obtain the first possible resonance frequency of a copper sample with different thicknesses is shown in Table 2.

Table 2. Numerical example based on Equation (1) for the first resonance frequency of a 1 mm and 20 mm thick copper sample.

Thickness	Wave Speed	First Resonance (n = 1)
1 mm 20 mm	$2300\,\mathrm{m\,s^{-1}}$ [39]	1.15 MHz 58 kHz

As can be seen, the first resonance frequency of a 1 mm thick copper sheet would be around 1.15 MHz. For the 20 mm thick copper sample, the first resonance frequency where EMAR theoretically works is around 58 kHz. Interestingly, even in related work where the samples are thicker (e.g., 10 mm), EMAR is still operated in the MHz frequency range, although theoretical operation below 1 MHz would be possible. Operating EMAR below 1 MHz offers additional advantages in terms of its applicability:

- Reduced supply voltage: As the EMAT consists of a coil, the overall impedance, and therefore, the required excitation voltage, is lower when the operation frequency is decreased. Indeed, as will be demonstrated in this work, EMAR is successfully operated at voltages below 60 V.
- Improved human and electrical safety: The lower operating voltage minimizes electrical hazards to personnel, eases compliance with safety regulations, and facilitates adaptation to standards for use in explosion-risk environments.

- Lower attenuation in the material: Depending on the ultrasonic frequency, the wave is naturally attenuated due to the material properties. In general, the attenuation in metals is lower at low frequencies and increases at higher frequencies [40]. That fact also promotes operation in the sub-MHz range.

In general, lower excitation frequencies make a possible realization of EMAR operation more comfortable in terms of necessary supply voltage and safety guidelines. Even though conventional high-frequency techniques may still be technically feasible, the practical considerations listed above illustrate why research into EMAR operation in the sub-MHz range is relevant for certain industrial applications.

Hence, this work provides a complementary study of the EMAR method when used at frequencies below 1 MHz. It addresses and analyzes the challenges and limitations of performing EMAR at lower frequencies compared with related work. As will be shown, the applicability of the EMAR method for thickness gauging encounters problems when used in the sub-MHz range. This work explains why these problems occur and identifies the conditions that must be met to successfully perform EMAR in this frequency range. In this regard, it is demonstrated that EMAR operation for thickness measurement in the sub-MHz range also achieves a relative accuracy better than 0.2 %.

For this work, copper blocks with a thickness of around 20 mm were selected as test specimens. As shown in Table 2, this thickness enables EMAR operation in the sub-MHz range. EMAR is then fundamentally investigated to estimate the thickness of the specimens based on the individual resonance frequencies.

The paper is structured as follows: In Section 2, the EMAT working principle is addressed by means of a certain EMAT design, and the EMAR method is explained in detail. Furthermore, in this section, the test specimens used for the experiments are introduced. Section 3 introduces the simulation framework for the EMAR study. Hereby, a system-based approach for simulating EMAR is presented. The corresponding simulation results are shown in Section 4. The actual measurement setup for the EMAR validation is presented in Section 5 and the measurement results are shown in Section 6. The thickness estimation based on the measurement results is presented in Section 7. Section 8 summarizes and discusses the results of this study.

2. Materials and Methods

This section presents the EMAT used in this work, explains the EMAR method and introduces the samples used in the experiments.

2.1. EMAT

The same EMAT is used for the ultrasonic wave excitation and detection without mechanical contact to the specimen. The EMAT working principle is well known and has been used in multiple applications [41–43]. There are three possible effects that can be used by an EMAT to generate mechanical waves in a specimen [12]:

- Magnetization
- Magnetostriction
- Lorentz forces

As shown later in this section, the specimens under test in this work are copper blocks, which are conductive materials and have a relative permeability of one. Hence, only the Lorentz force generation is present, and consequently, the other effects are negligible.

Lorentz forces are generated according to [12]

$$\vec{f_L} = \vec{J} \times \vec{B_0}. \tag{2}$$

Equation (2) states that the Lorentz force density and its direction $\vec{f_L}$ are calculated by the cross product of the current density $\vec{J}$ in the specimen and the static magnetic flux density $\vec{B_0}$. To generate a current density $\vec{J}$ inside the specimen, eddy currents can be used. In terms of an EMAT, this is achieved by applying a time-varying current through a coil. The coil is typically manufactured on a printed circuit board (PCB). The current through the coil generates a time-varying magnetic field, which induces the eddy current spatially beneath the coil area. The second requirement is a static magnetic flux density $\vec{B_0}$, typically provided by a set of permanent magnets.

In order to receive the ultrasonic wave, the EMAT is considered a velocity sensor. The ultrasonic wave to be received, mechanically excites the particles inside the material and forces them to move with the velocity $\vec{v}$. In the presence of a static magnetic flux density, a time-varying current density [12]

$$\vec{J_m} = \sigma(\vec{v} \times \vec{B_0}) \tag{3}$$

is generated inside the specimen with its electrical conductivity σ. This current density further generates a magnetic field, which induces a proportional voltage inside the EMAT coil. Its magnitude is proportional to the velocity $\vec{v}$ of the ultrasonic movements inside the specimen.

EMAT Design

Depending on the EMAT design, the EMAT is capable of generating and receiving certain wave types. According to [12], when using EMAR for thickness measurement, shear waves should be used. To generate shear waves within the material, a so-called racetrack EMAT is used [28]. It consists of a printed racetrack coil and a set of permanent magnets above the coil. A top view of the printed racetrack coil schematic is shown in Figure 2a.

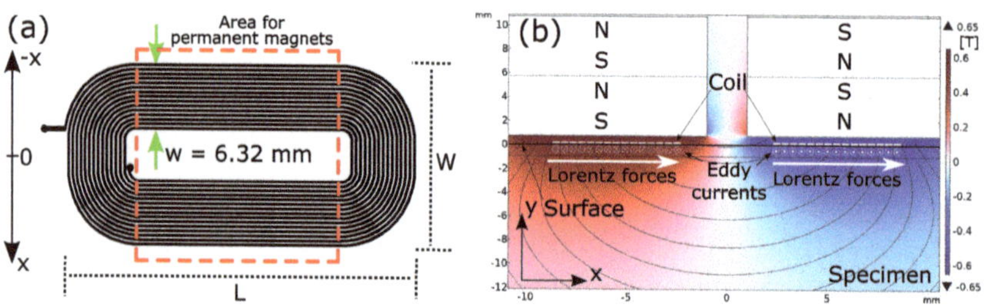

Figure 2. (**a**) Top view of racetrack coil configuration used for the EMAT. Each track spans a width of 6.32 mm. The red dotted square marks the area where the magnets are positioned. (**b**) 2D-Illustration of Lorentz force generation. A cross-section of the EMAT, including the racetrack coil and magnets on top, which provide the constant magnetic flux density, is shown. The color indicates the magnetic flux density $\vec{B}$ in y-direction.

To keep the EMAT handy for convenient use in the experiments, it has a length of L = 35 mm and a width W = 18 mm. Given these geometries and the design specifications of the PCB manufacturer, the EMAT ended up with N = 15 turns. They are spaced by 130 µm. In total, one track spans a width of w = 6.32 mm. This width is considered the track width. Both tracks are separated by a distance of 4 mm to primarily excite Lorentz forces parallel to the surface and thus enhance shear wave excitation. The red dashed area marks the area where the permanent magnets are put. The eddy currents and resulting Lorentz forces are generated within this area.

A cross-section of the EMAT with its permanent magnets, corresponding magnetic flux density field lines, eddy currents, and resulting Lorentz forces is shown in Figure 2b. As can be seen, the magnets are magnetized in different directions. The corresponding magnetic field lines in the vicinity of the coil are mainly normal to the surface of the specimen. Given the racetrack coil, the induced eddy currents directly beneath the surface point in opposite directions, and thus the resulting Lorentz forces will consequently point in the x-direction, and shear forces are generated. The magnetic field lines in the gap between the two tracks are mainly parallel to the surface and would also excite longitudinal waves. However, there are hardly any eddy currents induced in this gap. Therefore, the excitation of a possible longitudinal wave is weak compared with the shear wave excitation.

As the Racetrack EMAT is introduced for thickness measurement using EMAR, the next section briefly recaps the working principle of EMAR.

2.2. EMAR Working Principle

The EMAR method assumes a standing wave along the thickness of the sample [12]. This standing wave is formed by a generated wave traveling along the thickness and is constantly reflected at the top and bottom of the specimen. For this measurement method, the EMAT is suitable as it generates the wave without mechanical contact to the specimen, and therefore, the reflection of the wave is not influenced by the transducer. Figure 3 shows an illustration of the theoretical reflection principle of ultrasonic waves. In resonance mode, the EMAT excites another wave just at the same time the first reflected wave arrives at the upper boundary.

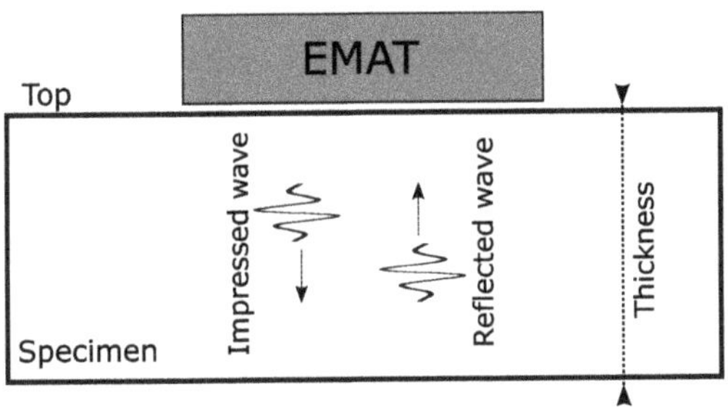

Figure 3. Depiction of EMAR working principle. The EMAT impresses a wave along the thickness of the specimen, which is reflected back to the top. When the timing is exact, the reflected wave is constructively interfering with another impressed wave from the EMAT.

If the timing is exact, the reflected wave from the bottom and the newly generated wave constructively interfere and form a higher resulting amplitude [12]. By repeating this procedure multiple times, the resulting ultrasonic wave amplitude increases further. The time points at which the wave has to be excited are a function of the ultrasonic wave speed inside the material, as well as the thickness of the block. The time it takes the wave to return to the upper surface is given by

$$t_{return} = \frac{2d}{c}, \tag{4}$$

with d the thickness of the block and c the ultrasonic velocity in the given material. If the excitation is repeated at the same time interval over and over again, a repetition frequency can be defined, and if the resonance condition is met, the amplitudes of each pulse add constructively, and therefore, the mechanical thickness resonance frequency can be defined. By further investigation, it turns out that each integer multiple of the resonance frequency satisfies the resonance condition, and thus, there is an infinite number of theoretical resonance frequencies given by

$$f_{res,n} = n\frac{1}{t_{return}} = n\frac{c}{2d}, \tag{5}$$

where n is any positive integer. If the ultrasonic wave speed is known and the resonance frequency is measured, then this method allows for determining the thickness of the block as

$$d = n\frac{c}{2f_{res,meas}}. \tag{6}$$

Instead of a pulse train, a continuous sine wave, which is often simpler to generate, can be used to excite the wave inside the specimen. Here, the resonance condition is met if the frequency of the sinusoidal excitation signal is equal to any of the resonance frequencies.

2.3. Test Samples

The test samples in this work are two copper blocks with a length of 100 mm and a width of 60 mm, but with different thicknesses. Figure 4 depicts the two specimens that are later used in the experiments.

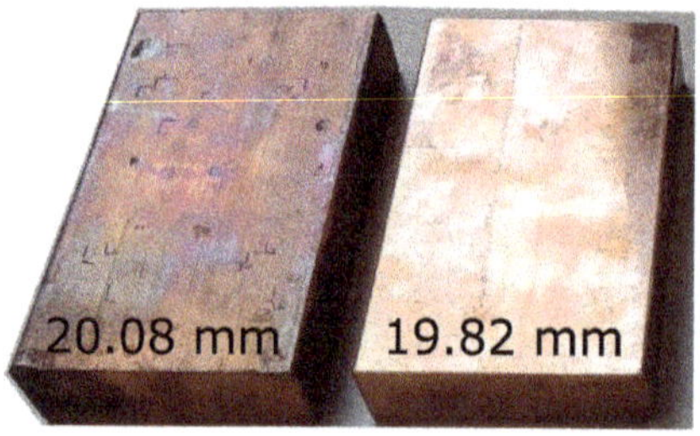

Figure 4. Test specimens used in this work. Two copper blocks are taken for the study. One has a thickness of 20.08 mm and the other one 19.82 mm, respectively. Those thickness values were determined with a caliper (10 µm resolution).

One has a thickness of 20.08 mm and the other has a thickness of 19.82 mm. These thickness values are measured with a caliper providing a resolution of 10 µm and are considered as true thickness values throughout the work.

In order to operate EMAR on these blocks, one has to compute the shear wave speed for the material. When the mechanical properties of the copper are known, the shear wave velocity c_S is computed by

$$c_S = \sqrt{\frac{E}{2\rho(1+\nu)}} \tag{7}$$

Here, E is the Young's Modulus, ρ is the density of the material and ν is referred to as Poisson's ratio. The mechanical parameters for copper vary in a certain range depending on the copper composition and its processing. Table 3 gives an overview of the parameter range found in the literature [44,45] and our choice of the parameters for the EMAR validation.

Table 3. Mechanical parameter range of copper found in the literature and chosen values in this work for the EMAR study.

Parameter	Range	Choice
Young's Modulus in GPa	110–138	133.5
Density in kg/m^3	8900–8960	8900
Poisson's ratio	0.3–0.34	0.3

As can be seen later in the measurements, the best thickness estimation is obtained when choosing the shear wave speed $c_s = 2401.92\,\frac{\text{m}}{\text{s}}$. Given Equation (7), this shear wave speed is obtained when choosing the copper material parameters as shown in Table 3. This wave speed value serves as the first estimation for the computation of the resonance frequencies in the copper blocks. For this first estimation, an exemplary copper sample with the chosen mechanical parameters and thickness of 20 mm is assumed. According to Equation (5), the resonance frequencies are derived as

$$f_{res,n} = n \cdot 60.05\,\text{kHz}, \tag{8}$$

with n being any positive integer. Given Equation (8), the first 16 resonance frequencies are lower than 1 MHz, which now allows for this fundamental study of EMAR being operated below 1 MHz.

3. Simulation Setup

In order to verify the EMAR with respect to the thickness estimation of a copper block, a 2D simulation model is set up in COMSOL Multiphysics (Version 6.1). The finite element (FE) model is shown in Figure 5.

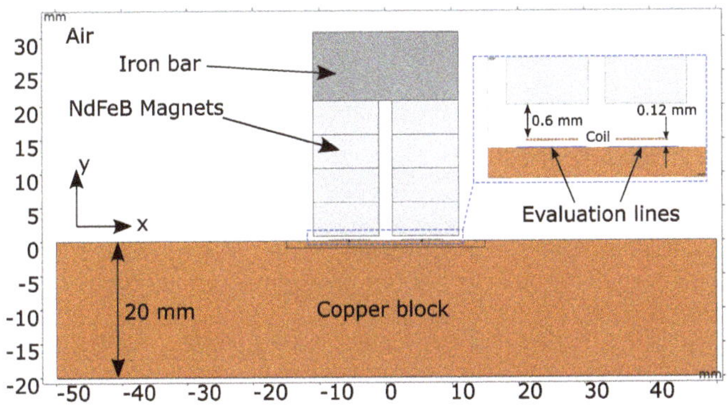

Figure 5. 2D-FE model. The EMAT is centered on top of the 20 mm copper sample. It consists of the racetrack coil and stacked NdFeB magnets magnetically shorted on top with an iron bar. The EMAT is lifted off by 0.12 mm from the copper surface. The surrounding domain is air.

As can be seen, it represents a cross-section of the EMAT on top of the copper block. The modeled racetrack EMAT consists of 15 turns, where each coil segment is 0.3 mm wide and has a height of 35 µm. On top of the coil, at a distance of 0.6 mm, which corresponds to the thickness of the PCB, the permanent magnets are placed. They are sintered neodymium (NdFeB-N42) magnets which have a remanent flux density of 1.31 T and provide the static magnetic field for the ultrasonic wave generation. The EMAT is lifted from the copper sample by 0.12 mm. This air gap height is later applied in the experiment as well by putting a single layer of tape at the border of the PCB. The air gap is mandatory to avoid any mechanical interaction between the EMAT and the block, as this would affect the standing resonance wave along the thickness. The copper block is modeled as a rectangle with a thickness of exactly 20 mm and a length of 100 mm. The chosen material parameters from Table 3 are assigned. The EMAT is centered on top of the copper sample. The EMAT and the copper block are surrounded by an air domain. Using COMSOL, the static magnetic field, the eddy current induction, the generation of the Lorentz forces, as well as the mechanical pulse propagation are simulated.

Furthermore, the mesh size has to be chosen properly, especially inside the copper domain, where the wave propagation takes place. For the EMAR verification, frequencies up to 1 MHz are considered. At this frequency, the physical wavelength of the ultrasonic wave is $\lambda = c_S/1\,\text{MHz} = 2.4\,\text{mm}$. According to modeling recommendations, the mesh size should be at most $\frac{\lambda}{10}$ to properly simulate the wave propagation. Thus, the mesh inside the copper sample is set to a maximum size of 240 µm.

Two evaluation lines are placed directly on the upper edge of the copper surface and beneath the left and right coil tracks. Along the lines, the velocity component in the x-direction is integrated and then divided by the length of the line to obtain the mean velocity in the x-direction at every time instance. This mean velocity below the coil tracks is responsible for generating a proportional voltage inside the EMAT coil as shown in Equation (3).

System Identification

The model introduced above is used to verify the EMAR method from 40 kHz to 1 MHz, as well as to exemplarily illustrate the wave propagation in the copper specimen. One way to validate the EMAR simulation is to impress a sinusoidal pulse train or continuous excitation current through the coil for a certain amount of time and then perform line evaluations of the velocity in the x-direction for the mechanical response. This can then be conducted for each excitation frequency. However, this approach is expensive in terms of simulation time and evaluation.

Another approach is to characterize the EMAT with the copper block as a linear system with its impulse response $h(t)$. The system input $x(t)$ is the current through the coil and the output $y(t)$ is the sum of the integrated velocity in the x-direction of the two evaluation lines. The velocity in this direction induces a proportional voltage in the EMAT, which is measured later on in the experiments. In order to compute the impulse response $h(t)$, a step response is used. Hereby, a current step at the input is applied, and the output in the time domain is evaluated. The simulation time for the system step response is 250 µs in steps of 1 ns. As will be shown later, this simulation time is sufficient for the validation of the EMAR simulation.

Once the step response is acquired, it is differentiated in the time domain to obtain the impulse response $h(t)$ of the system. The impulse response includes the mechanical description of the system over time and thus holds the resonance condition used for EMAR.

Given the simulated impulse response from the simulation, one can then apply any time signal at the input to compute the output of the system by the convolution of the input with the impulse response as

$$y(t) = h(t) * x(t). \tag{9}$$

The convolution operation is fast and can be performed in a program like Python or MATLAB (R2022b). Hence, with this approach, only one FEM-based simulation is required to generate the step response, and the EMAR method is verified within post-processing of the simulation. For EMAR, a sinusoidal pulse train with frequencies from 40 kHz to 1 MHz, in steps of $f_{res,1}/5$, are applied for 125 µs as input and the output for every excitation frequency is transferred into the frequency domain for generating the EMAR spectrum.

4. Simulation Results

As described above, the main simulation result from the COMSOL domain is the step response of the system $h(t)$. It is convoluted with a sinusoidal excitation with frequencies from 40 kHz to 1 MHz. The output of this convolution represents the amplitude of the copper particle velocity within the ultrasonic wave. An exemplary result of two convolutions at the first and 16th resonance frequency with a sinusoidal current (amplitude of 1 A) is shown in Figure 6.

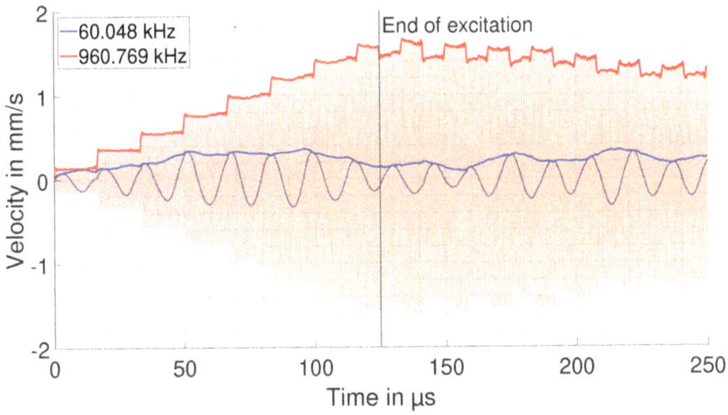

Figure 6. Convolution output at the first and 16th resonance, representing the amplitude of the copper particle velocity within the ultrasonic wave. Although the resonance condition is met, no significant constructive interference is seen for the first resonance frequency. For the 16th resonance frequency, the output adds up nicely every time the wave returns from the bottom surface until the end of the excitation.

Although both frequencies theoretically meet the EMAR condition, only the excitation at the 16th resonance frequency adds up nicely during excitation time and then remains high for the rest of the time. In the case of the first resonance frequency, the signal does not add up much, and thus, after the end of excitation, it is lower by a factor of 8 compared with the higher frequency. The reason for this is addressed later on in this section. In order to compute the EMAR spectrum from the convolution, the part of the signal from 125 µs to 250 µs (time range in which the excitation is already over) is transferred into the frequency domain for every excitation frequency and one obtains the desired EMAR spectrum. The spectrum from the simulation is shown in Figure 7.

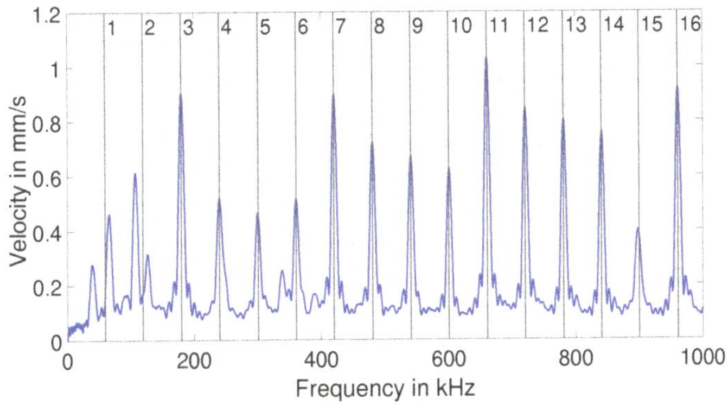

Figure 7. Computed EMAR resonances given by the simulated step response and further computation.

As can be seen, the spectrum shows that the simulated resonances match the calculated frequencies of Equation (8) for $n \geq 3$. These resonances confirm the system-based approach for simulating EMAR. However, it is also seen that the first and second resonance frequencies do not fit with the theoretical frequencies. This is actually the result of the velocity not adding up during excitation, as shown in Figure 6. Here, EMAR runs into problems when trying to use it for thickness estimation. And the reason for this problem is the weak directivity of the impressed wave towards the bottom of the block, as shown in the next section.

Directivity

To analyze the directivity, the simulated wave motion of the propagating wave is illustrated at a certain point in time. Hereby, the first and 16th resonance frequency are examined. Figure 8 shows the simulated standing wave motion (x-velocity) along the block thickness for the two frequencies at 15.85 µs. This point in time is chosen as the wave was already reflected back once from the bottom of the block and is propagating towards the top again.

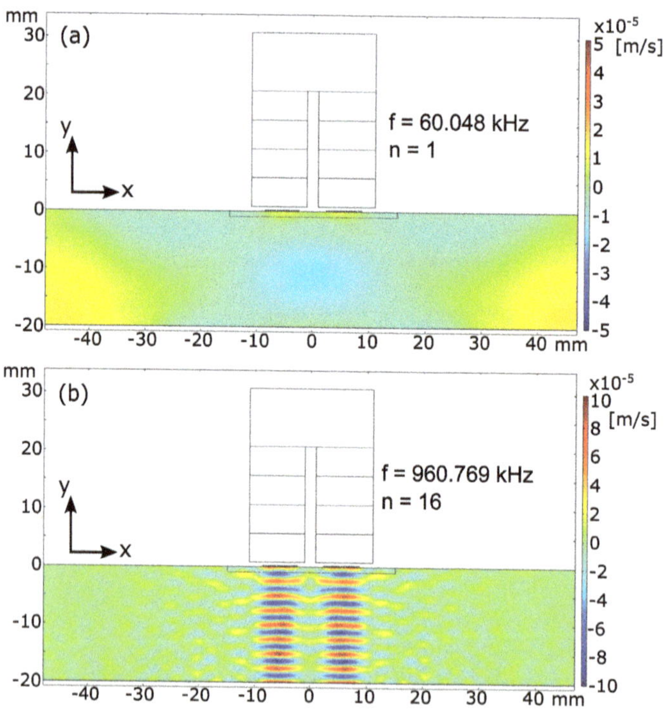

Figure 8. Simulated standing wave motion inside the copper block in terms of velocity in x-direction at 15.85 µs. (**a**) First resonance frequency. (**b**) 16th resonance frequency.

It can clearly be seen that the wave in Figure 8a does not propagate nicely below the EMAT. Indeed, a strong movement is also seen on the side walls, indicating the entire domain is mechanically excited. For this frequency, the impressed wave does not stay spatially beneath the EMAT and thus no constructive interference builds up. A completely different behavior is seen in Figure 8b. Here, the wave movement is more focused towards the bottom and stays spatially beneath the EMAT. In this case, the constructive interference takes place and a standing wave along the thickness builds up.

The simulation results already clearly show that the impressed wave hardly remains spatially below the EMAT when attempting to perform EMAR at very low frequencies. Instead, it mechanically excites the entire test specimen. As a result, the so-called eigenmodes of the blocks are excited, which leads to a shift in the actual thickness resonance frequency and generates additional resonance frequencies. This is especially seen for frequencies $< 150\,\text{kHz}$ as shown in Figure 7. These resonances are detected by the EMAT as well and thus falsify a potential thickness evaluation.

In order to verify this directivity behavior seen in the simulation, experiments in a laboratory environment are performed and compared with the outcome of the simulation.

5. Measurement Setup and Signal Processing

This section describes the measurement setup for thickness measurement with EMAR and the signal processing. The aim of the measurement experiments is to examine the applicability of the EMAR method for estimating the specimen thickness on the basis of the individual resonance frequencies. Furthermore, the results are compared with the outcome of the simulation. An illustrative overview of the measurement setup is seen in Figure 9.

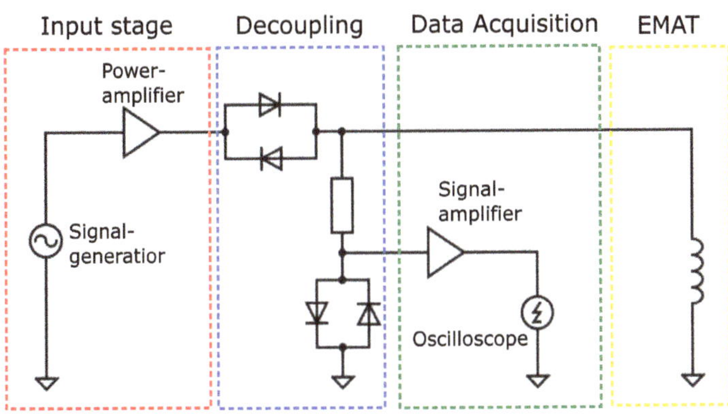

Figure 9. Schematic illustration of the measurement setup. The input stage provides the high current for driving the EMAT. The commonly used decoupling network allows for the EMAT to be used as transmitter and receiver. The received EMAT voltage is amplified and stored within the data acquisition block. The last part is considered the EMAT itself.

The setup consists of four parts. The first part is the input stage, consisting of a signal generator and a power amplifier. The input stage is responsible for providing the required signal form and current to drive the EMAT. The second part is the decoupling network. As shown, it decouples the excitation signal from the input to the signal amplifier. In this way, the EMAT can be used as a transmitter and receiver. The third part is the data acquisition, where the EMAT signal generated from the ultrasonic wave is amplified and then measured by a USB-oscilloscope. The last part is the EMAT itself, acting as a generator for the excitation of the ultrasonic wave and subsequently as a receiver to measure the ultrasonic wave movement.

A picture of the actual setup used in the laboratory environment and the EMAT on the copper block is shown in Figure 10.

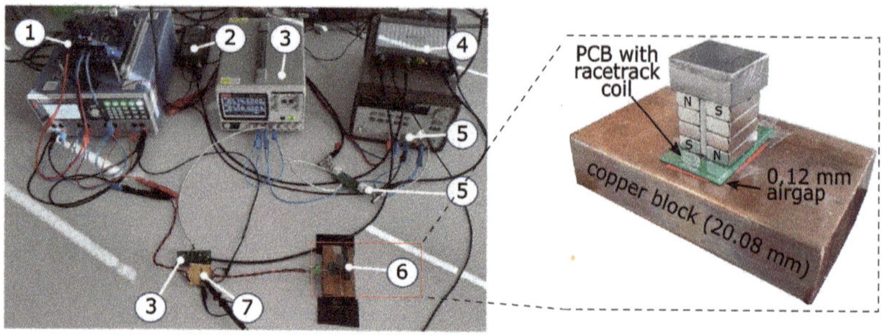

Figure 10. Measurement setup used in laboratory environment. 1: Power amplifier with power supply; 2: Current probe; 3: Amplifier for the measured EMAT signal and power supply; 4: Picoscope (Signal generator and data acquisition); 5: Pre-amplifier and power supply; 6: EMAT on copper block; 7: Decoupling network.

For the power amplifier, the APEX PA05 (Apex Microtechnology, Tucson, AZ 85741, USA) is used. This amplifier can drive output currents up to 20 A. The decoupling network is built on a PCB, and the signal amplifier is built as a non-inverting amplifier with a gain of 100. The EMAT is placed at the center of the copper block and lifted 0.12 mm from the surface of the block by means of a tape at the border of the PCB. This air gap height is the same as in the simulation. Given the EMAT on the copper block, an ohmic part of $R = 2.6\,\Omega$ and an inductive part of $L = 0.62\,\mu\text{H}$ of the racetrack coil were measured. The overall absolute value of the coil impedance is

$$|\underline{Z(f)}| = \sqrt{R^2 + (2\pi f L)^2}, \tag{10}$$

with f as the excitation frequency. The sinusoidal current amplitude during the excitation phase throughout the experiments is measured for every excitation frequency, and the voltage amplitude at each frequency is computed with the absolute value of the impedance $|\underline{Z(f)}|$. Figure 11 depicts the measured current through the EMAT and the corresponding voltage over the excitation frequencies.

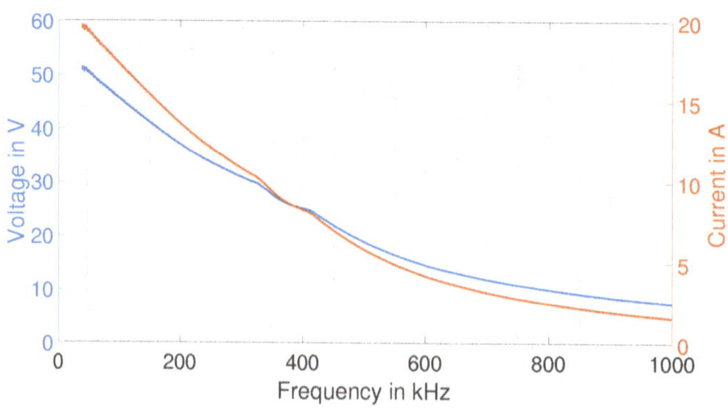

Figure 11. Measured current amplitude during the EMAT excitation and corresponding calculated voltage provided by the power amplifier.

As shown and mentioned in the introduction, the EMAR operation is performed with supply voltages lower than 60 V. This lower voltage is advantageous in practice, as it simplifies the requirements for power electronics and insulation and increases operational safety, while still providing a substantial amount of current through the EMAT for the successful excitation of thickness resonances as will be shown later. As excitation frequency increases, the voltage decreases due to limitations of the power amplifier. Consequently, the current also decreases.

Measurement Procedure

The EMAR method is verified for the frequency range of 40 kHz to 1 MHz in steps of 250 Hz. Within this frequency range, the first 16 resonance frequencies are expected to be excited within the specimen as follows. By means of the signal generator, a sine train with the desired excitation frequency is provided to the power amplifier, which drives the excitation current through the EMAT. The duration of the excitation is set to 600 µs. Depending on the excitation frequency, 24 to 600 pulses within the sine train are applied at the input. Figure 12a exemplary depicts the unfiltered output of the signal amplifier at an excitation frequency of 658 kHz for the described measurement procedure. The sampling rate of the Picoscope is set to 20 MHz.

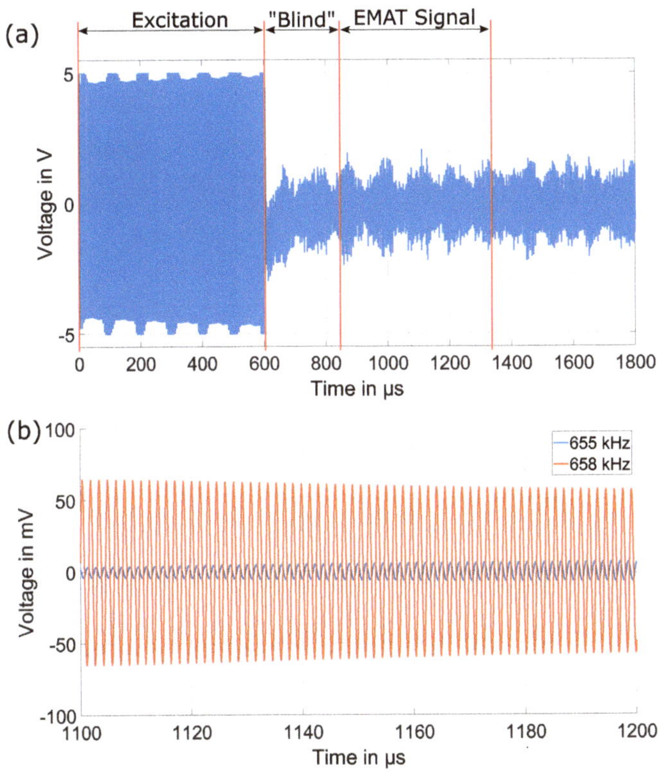

Figure 12. (**a**) Exemplary unfiltered output from the signal amplifier at excitation frequency of 658 kHz. (**b**) Zoom into the bandpass filtered EMAT time signals for excitation frequencies of 655 kHz and 658 kHz within the 500 µs evaluation window.

During excitation, the output of the amplifier is driven into saturation. After the excitation, there is a "blind" window of about 250 µs where no measurements are possible due to necessary signal settling. After 850 µs, the EMAT signal is measured. As can be seen, the EMAT signal is quite noisy, but additional signal post-processing, including filtering operations, will help to extract the necessary information for the EMAR operation.

For the post-processing, the measured EMAT signal is truncated to a length of 500 µs. With regard to the choice of this time window, various lengths were investigated, and the window position was also shifted. However, the previously given parameters (starting time and length) provide the best results in terms of signal amplitude. With the length of 500 µs, a resolution of 2 kHz is achieved in the frequency domain. As will be seen later, this resolution is sufficient to correctly resolve and determine the position of the resonance peaks. The signal within this time window is additionally bandpass filtered at the given excitation frequency. The filter has a bandwidth of 10 kHz. A zoom into the bandpass filtered EMAT time signals between 1100 µs and 1200 µs for exemplary frequencies 655 kHz and 658 kHz is shown in Figure 12b. These signals are measurements from the 20.08 mm copper block. As can be seen, the amplitude of the signal at 658 kHz is roughly 10 times or 20 dB higher compared with the signal at 655 kHz. That is because at 658 kHz, the 11th resonance frequency is mechanically excited within the copper block, as will be shown later. Within this time window, so-called coherent sampling [46] is performed by cutting out an integer number of periods for the given excitation frequency. From this sequence, the Fourier transform is computed, and the voltage value at the excitation frequency is extracted over the entire range of excitation frequencies and plotted for the contribution to the final EMAR spectrum.

6. Measurement Results

As described above, the aim of the measurement experiments and the corresponding results is to examine the applicability of the EMAR method for estimating the specimen thickness on the basis of the individual resonance frequencies of the EMAR spectrum.

The measured EMAR spectrum for the test samples is shown in Figure 13.

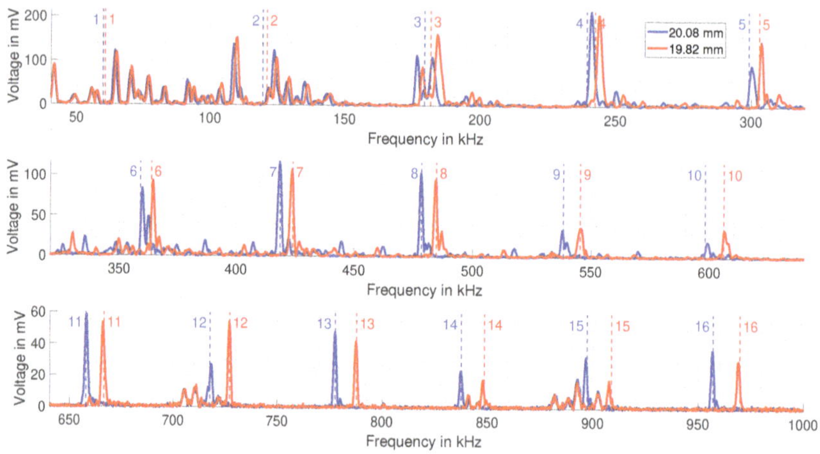

Figure 13. Measured EMAR spectrum in the frequency range from 40 kHz up to 1 MHz. In blue, the spectrum for the 20.08 mm block is shown and in red the spectrum for the 19.82 mm block is depicted. Furthermore, the predicted resonance frequencies for each thickness are included.

In blue, the spectrum of the 20.08 mm copper block is plotted, and in red, the spectrum of the 19.82 mm block is shown. The spectrum is divided into three single spectra for better illustration. In the first row, the frequencies from 40 kHz to 320 kHz are plotted. Within this frequency range, the first five resonance frequencies are contained. The second row, ranging from 320.25 kHz to 640 kHz, includes the resonance frequencies number 6 to 10. The third row contains the last six resonance frequencies from number 11 to 16. Each row also contains the predicted resonant frequencies for the 20.08 mm thick copper block and for the 19.82 mm block. The predicted resonances are numbered vertical dashed lines in the corresponding color for each specimen. The number marks the nth predicted resonance frequency. What can already be seen is that the maximum peak height of the measured resonances decreases for increasing frequency (Please note the different vertical axis scaling). This is due to the limitations of the power amplifier, leading to decreasing current at higher frequencies.

When analyzing each row shown in the measured spectrum in Figure 13, one can see in the first row that the predicted resonance frequencies poorly match the actual measured resonance frequencies. Especially for the resonance frequencies 1 to 3, multiple peaks around the predicted resonance frequencies are observed. These peaks arise from the fact that the directivity of the ultrasonic wave towards the bottom of the block is weak when trying to generate a standing mechanical wave along the thickness of the block, as shown in Figure 8. Thus, these peaks from 1 to 3 are actual mechanical excited resonances from the eigenmodes of the block and overlap with the actual thickness resonance. When evaluating resonance frequency 4 to 5, distinct resonance peaks form next to the predicted resonances, but still do not match properly.

In the second row, the predicted and measured resonances start to coincide. However, there are still multiple small resonances from the eigenmodes. These could still distort the actual position of the thickness resonance, especially when the magnitude is low, e.g., for resonance frequencies 9 and 10.

The third row shows the best match between the predicted and measured resonance frequency. Furthermore, it can be seen that the small peaks between the actual thickness resonances vanished, thus the directivity of the ultrasonic wave must be mainly towards the bottom and no longer to the side. There is electromagnetic interference around 710 kHz and at 900 kHz, as these peaks also occur in the spectrum when there is no mechanical excitation. Based on the measured spectrum, one can now estimate the thickness of the blocks.

7. Thickness Estimation

To estimate the thickness of the blocks by the EMAR method, the highest amplitude resonance in the vicinity of the nth predicted resonance frequency is searched. Given the found resonance, a Gaussian fit into the resonance peak is performed to estimate the frequency for the maximum of the resonance. An exemplary result of a measured resonance peak (n = 11) with its Gaussian fit is shown in Figure 14.

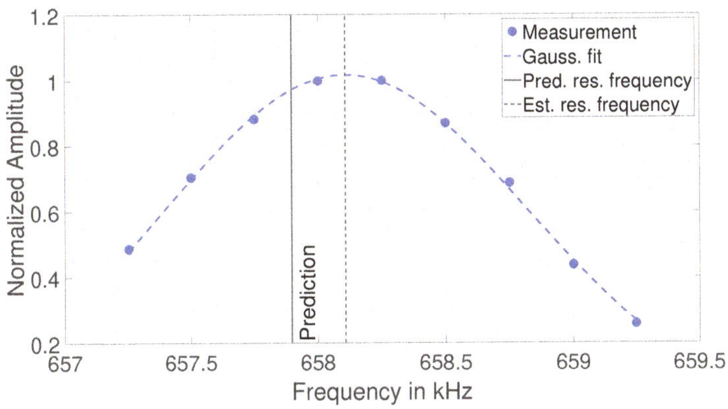

Figure 14. Measured resonance peak at the vicinity of the 11th resonance frequency. From the Gaussian fit, the resonance frequency of the peak is estimated (dashed black line). Also plotted is the predicted resonance frequency (black).

As can be seen, the Gaussian fit matches the measured resonance peak well. The peak's frequency is evaluated and used for the thickness estimation. Further, in this figure, the predicted resonance frequency is also plotted. It is clearly visible that the found resonance frequency is shifted compared with the actual predicted resonance frequency. It is interesting to note that all nine measuring points together form a "nice" resonance curve that is perfectly fitted by the Gaussian curve. Therefore, this shift in frequency is not due to an outlier, but must occur from some "systematic" behavior. Probably still the influence of a certain eigenmode, but it could also be some frequency-dependent material behavior affecting the actual shear wave speed compared with the mean velocity of $c_s = 2401.92\,\frac{\mathrm{m}}{\mathrm{s}}$.

The 16 evaluated frequencies based on the position of the maximum of the Gaussian fit for each block are denoted as $f_{\mathrm{res,meas}}$ and serve as input to compute the thickness according to

$$d_{\mathrm{est,n}} = n\frac{c}{2f_{\mathrm{res,meas}}}. \tag{11}$$

The method described above is also applied to the simulated EMAR spectrum, shown in Figure 7, and is plotted together with the measurements to compare them.

The result for the estimated thickness given the evaluated resonance frequencies from the measurements and the simulation is shown in Figure 15.

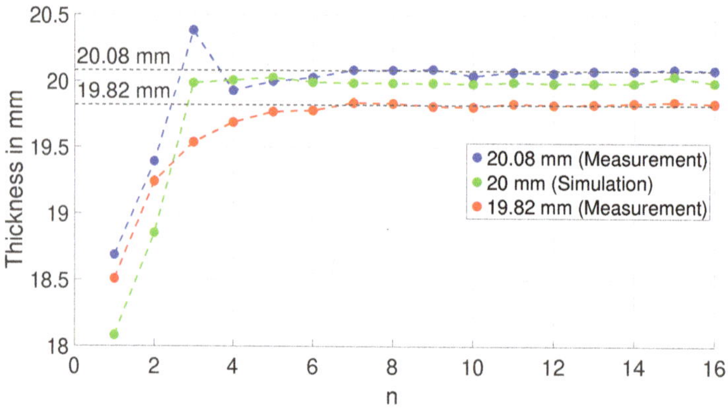

Figure 15. Estimated thickness from n evaluated resonance frequencies from measurement and simulation.

The measurement shows that the estimated thickness for both blocks significantly differs from the true thickness values for resonance frequencies n $<$ 6. As shown in the simulation results (Figure 8a), this is due to the poor directivity of the excited ultrasonic wave within this frequency range. For higher frequencies (n $>$ 6), the estimated thickness approaches the true value, as the directivity towards the bottom of the block increases. Similar to the measurement, the simulation also approaches the true thickness value of 20 mm once the third or higher frequencies are taken. Most probably, this is due to the fact that the third spatial dimension is missing in the simulation. Therefore, no additional eigenmodes from this third missing dimension interact with the thickness resonances.

The absolute value of the relative deviation from the true thickness for every resonance frequency is shown in Figure 16.

As with the measurements shown, the relative deviation decreases exponentially from around 8 % to about 0.2 % for resonant frequencies 1 to 6. For higher resonance frequencies, the deviation stays below 0.2 %. The deviation of the simulation shows a similar behavior except that the deviation already remains below 0.2 % for resonance frequencies n $\geq$ 3.

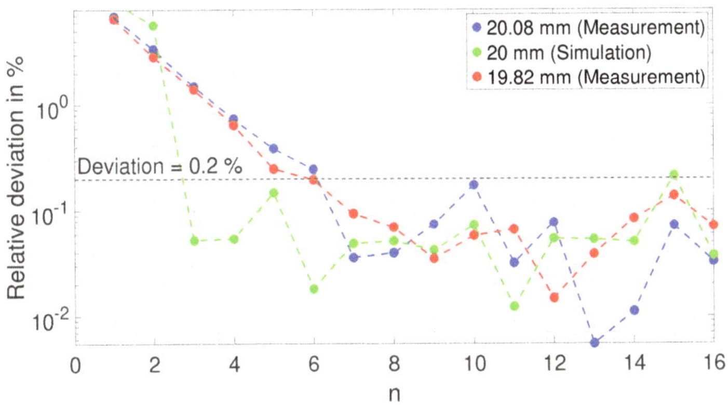

Figure 16. Absolute value of the relative deviation from the true thickness value for the measurements on the 20.08 mm and 19.82 mm copper block and for the simulation of the 20 mm block.

Further investigations show that the directivity of the propagating wave and the associated accuracy of the thickness measurement can be related to the EMAT geometry. Figure 17 depicts the ratio between the wavelength λ of the excited wave and EMAT track width w = 6.32 mm versus the estimated resonant frequencies for both blocks.

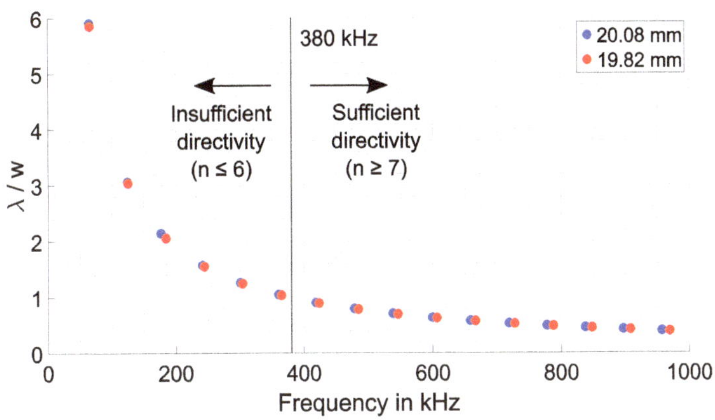

Figure 17. Ratio of wavelength λ over track width w versus the estimated resonance frequencies for the 20.08 mm and 19.82 mm copper block.

As can be seen, the ratio decreases with increasing frequency. The ratio is 1 at about 380 kHz. If the ratio is higher, which corresponds to the operation of EMAR at resonant frequencies 1 to 6, the thickness deviation is high due to the weak directivity of the wave, as shown in Figure 16. Increasing the operating frequency ($n > 6$) and consequently reducing the ratio λ / w to values below 1 increases the directivity, making EMAR suitable for thickness estimation.

8. Conclusions

This study demonstrates the successful application of EMAR for thickness measurement on 20 mm thick copper blocks for frequencies in the sub-MHz regime, achieving relative thickness deviations of less than 0.2 %. This shows that even for frequencies below 1 MHz, comparable accuracy to that reported in existing literature at megahertz frequencies can be achieved (see Table 1). Moreover, the operation at lower frequencies enables supply voltages below 60 V, improving safety and reducing equipment costs. In addition to these practical aspects, this work also reveals a critical design parameter not previously discussed in EMAR studies: The ratio of ultrasonic wavelength to EMAT track width (λ/w). We show that for $\lambda/w > 1$, wave directivity deteriorates, structural eigenmodes of the specimen become excited, and thickness estimation becomes unreliable. This observation explains why previous works generally operate at much higher frequencies, even when the sample thickness would theoretically allow lower-frequency operation. Overall, this study provides not only a practical low-voltage implementation of EMAR for thick samples, but also establishes a previously unreported design guideline ($\lambda/w < 1$) essential for maintaining measurement accuracy. These findings are expected to guide future EMAR system designs, particularly for safe and cost-effective industrial applications where slow wall thinning is monitored over long time periods and inspection speed is less critical.

Author Contributions: Conceptualization, A.S., B.S. and H.W.; methodology, A.S. and B.S.; software, D.A.; validation, A.S., D.A. and B.S.; investigation, A.S.; writing—original draft preparation, A.S.; writing—review and editing, D.A., B.S., A.H., G.K. and H.W.; supervision, H.W.; project administration, H.W. All authors have read and agreed to the published version of the manuscript.

Funding: This research received no external funding.

Institutional Review Board Statement: Not applicable.

Informed Consent Statement: Not applicable.

Data Availability Statement: The raw data supporting the conclusions of this article will be made available by the authors on request.

Acknowledgments: The financial support by the Austrian Federal Ministry for Digital and Economic Affairs, the National Foundation for Research, Technology and Development and the Christian Doppler Research Association is gratefully acknowledged. The study was supported by the Open Access Funding by the Graz University of Technology Fund.

Conflicts of Interest: Authors Andre Hochfellner and Gerald Klösch were employed by the company voestalpine Stahl Donawitz GmbH. The remaining authors declare that the research was conducted in the absence of any commercial or financial relationships that could be construed as a potential conflict of interest.

References

1. Billings, J.K. Noncontact ultrasonic gauging for industrial applications. *J. Acoust. Soc. Am.* **1986**, *79*, S60. [CrossRef]
2. Chen, X.; Li, J.; Wang, Z. Inversion Method in Pulsed Eddy Current Testing for Wall Thickness of Ferromagnetic Pipes. *IEEE Trans. Instrum. Meas.* **2020**, *69*, 9766–9773. [CrossRef]
3. Heinlein, S.; Mariani, S.; Milewczyk, J.; Vogt, T.; Cawley, P. Improved thickness measurement on rough surfaces by using guided wave cut-off frequency. *NDT E Int.* **2022**, *132*, 102713. [CrossRef]
4. Kumar, N.P.; Patankar, V.; Kulkarni, M. *Ultrasonic Gauging and Imaging of Metallic Tubes and Pipes: A Review*; International Atomic Energy Agency (IAEA): Vienna, Austria, 2020.
5. Li, F.; Bell, A.; Damjanovic, D.; Jo, W.; Ye, Z.G.; Zhang, S. Recent Advances in Piezoelectric Materials for Electromechanical Transducer Applications. *IEEE Trans. Ultrason. Ferroelectr. Freq. Control* **2022**, *69*, 2999–3002. [CrossRef]
6. Cegla, F.B.; Cawley, P.; Allin, J.; Davies, J. High-temperature (>500°C) wall thickness monitoring using dry-coupled ultrasonic waveguide transducers. *IEEE Trans. Ultrason. Ferroelectr. Freq. Control* **2011**, *58*, 156–167. [CrossRef] [PubMed]
7. Sait, S.; Abbas, Y.; Boubenider, F. Estimation of thin metal sheets thickness using piezoelectric generated ultrasound. *Appl. Acoust.* **2015**, *99*, 85–91. [CrossRef]
8. Zheng, S.; Zhang, S.; Luo, Y.; Xu, B.; Hao, W. Nondestructive analysis of debonding in composite/rubber/rubber structure using ultrasonic pulse-echo method. *Nondestruct. Test. Eval.* **2021**, *36*, 515–527. [CrossRef]
9. Song, J.; Guo, D.; Jia, J.; Tu, S. A new on-line ultrasonic thickness monitoring system for high temperature pipes. *Int. J. Press. Vessel. Pip.* **2022**, *199*, 104691. [CrossRef]
10. Zarei, A.; Pilla, S. Laser ultrasonics for nondestructive testing of composite materials and structures: A review. *Ultrasonics* **2024**, *136*, 107163. [CrossRef]

11. Lefevre, F.; Jenot, F.; Ouaftouh, M.; Duquennoy, M.; Ourak, M. Laser generated guided waves and finite element modeling for the thickness gauging of thin layers. *Rev. Sci. Instrum.* **2010**, *81*, 034901. [CrossRef]
12. Hirao, M.; Ogi, H. *Electromagnetic Acoustic Transducers*; Springer: Tokyo, Japan, 2017. [CrossRef]
13. Zhai, G.; Li, Y.; Qin, Y.; Liu, Y. Design Method of Multiwavelength EMATs Based on Spatial Domain Harmonic Control. *IEEE Trans. Ultrason. Ferroelectr. Freq. Control* **2021**, *68*, 2259–2270. [CrossRef]
14. Rueter, D.; Morgenstern, T. Ultrasound generation with high power and coil only EMAT concepts. *Ultrasonics* **2014**, *54*, 2141–2150. [CrossRef]
15. Siegl, A.; Schweighofer, B.; Bergmann, A.; Wegleiter, H. An Electromagnetic Acoustic Transducer for Generating Acoustic Waves in Lithium-Ion Pouch Cells. In Proceedings of the 2022 IEEE International Instrumentation and Measurement Technology Conference (I2MTC), Ottawa, ON, Canada, 16–19 May 2022; pp. 1–6. [CrossRef]
16. Li, Y.; Liu, Z.; Miao, Y.; Yuan, W.; Liu, Z. Study of a spiral-coil EMAT for rail subsurface inspection. *Ultrasonics* **2020**, *108*, 106169. [CrossRef]
17. Pei, N.; Zhao, B.; Bond, L.J.; Xu, C. Analysis of the directivity of longitudinal waves based on double-fold coil phased EMAT. *Ultrasonics* **2022**, *125*, 106788. [CrossRef] [PubMed]
18. Yang, R.; Li, Z.; Wang, S.; Jiang, C. Analytical modelling and analysis of the meander-line coil EMATs. *Ultrasonics* **2025**, *146*, 107493. [CrossRef] [PubMed]
19. Lunn, N.; Dixon, S.; Potter, M. High temperature EMAT design for scanning or fixed point operation on magnetite coated steel. *NDT E Int.* **2017**, *89*, 74–80.
20. Zhai, G.; Liang, B.; Li, X.; Ge, Y.; Wang, S. High-temperature EMAT with double-coil configuration generates shear and longitudinal wave modes in paramagnetic steel. *NDT E Int.* **2022**, *125*, 102572. [CrossRef]
21. Siegl, A.; Leithner, S.; Schweighofer, B.; Wegleiter, H. Excitation of Mechanical Resonances in the Stationary Ring of a Mechanical Seal by a Continuously Operated Electromagnetic Acoustic Transducer. *Sensors* **2023**, *23*, 1015. [CrossRef]
22. Zuo, P. Underwater quantitative thickness mapping through marine growth for corrosion measurement using shear wave EMAT with high lift-off performance. *Ultrasonics* **2024**, *143*, 107426. [CrossRef]
23. Han, S.W.; Cho, S.H.; Kang, T.; Moon, S.I. Design and test of electromagnetic acoustic transducer applicable to wall-thinning inspection of containment liner plates. *Trans. Korean Soc. Press. Vessel. Pip.* **2019**, *15*, 46–52.
24. Rieger, K.; Erni, D.; Rueter, D. Examination of the Liquid Volume Inside Metal Tanks Using Noncontact EMATs From Outside. *IEEE Trans. Ultrason. Ferroelectr. Freq. Control* **2021**, *68*, 1314–1327. [CrossRef]
25. Shi, Y.; Tian, S.; Jiang, J.; Lei, T.; Wang, S.; Lin, X.; Xu, K. Thickness Measurements with EMAT Based on Fuzzy Logic. *Sensors* **2024**, *24*, 4066. [CrossRef]
26. Huang, S.; Zhao, W.; Zhang, Y.; Wang, S. Study on the lift-off effect of EMAT. *Sens. Actuators A Phys.* **2009**, *153*, 218–221. [CrossRef]
27. Abderahmane, A.; Clausse, B.; Lhémery, A.; Daniel, L. Mechanisms of elastic wave generation by EMAT in ferromagnetic media. *Ultrasonics* **2024**, *138*, 107218. [CrossRef]
28. Hirao, M.; Ogi, H. Electromagnetic acoustic resonance and materials characterization. *Ultrasonics* **1997**, *35*, 413–421. [CrossRef]
29. Dixon, S.; Edwards, C.; Palmer, S. High accuracy non-contact ultrasonic thickness gauging of aluminium sheet using electromagnetic acoustic transducers. *Ultrasonics* **2001**, *39*, 445–453. [CrossRef] [PubMed]
30. Yuan, X.; Li, W.; Deng, M. Non-contact characterization of material anisotropy of additive manufacturing components by electromagnetic acoustic resonance technique. *Meas. Sci. Technol.* **2023**, *35*, 026001. [CrossRef]
31. Yoshida, M.; Asano, T. A new method to measure the oxide layer thickness on steels using electromagnetic acoustic resonance. *J. Nondestruct. Eval.* **2003**, *22*, 11–21. [CrossRef]
32. Liu, T.; Pei, C.; Cheng, X.; Zhou, H.; Xiao, P.; Chen, Z. Adhesive debonding inspection with a small EMAT in resonant mode. *NDT E Int.* **2018**, *98*, 110–116. [CrossRef]
33. Kawashima, K.; Wright, O.B.; Hyoguchi, T. High frequency resonant electromagnetic generation and detection of ultrasonic waves. *Jpn. J. Appl. Phys.* **1994**, *33*, 2837. [CrossRef]
34. Hobbis, A.; Aruleswaran, A. Non-contact thickness gauging of aluminium strip using EMAT technology. *Nondestruct. Test. Eval.* **2005**, *20*, 211–220. [CrossRef]
35. Chen, W.; Lu, C.; Li, X.; Shi, W.; Zhou, Y.; Liu, Y.; Zhang, S. A novel laser-EMAT ultrasonic longitudinal wave resonance method for wall thickness measurement at high temperatures. *Ultrasonics* **2024**, *141*, 107340. [CrossRef]
36. Li, Y.; Cai, Z.; Chen, L. Detection of Sloped Aluminum Plate Based on Electromagnetic Acoustic Resonance. *IEEE Trans. Instrum. Meas.* **2022**, *71*, 6000812. [CrossRef]
37. Yusa, N.; Song, H.; Iwata, D.; Uchimoto, T.; Takagi, T.; Moroi, M. Probabilistic evaluation of EMAR signals to evaluate pipe wall thickness and its application to pipe wall thinning management. *NDT E Int.* **2021**, *122*, 102475. [CrossRef]
38. Cai, Z.; Sun, Y.; Lu, Z.; Zhao, Q. Research on Identification and Detection of Aluminum Plate Thickness Step Change Based on Electromagnetic Acoustic Resonance. *Magnetochemistry* **2023**, *9*, 86. [CrossRef]

39. Engineering ToolBox. Solids and Metals—Speed of Sound. Available online: https://www.engineeringtoolbox.com/sound-speed-solids-d_713.html (accessed on 4 October 2024).
40. Ono, K. A Comprehensive Report on Ultrasonic Attenuation of Engineering Materials, Including Metals, Ceramics, Polymers, Fiber-Reinforced Composites, Wood, and Rocks. *Appl. Sci.* **2020**, *10*, 2230. [CrossRef]
41. Khalili, P.; Cegla, F. Excitation of Single-Mode Shear-Horizontal Guided Waves and Evaluation of Their Sensitivity to Very Shallow Crack-Like Defects. *IEEE Trans. Ultrason. Ferroelectr. Freq. Control* **2021**, *68*, 818–828. [CrossRef]
42. Sarris, G.; Haslinger, S.G.; Huthwaite, P.; Lowe, M.J.S. Fatigue State Characterization of Steel Pipes Using Ultrasonic Shear Waves. *IEEE Trans. Ultrason. Ferroelectr. Freq. Control* **2023**, *70*, 72–80. [CrossRef] [PubMed]
43. Siegl, A.; Schweighofer, B.; Wegleiter, H. A Sensor Model to Simulate the Excitation and Propagation of Lamb Waves in Lithium-Ion Pouch Cells. *IEEE Sens. Lett.* **2023**, *7*, 6004504. [CrossRef]
44. AmesWeb. Modulus of Elasticity for Metals. Available online: https://amesweb.info/Materials/Modulus-of-Elasticity-Metals.aspx (accessed on 4 October 2024).
45. Koti, V.; George, R.; Shakiba, A.; Murthy, K. Mechanical properties of copper nanocomposites reinforced with uncoated and nickel coated carbon nanotubes. *FME Trans.* **2018**, *46*, 623–630. [CrossRef]
46. Oppenheim, A.; Schafer, R.; Buck, J. *Discrete-Time Signal Processing*; Prentice Hall international editions; Prentice Hall: Saddle River, NJ, USA, 1999.

sensors

MDPI

Article

Development of an In-Situ Multifrequency Electromagnetic Sensor for Real-Time Microstructure Monitoring in a Continuous Annealing Furnace

John W. Wilson [1,*], Mohsen A. Jolfaei [2], Lei Zhou [2], Carl Slater [2], Claire Davis [2] and Anthony J. Peyton [1]

1 Department of Electrical and Electronic Engineering, University of Manchester, Manchester M13 9PL, UK; a.peyton@manchester.ac.uk
2 Warwick Manufacturing Group, University of Warwick, Coventry CV4 7AL, UK; mohsen.jolfaei@warwick.ac.uk (M.A.J.); lei.zhou@warwick.ac.uk (L.Z.); c.d.slater@warwick.ac.uk (C.S.); claire.davis@warwick.ac.uk (C.D.)
* Correspondence: john.wilson@manchester.ac.uk

Abstract: The continuous annealing process is widely used in the production of advanced high-strength steels. However, to tightly regulate the mechanical properties of the steel, precise control of processing parameters is needed. Although some techniques are available to monitor the mechanical properties of the steel on entry and exit to the furnace, monitoring the evolving microstructure of the steel through installation of sensors in the annealing line is extremely challenging due to the high temperature, high speed of the steel strip and limited space in the furnace. This study presents the development and validation of a multifrequency electromagnetic sensor system for real-time monitoring of microstructural transformations in steel during thermal cycling, intended for deployment in a continuous annealing line. Experiments were conducted on austenitic stainless steel to study the signal response to an increase in resistivity without a change in magnetic permeability. Pure nickel was tested to investigate the response to a change in magnetic permeability and the ferromagnetic-to-paramagnetic transition at its Curie temperature. A ferritic stainless steel was also tested to assess the performance of the system for high-temperature ferromagnetic materials and a higher-temperature ferromagnetic-to-paramagnetic transition. The tests indicate a strong response to material resistivity and permeability changes, with complementary information from different frequencies. Test results are supplemented by a finite element modelling study into the effect of a change in frequency and permeability on sensor response, with a discussion on the implications of experimental and modelling results for future applications. The results show that the developed system has the potential to characterise thermally induced changes in steels, establishing proof of concept for non-destructive, high-temperature electromagnetic sensing in steel processing and setting the foundation for further industrial deployment in phase and recrystallisation monitoring.

Keywords: electromagnetic sensor; non-destructive testing; steel; magnetic properties; annealing; high temperature

1. Introduction

Continuous annealing furnaces provide a controlled thermal environment for adjusting steel microstructure post cold-rolling. The continuous annealing process (CAP) subjects the cold-rolled material to a controlled heating and cooling sequence to improve its properties, such as ductility and strength, by passing the material through a series of gas or induction furnaces and cooling sections. Unlike other industrial annealing processes, CAP

 https://doi.org/10.3390/s25165158

is a non-equilibrium, rapid process that lasts for approximately 60–120 s and is therefore strongly influenced by the kinetics of the metallurgical process of interest (for example, austenitisation, recrystallisation, phase transformation) [1]. CAP enables precise control of heating and cooling rates in a high-throughput manner, making it ideal for the production of advanced steels such as dual-phase and interstitial free grades [2,3].

Given the need for precise process control to produce high-value steels in CAP, in situ characterisation methods that can reliably track evolving microstructures to ensure mechanical property targets are much sought after. It is, however, a challenging environment with high temperatures, limited space, fast-moving steel strip and a large lift-off (distance between sensor and strip) to contend with. Current techniques for the determination of continuous annealing process parameters involve a combination of mathematical, physical and microstructural models to inform process parameters along with some online measurements from strip speed and tension and a limited number of pyrometers for temperature measurement, [4]. Additional information can be gained from employing a system such as IMPOC, which uses remanent magnetisation to obtain the material's tensile strength and yield strength during production [5]. However, IMPOC is typically only employed on entry or exit from the line, not during the annealing process, as it cannot be used within a high-temperature environment. One system is available that permits the steel structure to be determined within the annealing process. It uses X-ray diffraction to define the crystalline phase fraction, and hence the austenite content of the material [6]. However, it is not widely deployed due to line-of-sight constraints, high-temperature interference, and the need for radiation shielding.

Electromagnetic (EM) sensors have demonstrated significant potential in characterising the microstructure of steel, both in offline and online applications [7]. EM techniques provide valuable insights into heat-induced microstructural changes in metals, offering a complementary approach by leveraging changes in magnetic properties and electrical resistivity to deliver real-time information on microstructural evolution. In the lab, EM sensors have been used to study phenomena such as recovery and recrystallisation [8], mechanical properties [9] and magnetic anisotropy [10]. In [8] changes in the microstructural state of an interstitial free steel caused by recovery and recrystallisation during heat treatment are detected using an EM sensor within the furnace. The measurement coils are formed around a ceramic cylinder, with the sample placed in the cylinder for measurement. This configuration eliminates problems with lift-off as the entire sample is within the measurement coil, so it can give very accurate results. However, it cannot be used on open plates or strip samples, so it is not suitable for industrial installations.

The use of multifrequency excitation in electromagnetic non-destructive testing is well established, with publications dating back to the 1960s [11,12]. Utilising the skin depth effect [13], the material under inspection can be probed at several depths simultaneously [14], offering some key advantages over single-frequency applications [15]. Defects in metals can be analysed in more detail, allowing the characterisation of flaws with more complex shapes [16] and improving defect characterisation capabilities in the presence of variations in conductivity, permeability, geometry and probe lift-off by subtracting the characteristic signals of these variations [17]. Multifrequency testing is also advantageous where the material property of interest varies with depth, for example, in laser hardened steel [18], for the characterisation of residual stresses [19] and for the identification of bimetallic coins [20].

Some EM measurement systems have been developed for industrial applications: IMPOC, HACOM, and 3MA are already widely used in online settings like strip steel production with signals correlated with mechanical properties [21]. For real-time microstructure monitoring, the EMspec™ system has been developed to track phase transformations during hot strip steel processing, where the sensor is located close to the hot strip but is water-cooled and therefore does not need to be designed from high-temperature resistant materials [22]. However, for CAP installation, the sensor assembly needs to be located in the hot zone and exposed to temperatures in the range of 650–850 °C. This is not possible for the current commercial systems, as they are complex constructions employing components such as ferrite cores, which would not survive continued exposure to high temperatures. For CAP installation a simplified approach is needed, with only the sensing coils and minimal supporting structure located in the hot zone and all other system components located externally to the furnace.

This study details the development of a multifrequency EM measurement system for deployment in a continuous annealing furnace. The primary aim of the work is to demonstrate the capability and robustness of the sensor system in tracking temperature-dependent changes in permeability and conductivity under high-temperature conditions. Although transformation phenomena are not directly studied, different materials are tested to investigate different aspects of the sensor response, with the scope of the paper centred on sensor design and its functional viability within CAP environments.

2. Materials and Methods

2.1. Experimental Test Rig

Figure 1a shows a diagram of the annealing simulator test rig built by the University of Warwick. A Carbolite Gero (Hope Valley, UK) HTRV vertical tube furnace is mounted in a type 304 stainless steel frame with a winch positioned above it to allow the sample holder (Figure 1b) to be moved in and out of the furnace using a wire rope. The tube furnace has a maximum temperature of 1800 °C. The user has the option of slowly cooling the sample in the furnace, lowering the sample out of the furnace for faster cooling in air or forced cooling using the compressed air jets. The sample holder was laser cut from a sheet of 4 mm austenitic stainless steel and allows the measurement coils to be clamped in place, close to the sample surface, with the sample mounted in the centre of the sample holder, with short lengths of thermocouple wire holding it in place.

Figure 2a shows the coil assembly. A commercially produced induction heating coil [23] is used for excitation. The coil is made from 3.5 turns of 4 mm solid copper rod and was re-sleeved using alumina silica fibre sleeving for increased temperature resistance. It also provides a readily available, rigid structure to act as a former for the receive coil. A 12-turn receive coil, made from 0.8 mm chromel wire taken from a K-type thermocouple extension cable, was wrapped around the outside of the transmit coil and secured using thermocouple sleeving. A photograph of the sample holder with sample and coil assembly is shown in Figure 2b.

A block diagram of the EM measurement system developed at the University of Manchester is shown in Figure 3. The measurement system consists of a custom-built control box, which provides a multifrequency excitation voltage to a Kepco (Flushing, NY, USA) 400-Watt Bipolar Operational Power supply (BOP). The output current of the BOP is proportional to the input voltage; therefore, the excitation current remains the same throughout the test, irrespective of thermally induced changes in transmit coil resistance. The receive coil is interfaced with the control box, which amplifies the coil voltage and records the signal using a National Instruments (Austin, TX, USA) data acquisition device.

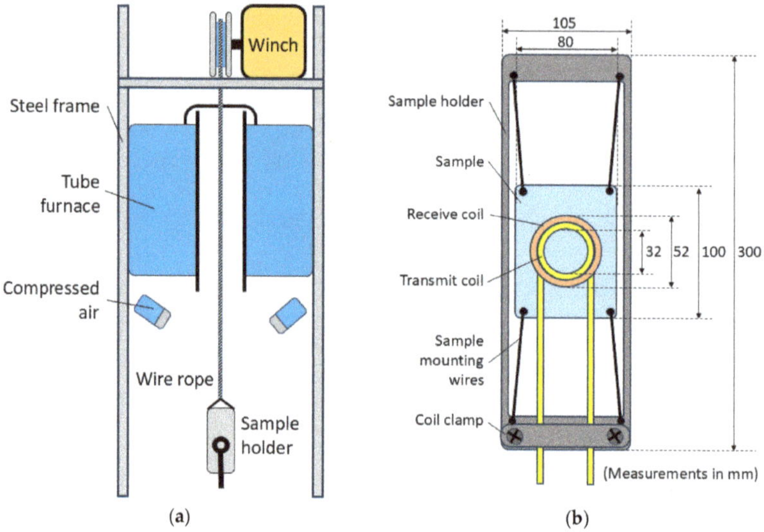

Figure 1. Diagram of the experimental rig (**a**) and diagram of the sample holder and coil (**b**).

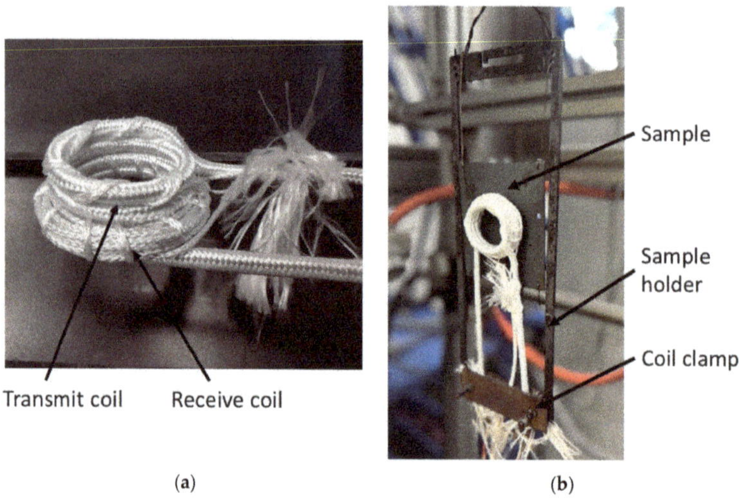

Figure 2. Photograph of the sensor assembly (**a**) and photograph of the sample holder and coil assembly (**b**).

K-type thermocouples were used to record the temperature of the coil and the sample at various positions throughout the tests. Data was acquired simultaneously from the EM measurement system and the K-type thermocouples using a Pico Technology (St Neots, UK) temperature data logger. Temperature and EM signals were recorded throughout the experiments, with the recorded temperature and time interpolated and synchronised to align the temperature with the corresponding EM data.

The EM measurement system can apply a range of excitation frequencies simultaneously. LabVIEW (Version 21.0) software is used to generate the multifrequency excitation signal, while extracting amplitude and phase values from the receive coil signal for each frequency using a windowed fast Fourier transform. Amplitude and phase values are plotted in real time and recorded. Data is sampled at a rate of 1 MS/s with one frame of data acquired every 100 ms.

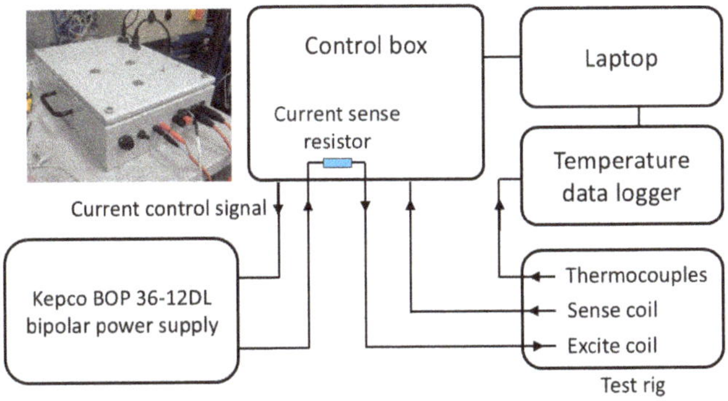

Figure 3. Block diagram of the EM measurement system.

2.2. Excitation Parameter Selection

A test was set up to determine the optimal excitation parameters for the tests using plates made from ferrite powder in a ceramic or polymer medium. This type of material was selected as it has a clearly defined magnetic permeability, along with a very high resistivity (>200 Ω/cm), which allows the assessment of the frequency response of the system to a variation in magnetic permeability without a significant contribution from electrical resistivity. The ferrite plates have known initial relative permeabilities of 230 (Low μ) [24], 650 (Med μ) [25] and 3000 (High μ) [26] and were tested with a lift-off of 3 mm. A frequency sweep from 100 Hz to 80 kHz was applied to the three samples, and an air measurement (response with no sample present) was subtracted. The amplitude response (Figure 4a) shows clear discrimination between the three samples, with the higher permeability sample producing a higher amplitude response throughout the frequency spectrum. Conversely, the phase response (Figure 4b) is very weak and not a good indicator of sample permeability. For this reason, the phase response is not considered here and has been omitted from the results in the remainder of this paper. It is worth noting that other EM systems such as EMspec [22] use a differential pair of receiver coils, where one receiver is positioned in the vicinity of the sample strip and another receiver further away. Such differential systems can provide a valuable phase response; however, they can be more sensitive to mechanical vibration. This study uses a simple transmitter—receiver configuration to demonstrate potential, but there is clearly scope for future refinement exploiting both phase and amplitude information, but that was not the purpose of this study.

It can be seen from the amplitude plot (Figure 4a) that the response at lower frequencies is very weak, and closer examination of the plot indicates that any frequency below 500 Hz would be unlikely to yield coherent results; therefore, 500 Hz was selected as the lowest excitation frequency. The amplitude response shows a peak at approximately 40 kHz, due to the roll-off for the frequency response of the amplifier. This frequency also gives the greatest difference between the three samples; consequently, 39.7 kHz was chosen as the highest excitation frequency. Two more intermediate frequencies were selected, resulting in the following frequency spectrum, with applied current weighting shown in brackets: 0.5 kHz (60%), 4.7 kHz (20%), 21.1 kHz (10%) and 39.7 kHz (10%). The current weighting is used to partially compensate for the weaker signal at lower frequencies; it is not intended to bring all the signals to exactly the same level.

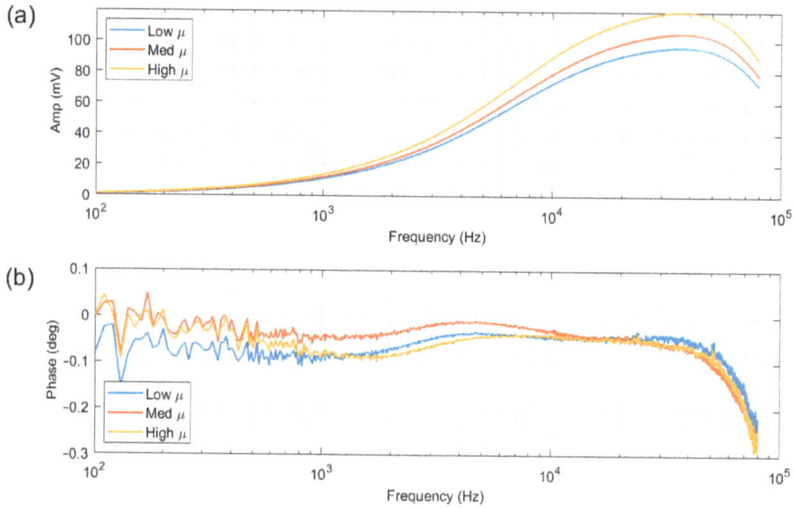

Figure 4. Amplitude (**a**) and phase (**b**) readings resulting from a frequency sweep from 100 Hz to 80 kHz with ferrite plates with low, medium (med), and high permeability values.

Figure 5 shows the resultant multifrequency transmit current and receive voltage for the system. The x-axis shown here corresponds to one cycle of the 500 Hz waveform, though 50 cycles are acquired for FFT analysis during testing. It can be seen from Figure 5a that the excitation uses the full ± 12 A current range of the power amplifier, with a clear contribution from the 500 Hz component. The amplitude of signal from the receive coil shown in Figure 5b is approximately ±350 mV, although a gain of 10 is applied to this before data acquisition. For the received voltage, the higher frequencies are dominant.

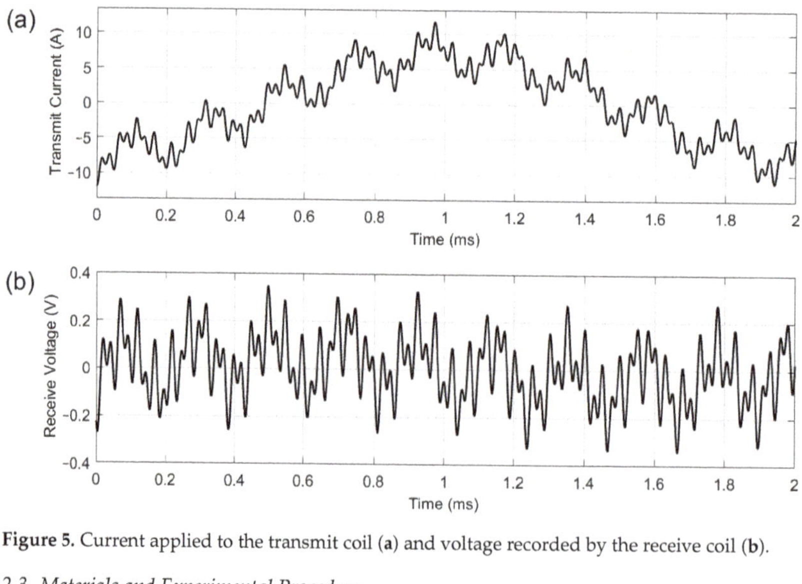

Figure 5. Current applied to the transmit coil (**a**) and voltage recorded by the receive coil (**b**).

2.3. Materials and Experimental Procedure

Three different measurements were carried out on the test rig using the same sample size with some variation in thickness. Although the sample holder can accommodate a sample measuring 225 mm × 80 mm, a smaller size of 100 mm × 80 mm was chosen for these tests to ensure uniform heating. Firstly, a 3 mm thick ceramic plate was used as a dummy sample to assess the level of background signal in the absence of a metal sample and provide data for background signal subtraction. For this test the sample was loaded into the furnace; the furnace was heated to 900 °C for austenitic steel and ferritic stainless calibration whilst 600 °C was used for nickel calibration test. Once the coil reached the target temperature, the sample holder was lowered out of the furnace and air cooled. The heating and cooling cycle for the test to 900 °C test is shown in Figure 6a.

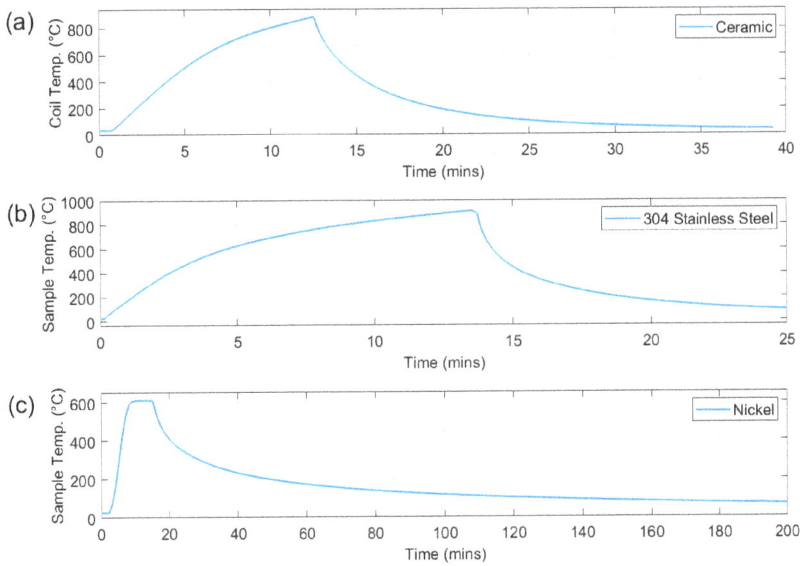

Figure 6. Heating and cooling cycles for ceramic plate (**a**), type 304 stainless steel (**b**), and nickel (**c**).

The second test used a 1 mm thick type 304 austenitic stainless-steel plate. As this material is paramagnetic, it provides an assessment of the response of the system to a change in resistivity without an accompanying change in magnetic permeability. For the 304 stainless steel the furnace was heated to 400 °C; the sample was loaded into the furnace; the furnace was then heated to 900 °C with an average sample heating rate of 0.76 °C/s; once the sample temperature reached 900 °C/s the sample holder was lowered out of the furnace and allowed to air cool to <100 °C. See Figure 6b.

A 1 mm thick 99% pure nickel plate was employed in the third test to assess the response of the system to changes in resistivity and magnetic permeability. Nickel was chosen as it has a relatively low Curie temperature (Tc) at 353–360 °C, so it presents the opportunity to also observe the accuracy of the system to characterise the ferromagnetic to paramagnetic transition at Tc. For this test the furnace was heated to 200 °C; the sample was loaded into the furnace; the furnace temperature was increased to 600 °C with an average heating rate of 1.67 °C/s; once the sample temperature reached 600 °C, the furnace was switched off, allowing the sample to cool to <100 °C. See Figure 6c. The slow furnace cooling rate was chosen to assess the system's ability to accurately determine the Tc of nickel during cooling.

The fourth test used a 0.9 mm thick type 430 ferritic stainless steel plate to assess the performance of the system when presented with a sample with a higher permeability and a higher temperature Curie point. This is more representative of the high-temperature behaviour that might be encountered in continuous annealing. The heating and cooling routine was identical to the test for the austenitic stainless steel shown in Figure 6b.

Type K thermocouples were spot-welded to the sample surface in the centre of the coil (middle of the sample) and directly opposite. A type K thermocouple was also embedded into the coil assembly to measure the temperature of the coil. It should be noted that this thermocouple did not contact the copper coil directly; it was sandwiched between the turns of the coil outside of the insulation. For the ceramic dummy sample, EM data is plotted against the coil temperature. For the stainless steel and nickel, the temperature of the surface of the material at the centre of the coil is used.

3. Finite Element Modelling

In order to gain insight into the interaction between the field and sample at the chosen frequencies, a finite element model was constructed. A two-dimensional, axisymmetric model was developed using COMSOL Multiphysics (Version 6.3). The exterior boundaries were set as magnetic insulation, and the interior boundaries were set as continuity. The complete mesh consists of 61,501 domain elements and 1240 boundary elements. The geometry of the sensor and sample in the model is shown in Figure 7a. Figure 7b shows the magnetic flux density distribution by rotating the modelled plane by 360 degrees.

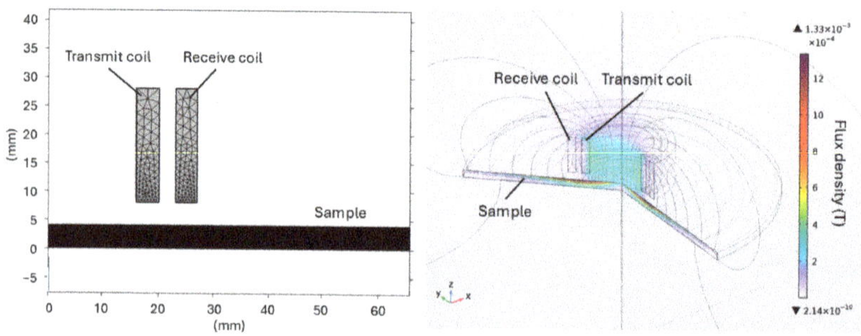

(a) (b)

Figure 7. Two-dimensional, axisymmetric FEM model with geometry of the sensor and sample (**a**) and magnetic flux density distribution by rotating the modelled plane 360 degrees (**b**). $\mu = 230$, frequency = 500 Hz, current = 2 A.

Figure 8 illustrates how permeability and frequency affect the depth of magnetic field penetration into the steel plate. As either permeability or frequency increases, the skin depth becomes shallower, the induced field becomes more concentrated under the coil, and the maximum flux density at the sample surface increases. Hence, the higher frequencies offer a more localised measurement and a stronger response at the expense of sample penetration.

As shown in Figure 9a, the modelling shows a monotonic increase in the receive coil inductance with increasing permeability across all frequencies. As frequency increases, so does the sensitivity to permeability changes, indicating that the highest frequency should result in the greatest sensitivity to a temperature-induced change in permeability in a ferromagnetic material. Conversely, the 500 Hz plot is relatively flat across all permeability values, indicating a poor correlation with permeability at this frequency. Figure 9b shows the receive coil inductance for a fixed permeability and a change in resistivity. It can be seen from the plot that the higher frequencies have the highest sensitivity to the change in resistivity, with 21.1 kHz and 39.7 kHz showing a very similar overall change.

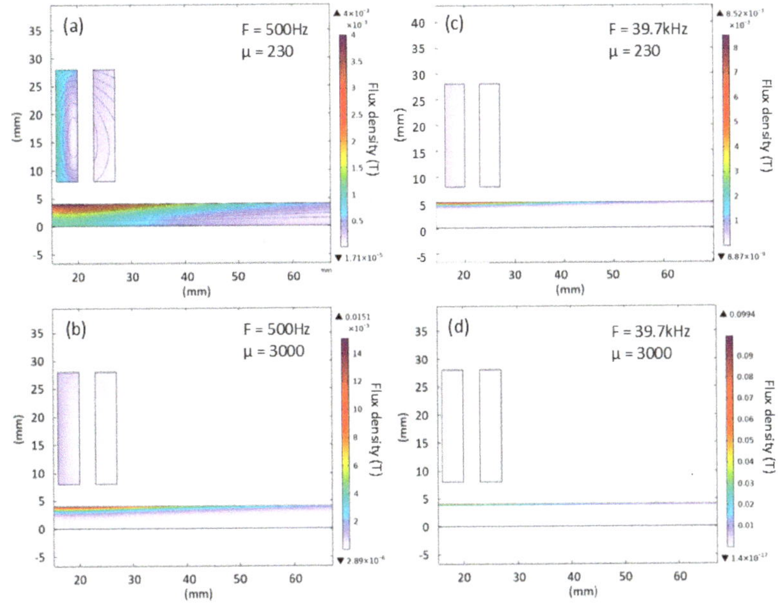

Figure 8. Flux density in the sample at low (**a**,**b**) and high (**c**,**d**) frequency excitation for two different permeability values.

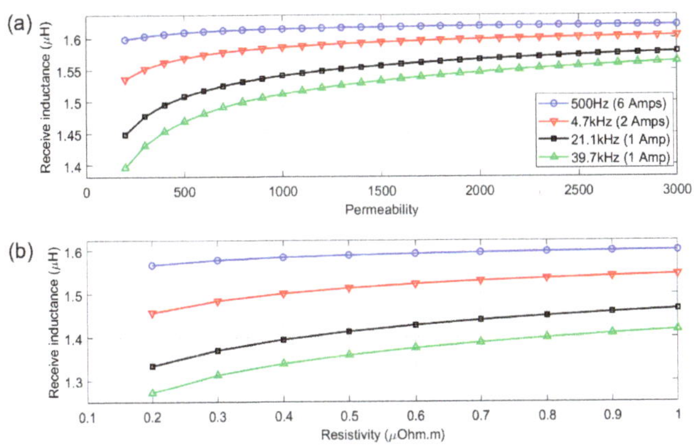

Figure 9. Magnitude of receive coil inductance for the chosen excitation frequencies and corresponding currents used in the experimental work for a variation in relative magnetic permeability (**a**) and resistivity (**b**). Fixed permeability of 500 used for (**b**).

4. Experimental Results

4.1. Ceramic Dummy Sample (Air Measurement)

Figure 10a,b show the results of the test with the ceramic dummy sample. Figure 10a shows the absolute amplitude values, whereas Figure 10b shows the change in amplitude from the start of the test to allow comparison of the different frequencies on the same plot. The remainder of the figures in this paper are shown in the same format.

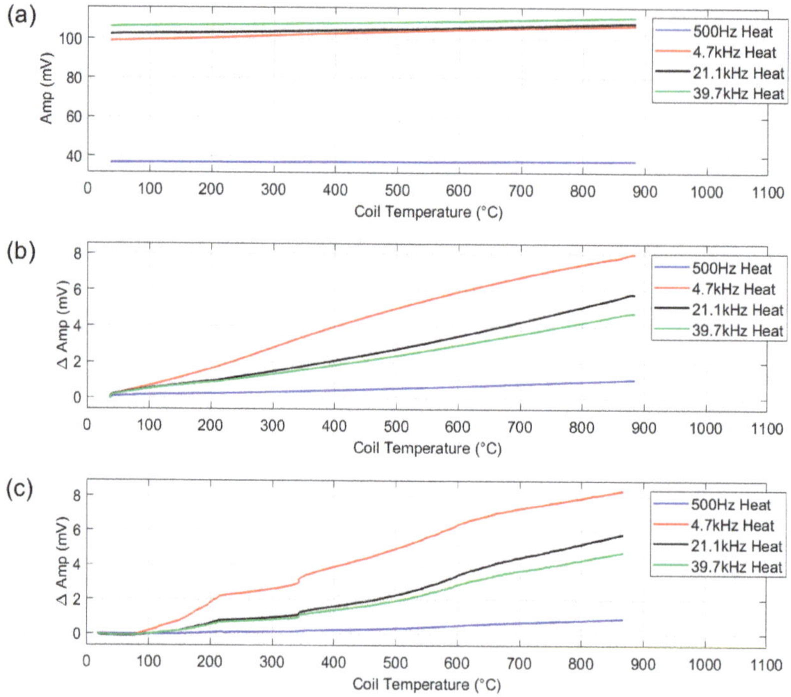

Figure 10. Multifrequency amplitude measurement for ceramic dummy sample in sample holder—absolute amplitude (**a**), multifrequency amplitude measurement for ceramic dummy sample in sample holder—change in amplitude (**b**) and multifrequency amplitude measurement for coil in furnace without sample holder (**c**). Represents the background signal of the system at the four chosen frequencies.

As the ceramic sample is not electromagnetically reactive, the measurement with the ceramic sample can be considered an air measurement. The ideal response for an air measurement in this scenario would be zero change in signal for the increase in temperature, indicating zero background field. However, a small change in amplitude is evident for all frequencies. It can be seen from the plot that the increase in amplitude on heating is almost linear with respect to coil temperature (x-axis), with 500 Hz showing the lowest overall change in amplitude, 4.7 kHz the highest and the other two higher frequencies lie between the two.

To assess if the interaction between the applied field and the austenitic stainless steel sample holder could be the source of the background signal, the test was repeated with the coil winched into the furnace without the sample holder, suspended from ceramic fibre cable, see Figure 10c. As the coil was not held in position by the sample holder in this test and could move freely in the furnace, the results are noisier; however, the trends in the data in Figure 10b,c are the same, indicating that the sample holder is not the source of the background signal.

The system is current-driven; therefore, the applied current does not vary with changes in the resistivity of the coil, so this can also be ruled out as the source of the background signal. As these factors have been eliminated, the background signal measured in this test must be caused by the generation of eddy currents in the 4 mm thick copper coil. It is expected that this background signal could be removed with future improvements in the system such as a balance differential coil configuration.

4.2. Austenitic Stainless Steel

Figure 11a shows the amplitude response to a sample made from type 304 austenitic stainless steel. This is nominally a fully paramagnetic material; therefore, magnetic permeability should remain constant, while resistivity increases with temperature, increasing the measured signal amplitude.

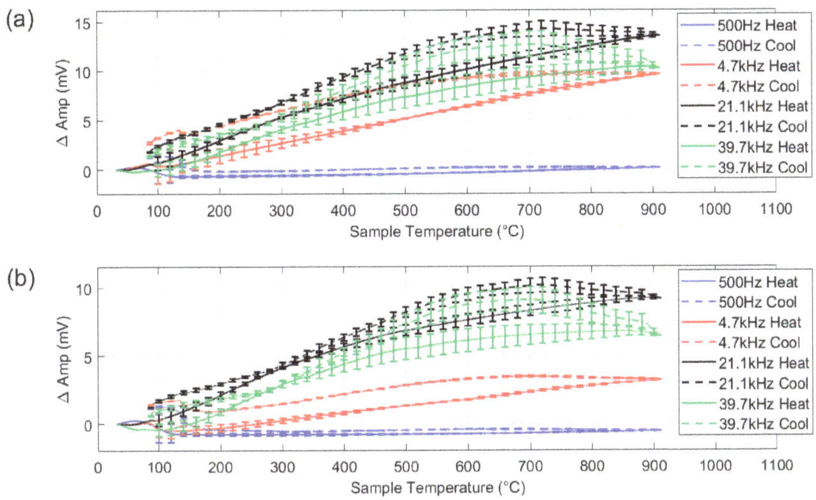

Figure 11. Multifrequency amplitude measurement for austenitic stainless steel (**a**), with background signal subtracted (**b**). With 1SD error bars.

Examination of Figure 11a shows that at the lowest frequency (500 Hz), the change in amplitude is minimal, indicating limited sensitivity to the thermally induced change in resistivity. As frequency increases, the sensor becomes increasingly responsive: the higher frequencies show a notable rise in amplitude and with temperature. On cooling, the signal amplitude increases slightly before decreasing. Examination of the material by electron backscatter diffraction showed trace amounts of retained martensite, which indicates that the material was weakly ferromagnetic before heating and after cooling, so the response here cannot be seen as purely paramagnetic. However, these effects are small in magnitude and secondary to the main focus of this work.

Figure 11b shows the response from the stainless steel with the air measurement shown in Figure 10b subtracted. Both data sets include a coil temperature measurement. The coil temperature measurements for the austenitic steel were compared to the coil temperature measurements for air, and the corresponding air amplitude values were subtracted. It can be seen from the plot that the air subtraction causes an overall reduction in amplitude, around 50% decrease for the two higher frequencies. As the austenitic steel sample has a relatively low room temperature conductivity (1.39 MS/m in comparison with 58.5 MS/m for the copper coil [27]), the signal from the sample is relatively weak, so the background signal from the copper coil tends to dominate, distorting the austenitic steel response, especially on cooling.

4.3. Pure Nickel

Figure 12a,b show the electromagnetic response to the heating and cooling of the nickel sample with and without background signal subtraction. Comparison of the two plots shows that background signal subtraction has very little impact, with some small shifts in the overall signal level at higher temperatures. As nickel has a Tc of 353–360 °C, a strong response to the ferromagnetic to paramagnetic transition is expected in this range. As with the austenitic steel (Figure 11), the higher frequencies give the sharpest response, this is attributed to increased sensitivity to near-surface properties due to a shallower penetration depth, as illustrated by the simulation results, though the 500 Hz plot does clearly reflect the Curie transition on heating and cooling.

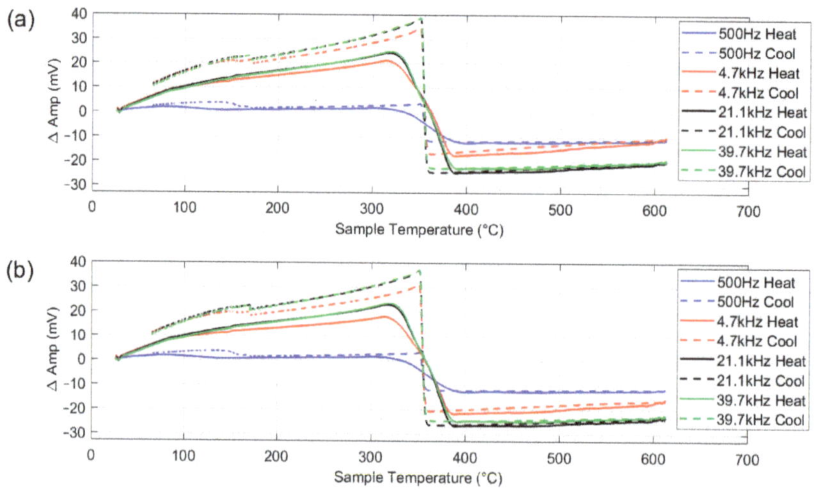

Figure 12. Multifrequency amplitude measurement for pure nickel (**a**), with background signal subtracted (**b**).

The initial increase in amplitude at lower temperatures can mainly be attributed to the increase in magnetic permeability with temperature, with some contribution from the increase in resistivity, as with the austenitic steel discussed in Section 4.2. As temperature rises, the permeability increases, enhancing the interaction with the electromagnetic field, which results in a higher induced voltage and, consequently, a higher amplitude signal. This trend continues until the material approaches Tc, where magnetic permeability drops sharply as the nickel transitions to a paramagnetic state. After that, further temperature increases invoke a small increase in amplitude as a result of a further increase in resistivity. On cooling, the process is reversed.

Closer examination of the strongest response, the 39.7 kHz plot, shows that on heating, the change in amplitude corresponding to the Curie transition is somewhat blurred, starting at around 325 °C and finishing at around 392 °C. This is caused by the relatively fast heating of the sample, 2.18 °C/s between 300 °C and 400 °C, meaning that different parts of the sample are heating at slightly different rates and reaching Tc at different times. In contrast, on cooling, the change in amplitude corresponding to the Curie transition takes place between 360 °C and 351 °C, much closer to the expected transition temperature, reflecting the much slower cooling rate of 0.2 °C/s.

4.4. Ferritic Stainless Steel

Figure 13 shows the EM response for the 430 ferritic stainless steel sheet. As with the nickel, the higher frequencies give the strongest response. Examination of the heating part of the cycle for the 39.7 kHz plot shows that there is an initial rise in amplitude as permeability increases, followed by a drop in signal level as the temperature increases past 600 °C. This drop in signal level is due to parts of the sample reaching the Curie temperature, as the relatively high heating rate caused some temperature non-uniformity in the sample. After the temperature passes Tc for all the samples, the signal levels off, apart from a very small increase due to the continuing increase in resistivity. The Tc is more clearly seen on cooling as the low cooling rate results in less temperature gradients across the sample. The lower Tc value (approx. 700 °C) than seen in iron or low carbon steels (approx. 780 °C) is due to the high Cr content in the steel.

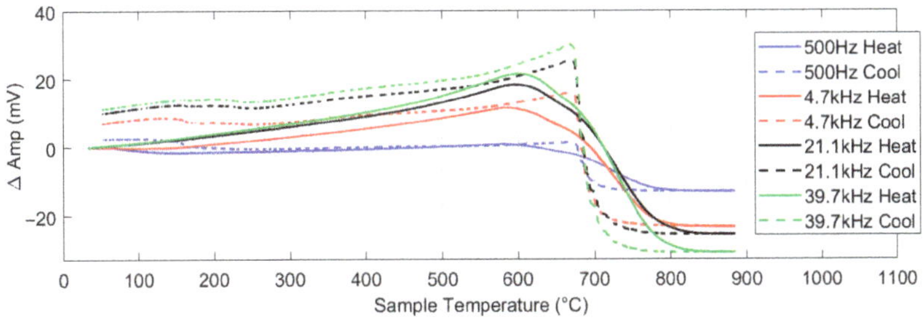

Figure 13. Multifrequency amplitude measurement for type 430 ferritic stainless steel with background signal subtracted.

5. Discussion

This paper outlines work to develop an electromagnetic sensor system for deployment in a continuous annealing furnace. The sensor system must be capable of in sítu characterisation of the material in the furnace, reliably tracking evolving microstructures and providing real-time feedback to ensure mechanical property targets are met. The sensor must be able to survive prolonged periods in an extreme environment at temperatures up to approximately 850 °C, depending on sensor positioning in the furnace. The system consists of three parts: the sensor assembly, custom-built hardware and software. In order to assess the system as a whole, a number of experiments were set up using an annealing simulator test rig which employs a vertical tube furnace for sample heating and cooling in conjunction with a winch to move the samples in and out of the furnace.

As can be seen in Figure 11, there is some variation, as indicated by the error bars, in the measured signal for repeated tests with the same sample. Some potential sources of error, including electrical noise from the furnace, measurement drift, variations in background signal and thermally induced changes in sensor geometry have been studied during the design stage and found to have negligible impact on the measured signal. The greatest potential source of error in this system is a variation in lift-off: the distance between the sensor and sample. Although, as shown in Figures 1 and 2, the position of the coil is fixed with respect to the frame, the sample is permitted some movement to allow for thermal expansion. Thus, some lift-off variation is possible. Tests with the ferrite powder samples showed that the signal level for all frequencies decreased by a factor of ≈ 6 for an increase in lift-off from 2 mm to 5 mm. Although it is unlikely that lift-off could increase this much during normal testing, it does illustrate the potential for lift-off-related error. Previous work has shown that through judicious sensor design, careful frequency selection, analysis of the signal phase and calibration procedures, a lift-off invariant multifrequency EM measurement system can be realised [21,22]. This will be a goal for the next iteration of this system.

The tests did not use samples representative of the types of materials processed in a continuous annealing furnace, rather the samples were selected to interrogate specific heat induced phenomena: a ceramic dummy sample to assess the background signal in the first test (Figure 10); an austenitic steel sheet to assess the response of the signal to a change in electrical resistivity only in the second test (Figure 11); a nickel sample to assess the capability of the system to accurately characterise the ferromagnetic to paramagnetic Curie transition in the third test (Figure 12); a ferritic stainless steel sample to assess the performance on the system when presented with a higher temperature ferromagnetic to paramagnetic Curie transition in a high permeability sample in the fourth test (Figure 13).

The test results show that although the system does exhibit a significant change in background signal, this is repeatable and can be successfully subtracted from the sample signal. The result for the nickel sample shows a very accurate Curie transition on cooling, however the signal corresponding to the Curie transition for heating is spread over a much wider temperature range, due to variations in temperature across the sample at the relatively high heating rate (compared to cooling rate) and the large diameter of the coil measuring a large sample area. The result for the ferritic stainless steel shows that the system is capable of characterising changes at high temperatures, as would be found in a CAP line. As with the nickel sample, there are some discrepancies between the measured temperature and the expected Curie transition temperature, again due to the size of the coil in comparison with the sample, combined with the fast heating. The presence of the coil adjacent to the sample also induced some temperature non-uniformity in the annealing simulator system. The EM measurement system is ultimately intended for deployment in a much larger furnace and on much larger samples (industrial CAP line with wide strip samples that continuously move past the sensor), therefore in practice, the coil will not encounter these large thermal gradients.

The range of frequencies employed in this work was primarily dictated by the bandwidth of the system, with modelling and experimental work used to assess the suitability of each frequency for material microstructure assessment. Figures 4, 8 and 9 show experiment and modelling results that illustrate the sensitivity of the system to materials with different permeabilities at different frequencies. The data shows a monotonic increase in magnitude with increasing permeability across all frequencies. This behaviour reinforces the experimental strategy: as steel undergoes phase transformation during annealing (changing permeability), the signal amplitude responds consistently. At higher frequencies (21.1 kHz and 39.7 kHz), the magnitude shows more distinct gradients with respect to permeability; these frequencies induce stronger, more localised signals, with a greater sensitivity to surface-level changes due to shallower penetration depths. In contrast, 500 Hz provides deeper penetration but less sensitivity to permeability changes; however, the greater penetration depth may make this frequency more suitable for bulk behaviour interrogation.

The combined insight from the experimental work and modelling will inform frequency selection for specific transformation tracking in the continuous annealing furnace application. For example, detecting ferrite–austenite transitions (which alter magnetic permeability) may benefit from multifrequency interpretation, using both deep and shallow probing characteristics. The high frequency magnitude shows the highest sensitivity to permeability changes, but if there is a surface effect on the material, low frequency data can be used to analyse through-thickness variation. Surface effects in steel can include phenomena such as decarburisation, where a lower carbon content at the surface can cause earlier transformation to ferrite on cooling than the bulk material and a higher permeability in that surface layer. If surface effects are not an issue, it may be judicious to discard the 500 Hz data, as it accounts for 60% of the applied current, and concentrate more energy in higher frequency bands.

In order to reach a higher environmental temperature, the sensor design is much simplified in comparison with other high-temperature EM measurement systems [5,21,22], leaving the signal more vulnerable to noise sources. For example, [21] uses a ferrite core to shape the applied field and increase maximum lift-off and a dummy sensor to subtract the background field. Although the same techniques cannot be applied to this system as the core would not survive, a similar approach can be adopted by employing a differential receiver coil configuration with associated signal processing to measure only the signal from the target steel strip. This would allow more sophisticated processing of the steel properties than attempted here. Isolating the signal from the target also allows further calibration and removal of variation in lift-off. Along with sensor improvements, further investigations will involve real-time microstructure change tracking for advanced steels such as dual-phase and interstitial free grades, with sensor output calibrated to represent real material properties such as magnetic permeability.

6. Conclusions

This study establishes a proof of concept for an EM test system capable of real-time detection of magnetic and structural transformations in metals during thermal cycling in a high-temperature furnace environment. An annealing simulator test rig was used to test a range of samples selected to interrogate different aspects of the sensor response. The test results show a correlation with the thermally induced change in conductivity in an austenitic stainless steel sample and a strong response to increasing permeability below Tc and the subsequent ferromagnetic—paramagnetic Curie transition in nickel and ferritic stainless steel. The sensor employs a simplified design in comparison to other sensors of this type, using a few turns of 4 mm thick copper rod for the transmit coil and high-temperature type K thermocouple wire for the receive coil, resulting in a robust temperature-resistant construction. The developed sensor assembly has proved to be tolerant to high-temperature deployment, surviving more than twenty tests where the sensor was exposed to thermal cycling up to 900 °C and one test where the sensor was heated to 715 °C and held at that temperature for five hours without significant damage. Future iterations of the sensor design will feature a differential receive coil configuration to aid in background signal subtraction and further investigation of the multifrequency phase response to mitigate the effects of lift-off variation on the measured signal. The eventual aim of the work is sensor deployment in an industrial continuous annealing line, providing real-time feedback for precise processing control in the production of high-value steels.

Author Contributions: Investigation, J.W.W. and M.A.J.; writing—original draft preparation, J.W.W., M.A.J. and L.Z.; writing—review and editing, C.D., A.J.P. and C.S.; software, L.Z.; funding acquisition, C.D., A.J.P. and C.S. All authors have read and agreed to the published version of the manuscript.

Funding: This work was supported by the UK Engineering and Physical Sciences Research Council (EPSRC) grant EP/W024713/1: High-temperature Electromagnetic Instrumentation for Metal Production (Hi-TEMP).

Institutional Review Board Statement: Not applicable.

Informed Consent Statement: Not applicable.

Data Availability Statement: Data supporting the conclusions of this article will be made available by the authors upon request.

Acknowledgments: The authors would like to thank Edward Hodierne for his support with the furnace experiments at the AMMC, University of Warwick.

Conflicts of Interest: The authors declare no conflicts of interest.

Abbreviations

The following abbreviations are used in this manuscript:

CAP	Continuous Annealing Process
EM	Electromagnetic
IMPOC	Impulse Magnetic Process Online Controller
HACOM	Harmonic Analysis Coil Online Measuring
3MA	Micromagnetic Multi-parameter, Microstructure and Stress Analysis
Tc	Curie Temperature
BOP	Bipolar Operational Power supply

References

1. Gorain, N.; Walunj, M.G.; Soni, M.K.; Kumar, B.R. Effect of continuous annealing process on various structure parameters of martensite of dual-phase steels. *Arch. Civ. Mech. Eng.* **2020**, *20*, 29. [CrossRef]
2. Deng, Y.; Li, Y.; Di, H.; Misra, R. Effect of heating rate during continuous annealing on microstructure and mechanical properties of high-strength dual-phase steel. *J. Mater. Eng. Perform.* **2019**, *28*, 4556–4564. [CrossRef]
3. Bi, Y.; Ge, R.; Yan, M.; Xiong, X. Ageing Susceptibility of Continuously Annealed Low-Carbon-Steel Strips. *Metals* **2022**, *12*, 1285. [CrossRef]
4. Sommers, U.; Carsten, A.K.; Brühl, F.; Sasse, C. Intelligent Furnace Technology for Annealing and Galvanising Lines. In Proceedings of the 21st Seminário de Automação e TI, Part of the ABM Week, São Paulo, SP, Brazil, 2–6 October 2017.
5. Van Den Berg, F.; Fintelman, D.; Yang, H.; Mocci, C.; Vannucci, M.; Colla, V. The Use of Advanced Data Analytics to Monitor Process-Induced Changes to the Microstructure and Mechanical Properties in Flat Steel Strip. In *Impact and Opportunities of Artificial Intelligence Techniques in the Steel Industry: Ongoing Applications, Perspectives and Future Trends*; Springer International Publishing: Berlin/Heidelberg, Germany, 2021; pp. 78–91. [CrossRef]
6. Sommers, U.; Klinkenberg, C.; Daube, T.; Brühl, F.; Sasse, C.; Lhoest, A.; Pensis, O.; Crutzen, J.P.; Horst, H. X-CAP—Closed-loop control of the AHSS annealing process via X-ray phase fraction measurement. In Proceedings of the ESTAD 2019, Duesseldorf, Germany, 24–28 June 2019.
7. Davis, C.; Strangwood, M.; Peyton, A. Overview of non-destructive evaluation of steel microstructures using multifrequency electromagnetic sensors. *Ironmak. Steelmak.* **2011**, *38*, 510–517. [CrossRef]
8. Zhou, L.; Wu, F.; Hall, R.; Davis, C. Electromagnetic sensors for in-situ dynamic microstructure monitoring of recovery and recrystallisation in interstitial free steels. *J. Magn. Magn. Mater.* **2022**, *551*, 169187. [CrossRef]
9. Aghadavoudi-Jolfaei, M.; Shen, J.; Smith, A.; Zhou, L.; Davis, C. Non-destructive measurement of microstructure and tensile strength in varying thickness commercial DP steel strip using an EM sensor. *J. Magn. Magn. Mater.* **2019**, *473*, 477–483. [CrossRef]
10. Jolfaei, M.A.; Liu, J.; Zhou, L.F.; Van Den Berg, F.; Davis, C. Non-destructive evaluation of magnetic anisotropy associated with crystallographic texture of interstitial free steels. *J. Magn. Magn. Mater.* **2023**, *568*, 170374. [CrossRef]
11. Libby, H.L. *Broadband Electromagnetic Testing Methods. Part IV. Multiparameter Test Principles (No. BNWL-953)*; Battelle Pacific Northwest Labs.: Richland, WA, USA, 1969.
12. Plakhotnyuk, A.N. Nondestructive testing of work-hardening depth in ferromagnetic components by a multifrequency electromagnetic method. *Sov. Phys. J.* **1968**, *11*, 141–144. [CrossRef]
13. Sophian, A.; Tian, G.Y.; Taylor, D.; Rudlin, J. Electromagnetic and eddy current NDT: A review. *Insight* **2001**, *43*, 302–306.
14. Blitz, J.; Peat, T.S. *The Application of Multi-Frequency Eddy Currents to Testing Ferromagnetic Metals*; NDT International: West Chester, PA, USA, 1981; Volume 14, Part 1, pp. 15–17.
15. Libby, H.L. *Multiparameter Eddy Current Concepts. Research Techniques in Nondestructive Testing*; Academic Press Inc. Ltd.: London, UK, 1970.
16. Avila, J.R.S.; Chen, Z.; Xu, H.; Yin, W. A multi-frequency NDT system for imaging and detection of cracks. In Proceedings of the 2018 IEEE International Symposium on Circuits and Systems (ISCAS), Florence, Italy, 27–30 May 2018; IEEE: Piscataway, NJ, USA, 2018; pp. 1–4.
17. Liu, Z.; Tsukada, K.; Hanasaki, K. One-dimensional Eddy Current Multi-frequency Data Fusion: A Multi-resolution Analysis Approach. *Insight* **1998**, *40*, 286–289.
18. Sidi-Ahmed, K.; Maouche, B.; Gabi, Y.; Alloui, L.; Straß, B.; Wolter, B.; Feliachi, M. Numerical simulations and experimental investigation of laser hardening depth investigation via 3MA-eddy current technique. *J. Magn. Magn. Mater.* **2022**, *550*, 169046. [CrossRef]
19. Hristoforou, E.V. Permeability sensors for magnetic steel structural health monitoring. *Sensors* **2025**, *25*, 606. [CrossRef] [PubMed]

20. Munjal, R.; Sajjad, F.A.; Wendler, F.; Kanoun, O. Multifrequency inductive sensor system for classification of bimetallic coins. *IEEE Trans. Instrum. Meas.* **2020**, *70*, 1–9. [CrossRef]
21. Dickinson, S.J.; Binns, R.; Yin, W.; Davis, C.; Peyton, A.J. The Development of a Multifrequency Electromagnetic Instrument for Monitoring the Phase Transformation of Hot Strip Steel. *IEEE Trans. Instrum. Meas.* **2007**, *56*, 879–886. [CrossRef]
22. Shen, J.; Zhou, L.; Jacobs, W.; Hunt, P.; Davis, C. Real-time in-line steel microstructure control through magnetic properties using an EM sensor. *J. Magn. Magn. Mater.* **2019**, *490*, 165504. [CrossRef]
23. BGS Technic KG. BGS Induction Coil for Induction Heater. 2025. Available online: https://www.bgstechnic.com/en/Products/Electric-Cordless-Battery-Tools/Induction-Heater/Induction-Coil-for-Induction-Heater-32-mm-straight-type-for-BGS-2169-3390-3391.html (accessed on 15 March 2025).
24. Kemet. Flex Suppressor Datasheet. 2025. Available online: https://content.kemet.com/datasheets/KEM_FS8004_NSS.pdf (accessed on 10 June 2025).
25. Laird Technologies, Inc. MP1496-0M0 Ferrite Plate. 2019. Available online: https://www.laird.com/products/inductive-components-emc-components-and-ferrite-cores/ferrite-plates-and-disks/mp-series/mp1496-0m0 (accessed on 10 June 2025).
26. Fair-Rite Products Corp. Ferrite Plate (3595000541). 2023. Available online: https://fair-rite.com/product/plate-3595000541/ (accessed on 10 June 2025).
27. Mitchell, B.S. *An Introduction to Materials Engineering and Science for Chemical and Materials Engineers*; John Wiley & Sons: Hoboken, NJ, USA, 2003.

sensors

MDPI

Review

Low-Cost Sensors in 5G RF-EMF Exposure Monitoring: Validity and Challenges

Phoka C. Rathebe * and Mota Kholopo

Department of Environmental Health, Faculty of Health Sciences, Doornfontein Campus, University of Johannesburg, P.O. Box 524, Johannesburg 2006, South Africa; motaliok@gmail.com
* Correspondence: prathebe@uj.ac.za; Tel.: +27-11-559-6641

Abstract

The deployment of 5G networks has transformed the landscape of radiofrequency electromagnetic field (RF-EMF) exposure patterns, shifting from high-power macro base stations to dense networks of small, beamforming cells. This review critically assesses the validity, challenges, and research gaps of low-cost RF-EMF sensors used for 5G exposure monitoring. An analysis of over 60 studies covering Sub-6 GHz and emerging mmWave systems shows that well-calibrated sensors can achieve measurement deviations of ±3–6 dB compared to professional instruments like the Narda SRM-3006, with long-term calibration drift less than 0.5 dB per month and RMS reproducibility around 5%. Typical outdoor 5G FR1 exposure levels range from 0.01 to 0.5 W/m^2 near small cells, while personal device use can cause transient exposures 10–30 dB higher. Although mmWave (24–100 GHz) and Wi-Fi 7/8 (~60 GHz) are underrepresented due to antenna and component limitations, Sub-6 GHz sensing platforms, including software-defined radio (SDR)-based and triaxial isotropic designs, provide sufficient sensitivity for both citizen and institutional monitoring. Major challenges involve calibration drift, frequency band gaps, data interoperability, and ethical management of participatory networks. Addressing these issues through standardized calibration protocols, machine learning-assisted drift correction, and open data frameworks will allow affordable sensors to complement professional monitoring, improve spatial coverage, and enhance public transparency in 5G RF-EMF exposure governance.

Keywords: 5G RF-EMF monitoring; low-cost sensors; citizen science; calibration uncertainty; participatory monitoring

1. Introduction

The fifth generation of mobile communication technology (5G) has been rolled out rapidly worldwide. Since 2018, governments and telecommunication companies have promoted it as essential infrastructure for the digital age, enabling ultra-fast data transfer, low latency, and the connectivity needed for applications like autonomous vehicles, telemedicine, and the Internet of Things (IoT) [1]. However, this vision is based on a physical reality quite different from earlier generations: instead of a few large towers, 5G depends on a dense network of small-cell transmitters embedded in everyday environments, from lampposts and traffic lights to the sides of residential buildings [2,3]. For many citizens, the technology has become unavoidable, not just as a service but as an infrastructure woven into the fabric of urban life. As in past generations, this visibility has been met with unease. The introduction of 2G in the 1990s raised fears about brain tumors from mobile phones; 3G deployment triggered local protests over mast siting near schools;

and 4G reignited debates over cumulative exposure and ecological effects [4]. With 5G, however, public concern has intersected with a new information landscape. Social media platforms have amplified both misinformation and distrust, giving unprecedented reach to those questioning the independence of regulators or the adequacy of safety standards [5]. In this environment, fear cannot simply be explained away as ignorance. It reflects a deeper dissonance between institutional framing of safety and community perception of risk.

Regulators like the International Commission on Non-Ionizing Radiation Protection (ICNIRP) and IEEE define safety in terms of numbers: meeting exposure limits set to prevent immediate thermal effects [1]. However, communities view risk more subjectively, considering long-term uncertainty, precautionary ethics, and fairness in how infrastructure is allocated [6]. Independent researchers share this skepticism, highlighting the absence of long-term epidemiological studies and the need for a precautionary approach [7]. As a result, deploying 5G has become more than just a technical project, it is also a test of trust, transparency, and democratic accountability [8]. One of the clearest signs of this trust gap is how exposure is monitored. Professional RF-EMF monitoring efforts use high-quality equipment, but their coverage is limited. Measurements are costly, sporadic, and usually limited to a few sites in major cities [9]. While technically precise, these efforts cannot capture how exposure varies over time, space, and different usage patterns. Even in countries with advanced monitoring systems, large areas remain unmeasured, and because these campaigns are episodic, their data quickly become outdated. Another major issue is accessibility. The results are often hidden in technical reports that are released months or years after the measurements are taken or presented in formats that are hard for non-experts to understand [10]. In countries like Poland, where stricter exposure limits show how uneasy people are [11], this lack of transparency makes it seem like monitoring is just for compliance, not for the public to know. The irony is clear; the infrastructure is on every street, but reliable data about exposure is still mostly invisible to the public.

Low-cost sensors present a viable path forward. Although less accurate than professional equipment, these devices can be widely deployed, providing continuous and distributed data [9,12]. Citizen science initiatives in other fields offer strong examples: PurpleAir's crowdsourced air quality maps now compete with official monitoring in detail [13]; community radiation mapping projects after the Fukushima disaster offered real-time transparency in ways government systems could not [14,15]; and noise mapping efforts have transformed urban planning debates [16,17]. In telecommunications, grassroots projects in the 2000s mapped Wi-Fi and 4G coverage, foreshadowing the potential of citizen-led exposure monitoring. Today, various initiatives are emerging around RF-EMF. In Greece, the National Observatory of Electromagnetic Fields has developed a network of frequency-selective sensors, some designed to be accessible to the public, providing near-real-time maps of exposure levels [18]. In France, civic groups have distributed affordable meters to households worried about antenna placements, creating a participatory dataset that complements official monitoring efforts [19]. These initiatives demonstrate both the potential benefits and the challenges of citizen-led monitoring. While it can democratize data access, concerns about calibration, reliability, and its acceptance in regulatory processes still persist [13,20,21]. Importantly, democratization involves more than just measurement; it also focuses on equity. Professional monitoring usually gathers in well-funded urban areas, whereas affordable sensors enable rural or marginalized communities to gather and analyze data independently. This process of measurement becomes a political act, giving citizens the opportunity to challenge official accounts and take part in influencing the environment they are exposed to [22–24]. Recent empirical evaluations of affordable 5G RF-EMF sensors offer both promising insights and clear limitations. For example, a distributed Sub-6 Ghz sensor network in Belgium and the Netherlands [9] experienced an

https://doi.org/10.3390/s26020533

average calibration drift of about 0.4 dB per month during outdoor use. Laboratory tests of SDR-based prototypes showed deviations of 1.78 dB under controlled conditions and up to 5.26 dB in field tests when compared to the professional Narda SRM-3006 m [25]. Similarly, a triaxial 5G sensor by Van der Straeten et al. [26] achieved a mean deviation of 2.8 dB within the reference equipment's uncertainty margin, demonstrating that low-cost sensors can attain measurement accuracy within ±5 dB. However, quantitative assessments for mmWave (>24 GHz) bands are still largely lacking, with most research focusing on Sub-6 GHz frequencies. This highlights the need for comprehensive reviews that not only cover sensor development but also evaluate their calibration reliability, qualitative performance, and spatial coverage potential within participatory 5G monitoring frameworks.

This paper contends that inexpensive sensors offer a vital chance to bridge the gap between institutional assurances and public doubts regarding 5G RF-EMF exposure monitoring. By increasing coverage, fostering citizen involvement, and enhancing transparency, they can shift exposure monitoring from an exclusive, technocratic task to a more inclusive democratic process. However, this potential is not guaranteed. Without careful focus on validity, interpretation, and governance, low-cost monitoring may generate data that regulators dismiss or that could be exploited in misinformation efforts. Therefore, the stakes go beyond technical concerns; they also involve the legitimacy of institutions, equitable access to information, and the future of public trust in digital infrastructure governance. Recent studies have made notable progress in validating low-cost 5G RF-EMF sensors, including SDR-based prototypes [25], triaxial 5G exposure sensors [26], and reviews of assessment technologies [27]. However, these works mainly concentrate on device calibration and technical performance. This review complements that by offering a broader, interdisciplinary overview that combines metrological, methodological, and societal perspectives of low-cost sensing. It places technical findings into the context of citizen science involvement, data interoperability, and policy frameworks, such as the European SEAWAVE and GOLIAT projects. The key contribution of this paper is providing a comprehensive view of how affordable sensing technologies can enhance transparency, inclusivity, and standardization in 5G RF-EMF exposure monitoring, bridging the gap between technical validation and governance.

2. Methodology

This review critically examines the emerging role of low-cost sensors in monitoring RF-EMF exposure from 5G networks, highlighting their technical validity, deployment challenges, and research gaps. For this review, "low-cost" refers to devices priced under approximately $1000 USD, in contrast to professional-grade meters such as the Narda SRM-3006, which cost over $10,000. This difference reflects roughly a tenfold cost gap often used in recent RF-EMF and environmental sensing research [9,25], separating affordable, community-level sensors from high precision professional equipment. It examines both commercially available sensors and academic prototypes, including those based on software-defined radios (SDRs). The review mainly concentrates on technologies validated for sub-6 GHz 5G exposure monitoring, while also exploring emerging research into the mmWave (24–100 GHz) range. Since no single low-cost platform currently covers both frequency ranges, the study emphasizes existing developments and identifies mmWave sensing as a significant research gap. The scope encompasses technical performance, methodological issues, and the broader societal implications of using these tools. The literature search was conducted using Scopus, IEEE Xplore, PubMed, Web of Science, and Google Scholar, using keywords like "5G RF-EMF exposure," "low-cost sensors," "mmWave sensing," "beamforming measurement," and "citizen science EMF." Gray literature, including documents from ICNIRP, WHO, national regulators, and NGO-led initiatives, was also

 https://doi.org/10.3390/s26020533

reviewed. This review was updated to include studies published through 2025, ensuring coverage of the most recent advances in low-cost and hybrid rf-emf exposure assessment, including analyses of broadband and personal exposimeter systems [28]. The focus was on publications from 2015 onward, aligning with the start of 5G infrastructure deployment. The inclusion criteria encompassed studies reporting empirical RF-EMF exposure data collected with low-cost or consumer-grade sensors.

Along with 5G New Radio emissions, the reviewed literature also examined overlapping Wi-Fi technologies (IEEE 802.11 a/b/g/n/ac/ax/be and ah, corresponding to Wi-Fi 4–8 and HaLow), which primarily operate within the 2.4 GHz, 5 GHz, 6 GHz, and 60 GHz bands. These frequency ranges significantly overlap with Sub-6 GHz 5G allocations (3.3–4.2 GHz), making them detectable by most broadband or SDR-based low-cost sensors used in this study. Although Halow (802.11 ah, 0.9 GHz) and Wi-Fi 7/8 (6–60 GHz) extend beyond the sensors' calibrated ranges, their emissions can still influence overall broadband readings. To manage this overlap, the studies included sensors that were either frequently calibrated or colocated with reference equipment to differentiate Wi-Fi traffic from 5G downlink signals. Temporal sampling was also considered. Since Wi-Fi and 5G signals are bursty and load-dependent, exposure levels can fluctuate rapidly over milliseconds, whereas international standards average exposure over 6 min. Most affordable sensors record field strength at 1–10 Hz and employ moving averages or root-mean-square calculations to align with these reference intervals. Some prototypes buffer data or synchronize with network activity to detect transient peaks while maintaining averaged compliance measurements [9,25]. These design methods enable low-cost sensors to adapt to RF device duty cycles and estimate real-time exposure levels within an acceptable uncertainty range (usually $\pm$3–6 dB).

Theoretical studies on propagation that did not involve physical measurements were excluded. Additionally, health-related research was considered only if it evaluated exposure levels using low-cost data. For each selected study or initiative, technical details were gathered, including sensor type, frequency range, calibration methods, measurement uncertainties, and development context. These details helped categorize common challenges into three groups: metrological issues like calibration drift and limited frequency coverage; data quality concerns, such as lack of standardization and interpretability; and gaps in implementation and scalability. The analytical framework used critical synthesis and triangulation to identify convergences, such as widespread calibration difficulties, and divergences, especially in how different applications interpret and handle uncertainty. While aiming for comprehensive coverage, the study has several limitations. Rapid advancements in 5G and sensor technologies mean some challenges noted here might soon become obsolete. Additionally, there are few rigorous assessments of low-cost sensors operating in mmWave bands, leaving important gaps. Lastly, the literature tends to focus on promising prototypes, which may underrepresent failed or inconclusive efforts. This methodology offers a structured foundation for evaluating the validity and relevance of low-cost 5G RF-EMF sensing technologies in both scientific research and societal applications.

3. 5G and RF-EMF Fundamentals

3.1. *Key Technological Features of 5G*

The fifth generation of mobile networks (5G) represents a major redesign of wireless communication systems. It is not just a single technology, but a suite of innovations aimed at providing extremely high data speeds, low latency, and vast device connectivity. Unlike earlier generations, 5G uses two primary frequency ranges: Sub-6 GHz and millimeter-wave (mmWave) bands. The Sub-6 GHz spectrum, usually

https://doi.org/10.3390/s26020533

from 700 MHz to 3.8 GHz, improves on 4G's coverage and capacity. Meanwhile, mmWave frequencies (24–100 GHz) support multi-gigabit data rates but have shorter transmission distances and are more easily blocked by obstacles such as foliage and buildings [27]. To address this propagation issue, 5G utilizes Massive-Input Multiple-Output (MIMO) antenna arrays and beamforming, which actively direct radio beams towards users to optimize signal strength and spectral efficiency [29]. Features like carrier aggregation, which merges multiple frequency blocks to increase throughput, and Dynamic Spectrum Sharing (DSS), enabling the concurrent use of 4G and 5G on the same frequencies, further improve network versatility. These advances considerably change the exposure landscape. Transitioning from a few high-power macrocells to a dense network of low-power small cells creates highly variable, dynamic electromagnetic environments. Power densities fluctuate rapidly due to traffic, user movement, and beam steering. While this network densification is essential for meeting 5G performance targets, it also presents new challenges in measuring exposure, developing models, and effectively communicating with the public [30].

3.2. Principles of RF-EMF Exposure

International exposure guidelines for RF-EMF are primarily established by the International Commission on Non-Ionizing Radiation Protection [31] and the IEEE International Commission on Electromagnetic Safety [32]. Both organizations set exposure limits based on the lowest scientifically validated threshold for adverse health effects, determined from experimental and epidemiological studies, rather than solely focusing on preventing immediate thermal responses. In these frameworks, fundamental basic restriction is expressed as the whole-body-averaged Specific Absorption Rate (SAR), measured in watts per kilogram (W/kg). SAR indicates the energy absorption rate from electromagnetic fields by body tissues. For far-field exposures, the more practical reference levels, electric field strength (V/m) and power density (V/m^2), are derived from SAR limits through computational and experimental methods using anatomically realistic human models [33]. These conversions include conservative reduction factors to account for biological variability among individuals and uncertainties in modeling, ensuring that meeting the reference levels also ensures compliance with the SAR safety thresholds. The WHO's EMF Project coordinates efforts to review scientific evidence and promote global consistency in exposure assessment methods [34,35]. Currently, international safety standards recognize thermal effects as the only scientifically proven health impacts of RF-EMF exposure below regulatory limits [31,36]. While some research groups explore potential non-thermal biological effects, such as oxidative stress or DNA alterations, these results are inconsistent and not considered confirmed health risks by the ICNIRP, IEEE, or WHO. The ICNIRP Data Gap Report highlights these topics for further study but affirms that current exposure limits are sufficiently protective [37]. Nevertheless, ongoing public disclosure and independent reviews, like the EUROPAEM EMF Guidelines and the BioInitiative Working Group, support the need for continued scrutiny of long-term, low-intensity exposure [38,39]. These differing perspectives highlight the dual challenge for policymakers: to uphold traceable, evidence-based standards while transparently responding to evolving scientific knowledge and community concerns. Developing a credible and balanced regulatory framework will depend on sustained interdisciplinary dialogue and openness to new methodologies.

3.3. Existing Exposure Assessment Methods

Accurate RF-EMF exposure assessment relies on reliable measurement tools and standardized methods. Common professional instruments include isotropic E-field probes,

https://doi.org/10.3390/s26020533

spectrum analyzers, and frequency selective meters. Isotropic probes measure fields uniformly along three orthogonal axes, providing orientation-independent readings. Spectrum analyzers offer frequency-resolved data, enabling differentiation between technologies like Wi-Fi, 4G, and 5G New Radio (NR) [40]. Field assessments usually use one of three methodological approaches: driven-by surveys, stationary monitoring stations, and personal dosimeters [29]. Driven-by surveys involve mobile campaigns using equipment mounted on vehicles to capture spatial variations. These are effective for city-scale mapping but are limited to accessible routes and only provide temporal snapshots [41]. Stationary monitoring stations provide continuous long-term measurements at fixed locations, offering high temporal resolution but limited spatial coverage. Greece's National Observatory of Electromagnetic Fields exemplifies this with its nationwide 5G-inclusive network, enabling real-time public reporting [18]. Personal dosimeters are lightweight wearable devices that track individual exposure in daily life. They are vital for epidemiological research, but face calibration difficulties and biases caused by body shielding [29]. Recent research highlights the need for measurement techniques to evolve with the growing technical complexity of 5G. Beamforming, quick power changes, and dynamic spectrum management render traditional averaging insufficient [27]. Additionally, enclosed and reflective settings like vehicles and buildings create standing-wave patterns, which complicate standard probe calibration [42]. Although professional instrumentation remains the standard for accuracy, its cost and logistical requirements limit spatial and temporal coverage. As a result, researchers and regulators increasingly acknowledge the complementary role of low-cost, networked sensors to enhance official monitoring systems. Properly integrating these distributed data streams through calibration, metadata standards, and transparent governance provides a pathway towards more comprehensive and participatory exposure assessment within the 5G environment.

Table 1 compares professional, low-cost, and participatory approaches to RF-EMF exposure monitoring. It highlights a trade-off between accuracy and accessibility: professional systems offer precision but limited transparency and coverage, while citizen and low-cost methods increase reach and participation at the cost of standardization. Bridging these models is crucial for establishing a more credible and democratic framework for 5G exposure assessment. Beyond methodological differences, exposure characteristics vary notably between user equipment (such as smartphones, tablets, and IoT wearables) and mobile infrastructure (for example, base stations and small cells). Measurements taken in public environments typically show that ambient RF-EMF from mobile networks contributes only a small portion of total personal exposure, usually between 5 and 20% of the total dose. User devices tend to dominate short-term peaks, especially during uplink data transmissions. For example, Velghe et al. [29] observed that exposure from a user's own device can be 10–30 dB higher than environmental levels, depending on usage patterns and proximity. Similarly, Remirez-Vazquez et al. [41] reported that base-station power densities in dense urban areas average 0.01–0.1 W/m^2, while localized exposures from handheld devices can transiently reach 1–10 W/m^2 at close proximity to the body. These differences highlight the importance of distinguishing between various sectors—residential, occupational, transportation, and medical—and between user-side and network-side sources when interpreting or validating low-cost sensor data. This distinction ensures that environmental field measurements are not confused with personal device emissions, thereby enhancing scientific accuracy and public communication.

https://doi.org/10.3390/s26020533

Table 1. Comparison of RF-EMF Exposure Assessment Approaches.

Monitoring Approach	Typical Instruments/Methods	Accuracy and Calibration	Spatial/Temporal Coverage	Cost and Accessibility	Data Transparency and Public Engagement	Key Limitations	Sources
Professional Monitoring (regulatory/Research)	Isotropic E-field probes, spectrum analyzers, frequency-selective meters, dosimeters; drive-by surveys, stationary monitoring networks	High—laboratory-calibrated instruments with traceable standards	Limited spatially, high temporal precision (in fixed locations)	High cost, specialized expertise required	Low—data often aggregated or delayed; limited public access	Sparse geographic coverage; episodic campaigns; inaccessible raw data	[18,40,41]
Low-Cost Sensor Networks (Research/Institutional Pilot Projects)	Compact RF power sensors, SDR-based detectors, IoT integrated sensor nodes	Moderate—calibration against reference instruments required; sensitivity varies	High spatial, continuous temporal coverage possible	Low to moderate cost, scalable deployments	Moderate—public dashboards increasingly used (e.g., NOEF Greece, France ANFR trials)	Calibration drift, signal discrimination challenges, environmental noise	[27,42]
Participatory Citizen Science Monitoring	Low-cost handheld EMF meters, smartphone-based detectors, community sensor kits	Variability—depends on sensor type, calibration, user training	Very high spatial, temporal (depends on volunteer activity)	Low cost, open participation	High—fosters trust, data sharing, cocreation of knowledge	Limited accuracy; inconsistent protocols; potential data misuse	[27,29,41]

4. The Landscape of Low-Cost RF-EMF Sensors

4.1. Commercial Solutions

In recent years, a small but expanding market of low-cost RF-EMF sensors has emerged to supplement professional monitoring systems. These commercial options, often aimed at hobbyists or small organizations, offer affordable, continuous exposure measurement with minimal expertise needed [43]. Examples include portable handheld meters, low-power IoT sensor nodes, and reconfigured SDRs. Table 2 lists the main specifications of selected off-the-shelf sensors. These devices generally work across Sub-6 GHz frequencies and sometimes cover the 3.5 GHz 5G NR band (Further examples in Figure 1a–c). Their dynamic range is usually limited (typically 0.1–20 V/m), and their measurement uncertainty tends to be higher than that of professional isotropic probes, generally ±3–6 dB. While some manufacturers claim frequency selectivity, most sensors employ broadband detection, which makes source attribution more challenging. Prices range from under 300 euros for basic handheld meters to 1500–2000 euros for SDR-based multi-band detectors [25].

Table 2. Representative Low-Cost RF-EMF Sensors and Their Technical Characteristics. The table lists selected low-cost and research-grade sensors, along with one professional reference instrument (Narda SRM-3006, Pfullingen, Germany) included for comparative benchmarking.

Device/Platform	Approx. Cost (EUR)	Frequency Range (GHz)	Dynamic Range	Measurement Type	Claimed Accuracy/Uncertainty	Key Features and Notes	References
* Narda SRM-3006 (Professional Reference Instrument)	>10,000	0.1–40 (depends on probe)	~100 (manufacturer spec)	Frequency-selective isotropic probe	±1.5 dB (confirmed in metrology calibration)	Laboratory-calibrated reference instrument used by research institutions, regulators and for ICNIRP compliance; high portability; real-time spectral analysis	[44–48]
Adalm-Pluto SDR Analog Devices, Figure 1b)	~400	0.325–3.8	0.1–20	Broadband/SDR-based	±4–6 dB	Open source SDR platform frequently adapted in research for 5G Sub-6 GHz sensing	[25]

https://doi.org/10.3390/s26020533

Table 2. *Cont.*

Device/Platform	Approx. Cost (EUR)	Frequency Range (GHz)	Dynamic Range	Measurement Type	Claimed Accuracy/Uncertainty	Key Features and Notes	References
ExposureSure Node v3 (IoT sensor)	~800	0.7–6.0	0.2–30	Broadband/IoT node	±3 dB	Cloud-connected EMF sensor designed for distributed network deployment; limited frequency selectivity	[27]
GQ EMF-390 (Consumer Handheld, Figure 1a)	~300	0.1–8.0	0.1–20	Broadband handheld	±5–6 dB	Widely available consumer meter combining EMF, ELF, and RF modes; coarse spectral discrimination	Manufacturer specifications
Low-Cost Triaxial 5G Sensor (Research Prototype)	~1000–1200	3.3–4.2 (5G n77/n78)	0.06–30	Triaxial analog-to-digital design	±3.12 dB	Field-validated against SRM-30006; deviation~2.8 dB within reference uncertainty	[26]

* The Narda SRM-3006 is not a low-cost device but is included as a professional benchmark for comparison with affordable and prototype sensors.

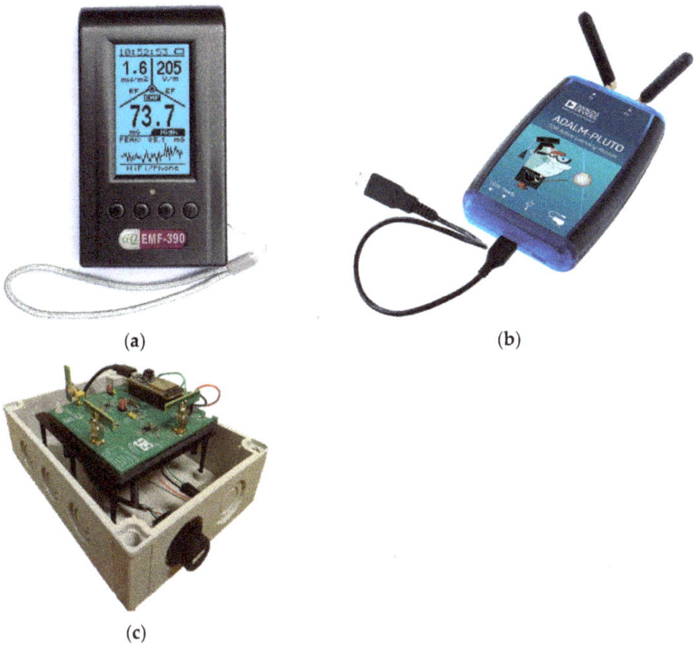

Figure 1. Representative Low-Cost and Research-Grade 5G RF-EMF Sensors. Examples of (**a**) GQ EMF-390 handheld meter (open access image), (**b**) Adla-Pluto SDR platform (image courtesy of manufacturer website and open access) and (**c**) Triaxial 5G RF-EMF research sensor [26].

While these commercial sensors can track overall exposure trends, their accuracy is still significantly lower than laboratory-grade equipment. Validation studies show average deviations of 2.5 to 5 dB when compared to professional monitors like the Narda SRM-3006, a difference that is often within the measurement uncertainty of the reference device [26].

4.2. Academic and Open-Source Prototypes

Alongside commercial innovation, academic and open-source communities have created various research prototypes that use SDRs, IoT chips, and inexpensive triaxial antennas [49,50]. These designs seek to balance cost-effectiveness with scientific openness. For

https://doi.org/10.3390/s26020533

example, researchers from Ghent University and The Hague University of Applied Sciences compared SDR-based sensors, such as the Adalm Pluto, with institutional prototypes like imec-WAVES and Smart Sensor Systems devices. They observed average deviations of only 1.78 dB in laboratory conditions and 5.26 dB in outdoor environments [25]. More recently, a low-cost triaxial 5G sensor designed to measure the n77 and n78 bands (3.3–4.2 GHz) showed a sensitivity of 0.06 V/m and an uncertainty of 3.12 dB, with an average deviation of just 2.8 dB compared to the SRM-3006 reference. This indicates a promising potential for large-scale deployment [26]. The open-source approach of many academic designs enables community validation and continuous improvement. Platforms like GitHub (v. 2.52.0) provide shared firmware, calibration scripts, and data repositories, promoting transparency that proprietary systems lack. This openness aligns with trends in citizen sensing areas like air quality and noise monitoring, where sharing calibration methods and ensuring reproducibility are key to building trust [51].

4.3. Citizen Science Initiatives

Low-cost sensors have also sparked a new wave of citizen science projects aimed at mapping RF exposure in daily environments. These initiatives build on the tradition of participatory mapping that was established earlier with Wi-Fi and 4G networks. An early example is the RF-EMF exposure sensing network in Antwerp, which used eleven fixed and five mobile low-cost sensors to track local exposure patterns. The system generated high-resolution spatial and temporal data, validated with professional equipment, demonstrating the viability of city-wide participatory EMF monitoring [12]. Similar projects have also emerged in France and Greece, where national observatories provide publicly accessible dashboards that combine official and citizen-generated data [18]. Citizen projects not only produce data but also build trust and encourage engagement, providing the public with a tangible way to understand and discuss exposure. Nonetheless, lessons from related environmental sensing fields reveal ongoing challenges: uneven data quality, inconsistent calibration, and unrealistic expectations of empowerment without proper data governance [52]. Achieving sustainable success requires consistent calibration, thorough metadata documentation, and ongoing feedback loops between citizens, scientists, and regulators.

4.4. Benchmarking Against Professional Equipment

Comparative validation is essential for confirming the reliability of low-cost sensors. Benchmarks conducted in labs and real-world settings regularly show accuracy differences that are significant but manageable. In a 2023 multi-lab comparison, low-cost 5G sensors differed from professional probes by an average of 1.78 dB and up to 5.8 dB in field tests [25]. Another validation in 2025 found deviations within 2.8 dB, which is within the uncertainty range of the commercial reference device [26]. Comparative validations offer specific insights into the quantitative performance of low-cost 5G sensors. Dprez et al. [25] compared four SDR-based prototypes and found mean deviations of 1.78 dB in controlled laboratory conditions and up to 5.26 dB in outdoor field trials, chiefly due to temperature drift and multipath interference. Similarly, Van der Straeten et al. [26] validated a triaxial 5G sensor (3.3–4.2 GHz) against the Narda SRM-3006 reference, observing an average deviation of 2.8 dB, well within the reference uncertainty (±3 dB). Aerts et al. [9] reported that in a distributed RF-EMF sensor network of 35 nodes, long-term calibration drift reached approximately 0.4 dB per month when uncorrected, highlighting the need for periodic co-location checks. These examples demonstrate that, when properly calibrated, Sub-6 GHz low-cost sensors can maintain total uncertainty below ±5 dB, which is acceptable for environmental monitoring though not yet sufficient for regulatory com-

https://doi.org/10.3390/s26020533

pliance. Although low-cost sensors show promising results, they face challenges such as signal drift, temperature sensitivity, and calibration decay over time [26,53,54]. Without frequent recalibration or reference checks, measurement uncertainty can exceed 5–7 dB, affecting comparability. Additionally, their frequency selectivity is limited: Sub-6 GHz sensors are becoming more reliable, but mmWave bands (>24 GHz) remain mostly out of reach for low-cost devices due to limitations in antennas and components [27]. When properly calibrated and transparently documented, low-cost sensors offer useful supplementary data to professional systems. They are especially valuable for detecting spatial exposure gradients, temporal changes, and public risk perceptions. Recent studies have extended this benchmarking work to incorporate intelligent mapping and hybrid measurement frameworks. For example, Najera et al. [28] compared broadband sensors and personal exposimeters using evolutionary programming optimization to develop high-resolution EMF exposure maps. Their results indicated that hybrid calibration between static broadband sensors and wearable exposimeters reduced spatial error by approximately 25% compared to standalone devices. Such research illustrates how algorithmic correction and data fusion can compensate for inherent uncertainties in low-cost sensors, aligning well with the discussion on machine-learning-based calibration in Section 5.1. Similarly, ongoing projects in Spain and Belgium demonstrate how integrating software intelligence into sensor networks can significantly enhance spatial accuracy and data usability. The way ahead is not about replacing reference-grade equipment but about integrating citizen-generated and institutional data within a well-organized, quality-controlled monitoring framework.

5. Core Challenges in Low-Cost 5G Sensing

5.1. Metrological and Technical Hurdles

Despite notable progress, inexpensive 5G RF-EMF sensors still encounter key metrological challenges that affect their long-term reliability and regulatory approval. The primary concern is calibration precision and the impact of temporal drift [25]. While initially calibrated against reference instruments, many sensors experience calibration loss over time caused by temperature changes, component aging, and electromagnetic interference [53,55,56]. Laboratory tests show deviations of 1.78–5.26 dB when compared to professional devices like the Narda SRM-3006, although these are generally within acceptable ranges, they are still vulnerable to drift during prolonged outdoor use [25,26]. Frequency coverage gaps further hinder their ability to accurately represent conditions. Most low-cost sensors work within the Sub-6 GHz bands but cannot detect millimeter-wave (mmWave) emissions (24–100 GHz), which are becoming more important for urban 5G networks [57,58]. Although experimental integrated-circuit transceivers have shown they can operate cost-effectively up to 39 GHz [59], no open-source or consumer sensor provides the same stability or beam tracking capabilities at these higher frequencies. Ongoing challenges include the limits of dynamic range and detection thresholds. Budget devices typically saturate at around 20 V/m and lose sensitivity below 0.05 V/m. Environmental factors in 5G settings—such as beam switching, Time Division Duplexing (TDD), and Dynamic Spectrum Sharing (DSS)—further complicate measurements [60]. While standard averaging intervals, like 6 min ICNIRP windows, can obscure transient peaks, accurately capturing these rapid fluctuations often requires higher sampling rates and more precise synchronization than low-cost analogue-to-digital converters generally offer [27]. Empirical studies consistently quantify calibration drift and measurement uncertainty in practical deployments. For example, the Adalm-Pluto SDR platform exhibited drift of up to 3 dB after six months of outdoor use [25], while temperature variations of ±10 °C caused a 0.5–0.8 dB bias [53]. In contrast, the triaxial 5G sensor by Van der Straeten et al. [26] achieved stability within ±0.2 dB over 24 h, showing that environmental compensation

https://doi.org/10.3390/s26020533

circuits can significantly improve consistency. Most consumer-grade broadband meters (e.g., GQ EMF-390) report total measurement uncertainty of ±5–6 dB, roughly double that of professional equipment (±2–3 dB). These quantified differences underscore the necessity of periodic reference calibration and drift correction for long-term reliability in citizen and research deployments. Recent developments in related sensing fields indicate that machine learning-based calibration may help reduce systematic biases and environmental drift. Neural networks and random forest models have already increased the accuracy of low-cost air-quality sensors by 30–50% over linear models [61]. Applying such adaptive calibration techniques to RF-EMF monitoring—potentially integrating ambient temperature, humidity, and device orientation—represents an open yet promising area for future research.

5.2. Data Quality and Interpretation Obstacles

Apart from technical accuracy, the reliability of low-cost FR-EMF datasets is compromised by inconsistent methods and varying contexts. The absence of standardized measurement protocols for averaging, sampling frequency, and calibration leads to datasets that cannot be easily compared across different projects and countries [52]. Additionally, without standardized metadata or uncertainty reports, raw field data risks being misinterpreted or misused in public discussions. Furthermore, the variability of 5G emissions over time and space is significantly higher than that of legacy systems. Techniques like adaptive beamforming and traffic-based power adjustments lead to exposure levels that can fluctuate by several decibels within seconds or between neighboring micro-cells [25]. This variability complicates efforts to obtain accurate average exposure measurements from limited datasets. Another challenge is the gap between environmental field strength and personal exposure. While fixed or handheld sensors record ambient conditions, individual absorption varies greatly based on body orientation, closeness to devices, and shielding from clothing or surroundings. Studies with dosimeters reveal that personal exposure from one's own mobile device can be one to two orders of magnitude higher than environmental measurements [29]. This disparity complicates epidemiological studies and public communication, as people often believe environmental readings directly indicate their personal dose. Standardizing data interpretation will thus need multi-modal methods that integrate affordable ambient sensors, personal dosimeters, and model-driven exposure assessments. Advances in spatial calibration techniques for sensor networks, like those used in air-quality monitoring, demonstrate that real-time bias adjustment and consistency across sites can be achieved with co-location learning models [62]. Such frameworks could also enhance citizen-driven RF-EMF networks.

5.3. Practical and Ethical Considerations

Deploying inexpensive sensors on a large scale also encounters practical and ethical challenges beyond measurement accuracy. Operational issues include unreliable power supply and connectivity; battery-powered sensors experience voltage drift that impacts sensitivity, while WiFi or LoRaWAN networks can cause latency and data losses [9,63–65]. Environmental factors such as heat, humidity, and rain reduce sensor durability, resulting in calibration issues or missing data [52]. The most pressing ethical concern is public misunderstanding. The bursty nature of 5G transmissions can lead to brief, high-amplitude peaks that exceed average exposure limits by several decibels, yet do not reflect genuine health risks. Without proper understanding, untrained users might interpret these raw, unaveraged peaks as hazardous, potentially spreading misinformation and leading to distrust [27]. Finally, privacy issues emerge due to the spatial granularity of dense sensor networks. Geolocated RF data can indirectly disclose device usage patterns, network

https://doi.org/10.3390/s26020533

congestion, or personal activities. Without explicit data-handling policies, community monitoring networks may breach privacy expectations or regulatory standards like the EU's GDPR. Ethical governance in citizen air-quality projects highlights the importance of anonymization, data aggregation, and community consent to ensure sustainable participatory sensing [66]. The development of affordable 5G sensing technologies depends not only on technological advancements but also on upholding epistemic rigor and ethical responsibility. To produce credible, useful, and socially accepted data, a coordinated approach is necessary, combining precise technical calibration, standardized methods for data interpretation, and effective public communication strategies. Table 3 summarizes the main technical, interpretive, and ethical challenges encountered in low-cost RF-EMF sensing. Although measurement accuracy is still below professional levels, new approaches, such as machine learning calibration, colocation validation, and participatory governance, indicate a developing ecosystem that strives to balance precision with accessibility.

Table 3. Summary of Core Challenges in Low-Cost 5G RF-EMF Sensing and Emerging Mitigation Strategies.

Challenge Domain	Key Issues	Typical Impact	Emerging Mitigation Strategies	Reference
Metrological and technical aspects	Calibration drift, frequency coverage gaps (especially mmWave); limited dynamic range; inability to capture transient 5G signals (beamforming, DSS)	Measurement uncertainty ±3–6 dB; field calibration drift ~0.3–0.5 dB/month; temperature bias up to 0.8 dB.	Periodic co-location reference probes; temperature-compensated circuits; ML-based calibration and drift correction	[9,25,26,53,67]
Data quality and interpretation	Lack of harmonized measurement protocols; inconsistent metadata; temporal variability; mismatch between ambient and personal exposure.	Non-comparable datasets; risk of public misinterpretation; limited integration into official systems.	Standardized metadata templates; multi-modal sensing (ambient + personal); spatial co-location and bias correction frameworks.	[29,52,59]
Practical and ethical considerations	Power supply and network instability; weather-related degradation; risk of public alarm from transient peaks; privacy and geolocation concerns.	Data loss; public distrust; potential privacy violations.	Durable enclosures and energy harvesting; automated time-averaging and filtering; anonymization and aggregation data publishing.	[27,66]

Table 3 highlights that developing affordable 5G RF-EMF sensing faces not only technical challenges but also epistemic and governance issues. Addressing calibration drift, protocol harmonization, and ethical data stewardship is crucial for these tools to transition from experimental prototypes to dependable additions to professional monitoring. The following section discusses the main research gaps in measurement science, data interoperability, and participatory governance that must be tackled to fully realize the potential of low-cost sensing for transparent and equitable 5G exposure assessment.

https://doi.org/10.3390/s26020533

6. Identifying Critical Research Gaps

6.1. Metrology and Sensor Design

The future of affordable 5G RF-EMF sensing depends on developing strict metrological standards comparable to professional tools. Although prototype devices show promising accuracy with deviations as low as 1.78 to 2.8 dB when compared to laboratory-grade probes [25,26], there is still no standardized validation or calibration protocol tailored for 5G's dynamic, beam-formed, and bursty signals. Currently, calibration methods rely on steady-state exposure in GTEM or anechoic chambers, which do not reflect real-world temporal modulation patterns. The field urgently needs calibration frameworks that address time-domain behavior, polarization diversity, and angular dependence specific to 5G New Radio (NR) emissions [27]. Expanding sensor frequency capabilities into the mmWave range (24–100 GHz) is equally urgent. Currently, low-cost sensors rarely go beyond 6 GHz, creating a gap in urban 5G deployments relying on FR2 bands. Recent advancements in low-cost intermediate-frequency transceivers built on 14 nm FinFET technology indicate that compact, flexible calibration systems for 39 GHz are now achievable [59,68]. The main engineering hurdle remains adapting these cost-effective innovations into mmWave isotropic probes for exposure monitoring. Another measurement challenge involves algorithmic error correction and vector fusion across different axes. While triaxial sensors are an advancement over single-axis devices, they still face issues like inter-axis coupling and inconsistent angular response. Using adaptive calibration matrices and machine-learning-based error fusion, similar to methods tested in low-cost air-quality sensors, could help reduce anisotropy and enhance isotropic accuracy [69]. Creating open, shareable calibration datasets and reference geometries would further foster a consensus in measurement standards.

6.2. Data Science and Modeling

While hardware advancements are essential, the current research gap lies in data fusion and modeling, which involves converting raw field strength data into scientifically valid and comparable exposure metrics. An immediate priority is combining low-cost data with high-quality professional measurements to improve the reliability of the entire network. Developing multilevel fusion frameworks that weight low-cost observations based on uncertainty or environmental factors is essential to avoid bias amplification when integrating heterogeneous data sources [70]. Another area of development involves creating algorithms that address angular dependence and variability in beamforming. Most low-cost sensors currently assume isotropic incident fields, leading to an underestimation of the anisotropic exposure patterns caused by 5G massive-MIMO and beam steering. Using simulation-based inverse modeling—integrating antenna radiation patterns with time-resolved power logs—could enable dynamic estimation of directional corrections [71,72]. Converting static measurements into realistic personal exposure metrics is still a complex challenge, both conceptually and computationally. Environmental readings indicate potential exposure but translating these into SAR or dose proxies involves modeling factors such as body posture, device proximity, and multipath signal propagation [29]. Combining ambient field maps with population mobility data could help generate probabilistic personal exposure estimates, which is an essential step for practical epidemiological applications. Finally, there is a lack of systematic uncertainty qualification pipelines. Few studies publish reproducible uncertainty budgets or use automated error propagation tools. Incorporating probabilistic modeling frameworks from air quality and radiation dosimetry could provide transparent uncertainty metrics for each observation. Additionally, advances in automated calibration methods, such as Random Forest and XGBoost models [69], should be applied to RF-EMF datasets. This would enable real-time bias correction and confidence estimation.

https://doi.org/10.3390/s26020533

6.3. Implementation and Policy Frameworks

The process of achieving societal and regulatory integration depends on effective implementation logistics and governance structures. At present, there is no agreed-upon approach for maintaining, recalibrating, or auditing large-scale low-cost sensing networks over time. Sustained field deployments need plans for regular recalibration, energy independence, and data synchronization to maintain measurement accuracy. Regulator acceptance of citizen-sourced exposure data is still an emerging area. Although environmental agencies are gradually acknowledging the importance of participatory air-quality data, there is no similar policy framework for EMF exposure reporting. Developing reliable calibration methods, open metadata standards, and interoperable APIs could facilitate the incorporation of verified citizen datasets into national observatories. Ethical public communication requires careful design. Open dashboards, if not well-explained, may cause anxiety by highlighting short-term changes without averaging or uncertainty markers. Using risk communication principles (such as color scales matching ICNIRP thresholds and clear metadata) can improve understanding and prevent alarm. Citizen air-sensing network frameworks focus on transparency, obtaining consent, and ensuring data anonymization [66]. A commonly overlooked issue is the lack of standardized reporting practices. Presently, there is no unified template defining key calibration details, sampling intervals, averaging durations, or uncertainty estimates for low-cost EMF sensors. Creating such standards, inspired by FAIR (Findable, Accessible, Interoperable, Reusable) data principles, would greatly improve reproducibility and support regulatory acceptance. Recent European research initiatives have advanced the standardization of RF-EMF exposure monitoring, especially under 5G conditions. The SEAWAVE (Systematic Review on EMF Exposure and 5G) project, supported by Horizon Europe, systematically reviews exposure metrics, biological effects, and epidemiological evidence related to 5G, aiding in the improvement of health risk assessment methods [73–75]. Additionally, the GOLIAT (Global System for Mobile Communication Exposure Assessment and Epidemiology) project develops harmonized exposure models, personal monitoring protocols, and cross-national epidemiological frameworks to connect population-level and environmental measurements [76,77]. These efforts lay an essential groundwork for incorporating participatory and low-cost sensing systems into scientifically sound and policy-relevant exposure assessments across Europe.

7. Discussion and Future Directions

7.1. Technical Context of 5G and Wi-Fi Standards in Exposure Monitoring

The studies analyzed indicate that affordable RF-EMF sensors need to accommodate the spectral and signaling characteristics of both 5G New Radio (NR) and WiFi (IEEE 802.11) technologies. 5G NR operates across two frequency ranges: Frequency Range 1 (FR1, 0.7–6 GHz) and Frequency Range 2 (FR2, 24–100 GHz), utilizing advanced features like massive-MIMO beamforming, dynamic spectrum sharing, time-division duplexing (TDD), and carrier aggregation. Environmental exposures in FR1 typically range from 0.01 to 0.5 W/m^2, with localized peaks reaching up to 2 W/m^2 near small cells [9,25]. Wi-Fi technologies (Wi-Fi 4 at 2.4/5 GHz, Wi-Fi 5 at 5 GHz, Wi-Fi 6/6E spanning 2.4–7 GHz, and HaLow at 0.9 GHz) emit intermittently, usually with duty cycles below 10%, resulting in highly variable instantaneous field strengths but relatively low average exposures [29,40]. Research indicates that low-cost Sub-6 GHz sensors can effectively monitor both 5G FR1 and Wi-Fi 4–6E, achieving accuracy levels of $\pm$3–6 dB. However, measurements at mmWave frequencies (FR2 above 24 GHz and WiFi 7/8 around 60 GHz) are still constrained by antenna design and component costs [26,27]. The most successful sensor setups combine triaxial isotropic antennas, SDR-based signal processing, and adaptive calibration to adapt to the dynamic temporal and spectral features of these standards. Incorporating this under-

 https://doi.org/10.3390/s26020533

standing of radio interference characteristics ensures that exposure assessments remain technically rigorous and that the discussion of results accurately reflects the operational realities of current wireless systems.

7.2. An Interdisciplinary Agenda

The future of low-cost 5G RF-EMF sensing relies on strong interdisciplinary collaboration. Engineers, data scientists, epidemiologists, and regulators each contribute a crucial part to the exposure-monitoring puzzle. Engineers are pushing hardware limits by creating triaxial isotropic sensors and mmWave-capable probes with less than 3 dB uncertainty [26,68]. Meanwhile, data scientists are designing machine learning models to correct biases, fill spatial gaps, and reconstruct transient 5G beam patterns [78]. Epidemiologists and exposure scientists need to work together to convert raw sensor data into metrics that are relevant for dose assessment, closing the gap between environmental measurements and individual exposure estimates [29]. Regulators and standards organizations (such as ICNIRP, IEC, and CENELEC) must then verify that participatory data align with safety limits and uncertainty considerations. Interdisciplinary collaboration will determine whether low-cost sensing can become a reliable supplement to official monitoring or if it remains a scattered citizen science effort lacking policy influence.

7.3. Differentiated Applications

The value of low-cost RF-EMF sensing depends on defining its use cases, as various applications require different performance standards [25]. For monitoring regulatory compliance, accuracy and traceability are essential. Devices need calibration traceable to national metrology standards and must maintain measurement uncertainty within ±2–3 dB [79]. For public engagement and risk communication, transparency, accessibility, and interpretability are more important than absolute accuracy [80]. Community dashboards, like those tested in Greece's National Observatory of EMF Exposure, present promising models for participatory data sharing [18], as long as they clearly show contextual uncertainty and averaging windows. In health research, ensuring data reliability, temporal detail, and contextual metadata is crucial. Longitudinal exposure studies should use harmonized field strength measurements with consistent averaging intervals (e.g., 6 min ICNIRP standards) and incorporate robust methods for uncertainty propagation [27]. By differentiating these application layers—compliance, communication, and research—the field can better align sensor design, calibration depth, and governance requirements.

7.4. Case Study Vignettes

Recent European initiatives demonstrate both the potential and constraints of affordable 5G sensing. In Belgium and the Netherlands, the Smart Sensor Systems (S^3R) network illustrates a hybrid citizen-academic approach, integrating affordable nodes with official oversight. These networks deliver spatial resolutions that surpass those of traditional campaigns, while deviations stay below 5 dB compared to certified instruments [25]. Greece's EMF Observatory at the national level combines data from both citizens and professionals in public dashboards, supported by calibration campaigns led by universities. Its success is due to a two-tier validation process, where citizen-collected data is cross-checked with data from certified sensors [18]. Laboratory benchmarking continues to validate field results. In controlled comparisons between the SRM-3006 professional meter and the triaxial low-cost 5G sensor by Ven der Straeten et. al, the triaxial low-cost 5G sensor showed deviations of only 2.8 dB, which falls within the commercial system's uncertainty range [26]. These findings demonstrate that, with thorough calibration and quality control, affordable sensors can effectively supplement professional infrastructure both scientifically and socially.

 https://doi.org/10.3390/s26020533

Qualitatively, these systems exhibit deviations of 2 to 5 dB from reference instruments and long-term drift rates under 0.5 dB per month. This confirms that modern low-cost sensors nearly match professional stability in real-world deployment.

In addition to the European citizen sensing initiatives discussed earlier, international research teams have achieved significant progress in high-resolution modeling of RF-EMF power density distributions. Manassas et al. [81] developed a machine learning framework that combines measurement data and GIS parameters to predict electromagnetic field exposure in complex urban environments, demonstrating how data-driven modeling can improve spatial estimation.

Similarly, Salem et al. [82] employed 3D ray-tracing and beamforming simulations to analyze uplink and downlink exposure in dense urban 5G networks, emphasizing the significant spatial variability in predicted power density and SAR across city environments. Ramirez-Vazquez et al. [41] conducted a systematic review and comparison of spot-measurement and mixed-method studies, illustrating how combined fixed and mobile measurements can effectively capture the variability of real-world RF exposure patterns. Moreover, comprehensive measurement-based assessments [83] using distributed drive-test networks have shown how combining stationary sensors with mobile surveys can produce detailed spatial maps of 5G downlink emissions. These high-precision modeling efforts deliver critical baseline data for understanding exposure heterogeneity but remain resource-intensive and are generally confined to research or regulatory contexts. In contrast, low-cost and participatory monitoring systems can provide complementary, distributed, and continuously updated exposure data at minimal cost, fostering transparency and community engagement.

Recent studies, including Najera et al. [28], show that combining personal dosimeters with static low-cost sensor nodes produces more accurate and stable EMF exposure maps over time than using either method alone. Their use of evolutionary programming to optimize sensor placement and calibration highlights a promising direction in low-cost sensing research. Incorporating such computational techniques into participatory networks could significantly improve spatial resolution and scientific reliability. This trend indicates that low-cost sensing is advancing from conceptual ideas to validated hybrid systems, supporting the practical potential discussed throughout the review. This synergy between top-down modeling and bottom-up sensing forms the methodological bridge this review aims to emphasize, highlighting the importance of integrating democratized data collection with advanced scientific modeling to achieve comprehensive 5G exposure assessment.

7.5. Standardization and Interoperability

The most significant factor for long-term impact is standardization and interoperability. At present, low-cost sensing exists within a fragmented ecosystem, with each research group using unique sampling frequencies, calibration protocols, and metadata schemas [27]. An international effort similar to CONSORT (for clinical trials) or PRISMA (for systematic reviews) could establish minimal reporting standards, including the calibration chain, frequency coverage, averaging interval, uncertainty budget, and data format. Without such frameworks, cross-study comparability and regulatory integration will remain limited. The efforts by the European Telecommunication Standards Institute (ETSI) and IEEE EMC committees already indicate a move toward standardizing RF-EMF measurement procedures for 5G small-cell environments [84]. However, extending these standards to low-cost, distributed networks will necessitate open metadata schemas and machine-readable calibration certificates. Insights from the FAIR data principles and citizen science codes of conduct highlight that transparency, traceability, and interoperability should be integrated from the start, rather than as afterthoughts.

 https://doi.org/10.3390/s26020533

8. Conclusions

The review examined over 60 studies on low-cost 5G and Wi-Fi RF-EMF exposure sensors, showing that most Sub-6 GHz systems have measurement deviations of ±3–6 dB compared to certified reference meters. Long-term outdoor deployments report calibration drift of less than 0.5 dB per month, while indoor tests show reproducibility within 5% RMs error. Typical ambient 5G FR1 exposure levels ranged from 0.01 to 0.5 W/m^2, rising to about 2 W/m^2 near small cells. Personal handset use can cause transient exposures 10 to 30 dB higher. These data demonstrate that affordable sensors, when properly calibrated, can approximate professional accuracy for community monitoring of 5G and Wi-Fi 4–6e fields. Nevertheless, measurement challenges persist for FR2 frequencies above 24 GHz and Wi-Fi 7/8 (60 GHz) due to antenna and component limitations. Future efforts should focus on mmWave calibration, standardized uncertainty assessment, and data-integration protocols to support transparent, large-scale exposure mapping.

Author Contributions: Conceptualization, M.K. and P.C.R.; methodology, M.K. and P.C.R.; validation, M.K. and P.C.R.; investigation, M.K. and P.C.R.; writing—original draft preparation, M.K. and P.C.R.; writing—review and editing, M.K. and P.C.R. All authors have read and agreed to the published version of the manuscript.

Funding: This study received no external funding.

Institutional Review Board Statement: Not applicable.

Informed Consent Statement: Not applicable.

Data Availability Statement: Not applicable.

Conflicts of Interest: The authors declare no conflicts of interest.

References

1. Bushberg, J.; Chou, C.; Foster, K.R.; Kavet, R.; Maxson, D.; Tell, R.A.; Ziskin, M.C. IEEE Committee on Man and Radiation—COMAR Technical Information Statement: Health and Safety Issues Concerning Exposure of the General Public to Electromagnetic Energy from 5G Wireless Communications Networks. *Health Phys.* **2020**, *119*, 236–246. [CrossRef] [PubMed]
2. Liu, Y.; Hu, Q.; Blough, D.M. Joint Link-Level and Network-Level Reconfiguration for mmWave Backhaul Survivability in Urban Environments. In Proceedings of the 22nd International ACM Conference on Modeling, Analysis and Simulation of Wireless and Mobile Systems, Miami Beach, FL, USA, 25–29 November 2019; Association for Computing Machinery: New York, NY, USA, 2019; pp. 143–151.
3. Meese, J.; Hegarty, K.; Wilken, R.; Yang, F.; Middleton, C. 5G and Urban Amenity: Regulatory Trends and Local Government Responses around Small Cell Deployment. *Digit. Policy Regul. Gov.* **2024**, *26*, 601–616. [CrossRef]
4. Russell, C.L. 5G Wireless Telecommunications Expansion: Public Health and Environmental Implications. *Environ. Res.* **2018**, *165*, 484–495. [CrossRef]
5. Rowley, J.; Mazar, H. Misunderstandings about Radiofrequency Electromagnetic Field Exposure and 5G Misinformation. In Proceedings of the 2021 IEEE International Conference on Microwaves, Antennas, Communications and Electronic Systems (COMCAS), Tel Aviv, Israel, 1–3 November 2021; pp. 219–221.
6. Wu, X.; Ma, L.; Low, D.; Sharma, S.; Papyshev, G. Beyond Precautionary Principle: Policy-Making under Uncertainty and Complexity. *Policy Des. Pract.* **2024**, *7*, 1–16. [CrossRef]
7. Frank, J. Electromagnetic Fields, 5G and Health: What about the Precautionary Principle? *J. Epidemiol. Community Health* **2021**, *75*, 562–566. [CrossRef]
8. Bouakkaz, S.; Suárez, L.; Cuppens, N.; Cuppens, F. Design of an Intelligent Trust Management Architecture for 5G Service Deployment. In Proceedings of the 11th International Conference on Information Systems Security and Privacy (ICISSP 2025), Porto, Portugal, 20–22 February 2025; pp. 310–317.
9. Aerts, S.; Vermeeren, G.; Van den Bossche, M.; Aminzadeh, R.; Verloock, L.; Thielens, A.; Leroux, P.; Bergs, J.; Braem, B.; Philippron, A.; et al. Lessons Learned from a Distributed RF-EMF Sensor Network. *Sensors* **2022**, *22*, 1715. [CrossRef] [PubMed]
10. Calvo Palomino, R. Towards Large-Scale and Collaborative Spectrum Monitoring Systems Using IoT Devices. Ph.D. Thesis, Universidad Carlos III de Madrid, Getafe, Spain, 2019.

 https://doi.org/10.3390/s26020533

11. Zmyślony, M.; Bieńkowski, P.; Bortkiewicz, A.; Karpowicz, J.; Kieliszek, J.; Politański, P.; Rydzyński, K. Protection of the Population Health from Electromagnetic Hazards—Challenges Resulting from the Implementation of the 5G Network Planned in Poland. *Med. Pr.* **2019**, *71*, 105–113. [CrossRef]
12. Vermeeren, G.; Van den Bossche, M.; Aminzadeh, R.; Aerts, S.; Leroux, P.; De Meyer, M.; Bergs, J.; Philippron, A.; Martens, L.; Joseph, W. RF EMF Exposure Sensing Network in a Smart-City Context. In Proceedings of the Joint Annual Meeting of the Bioelectromagnetics Society and the European BioElectromagnetics Association (BioEM2019), Montpellier, France, 23–28 June 2019; Bioelectromagnetics Society (BEMS): Columbia, SC, USA, 2019; pp. 239–243.
13. Riter, K.; Doraiswamy, P.; Clayton, A.C.; Rountree, K. *Solar Station for an Off-the-Grid Air Quality Sensor System*; RTI Press: Research Triangle Park, NC, USA, 2023.
14. Abe, Y. Temporal Citizen Science After Fukushima. *Int. J. Commun.* **2023**, *17*, 19.
15. Hultquist, C.; Oravecz, Z.; Cervone, G. A Bayesian Approach to Estimate the Spatial Distribution of Crowdsourced Radiation Measurements around Fukushima. *ISPRS Int. J. Geo-Inf.* **2021**, *10*, 822. [CrossRef]
16. Charvát, K.; Kepka, M. Crowdsourced Data. In *Big Data in Bioeconomy*; Springer: Berlin/Heidelberg, Germany, 2021. [CrossRef]
17. Seltzer, E.; Mahmoudi, D. Citizen Participation, Open Innovation, and Crowdsourcing: Challenges and Opportunities for Planning. *J. Plan. Lit.* **2013**, *28*, 3–18. [CrossRef]
18. Christopoulou, M.I.; Kyritsi, T.; Yalofas, A.; Koutounidis, D.; Karabetsos, E. 5G NR Launching in Greece: Preliminary in Situ and Monitoring Network Measurements of Electromagnetic Fields Exposure Levels at Rooftops. *Bioelectromagnetics* **2024**, *45*, 193–199. [CrossRef]
19. McCrory, G.; Veeckman, C.; Claeys, L. Citizen Science Is in the Air—Engagement Mechanisms from Technology-Mediated Citizen Science Projects Addressing Air Pollution. In Proceedings of the International Conference on Internet Science, Thessaloniki, Greece, 22–24 November 2017; Kompatsiaris, I., Cave, J., Satsiou, A., Carle, G., Passani, A., Kontopoulos, E., Diplaris, S., McMillan, D., Eds.; Springer International Publishing: Cham, Switzerland, 2017; pp. 28–38.
20. Campmier, M.J.; Gingrich, J.; Singh, S.; Baig, N.; Gani, S.; Upadhya, A.; Agrawal, P.; Kushwaha, M.; Mishra, H.R.; Pillarisetti, A.; et al. Seasonally Optimized Calibrations Improve Low-Cost Sensor Performance: Long-Term Field Evaluation of PurpleAir Sensors in Urban and Rural India. *Atmos. Meas. Tech.* **2023**, *16*, 4357–4374. [CrossRef]
21. Lassalle, M.W. The Present State of Low-Cost Air Quality Sensors in Japan and Their Accuracy. *Int. J. Environ. Stud.* **2024**, *81*, 1665–1683. [CrossRef]
22. Kenens, J.; Van Oudheusden, M.; van Hoyweghen, I.; Mizushima, N. Nonscalability of "Citizen Science" in Post-Fukushima Japan: Unpacking Articulations of Citizen Radiation Measuring Organizations. *Public Underst. Sci.* **2022**, *31*, 507–523. [CrossRef] [PubMed]
23. Kimura, A.H. Citizen Science in Post-Fukushima Japan: The Gendered Scientization of Radiation Measurement. *Sci. Cult.* **2019**, *28*, 327–350. [CrossRef]
24. da Schio, N. The Empowering Virtues of Citizen Science: Claiming Clean Air in Brussels. *Engag. Sci. Technol. Soc.* **2022**, *8*, 29–52. [CrossRef]
25. Deprez, K.; Colussi, L.; Korkmaz, E.; Aerts, S.; Land, D.; Littel, S.; Verloock, L.; Plets, D.; Joseph, W.; Bolte, J. Comparison of Low-Cost 5G Electromagnetic Field Sensors. *Sensors* **2023**, *23*, 3312. [CrossRef]
26. Van der Straeten, J.; Van Bladel, H.; Deprez, K.; Joseph, W.; Vermeeren, G. Design and Validation of a Low-Cost Triaxial 5G RF-EMF Exposure Sensor. *IEEE Sens. J.* **2025**, *25*, 16050–16060. [CrossRef]
27. Korkmaz, E.; Aerts, S.; Coesoij, R.; Bhatt, C.R.; Velghe, M.; Colussi, L.; Land, D.; Petroulakis, N.; Spirito, M.; Bolte, J. A Comprehensive Review of 5G NR RF-EMF Exposure Assessment Technologies: Fundamentals, Advancements, Challenges, Niches, and Implications. *Environ. Res.* **2024**, *260*, 119524. [CrossRef]
28. Nájera, A.; Sánchez-Montero, R.; González-Rubio, J.; Guillén-Pina, J.; Chocano-del-Cerro, R.; López-Espí, P.-L. Comparison Between Broadband and Personal Exposimeter Measurements for EMF Exposure Map Development Using Evolutionary Programming. *Appl. Sci.* **2025**, *15*, 7471. [CrossRef]
29. Velghe, M.; Aerts, S.; Martens, L.; Joseph, W.; Thielens, A. Protocol for Personal RF-EMF Exposure Measurement Studies in 5th Generation Telecommunication Networks. *Environ. Health* **2021**, *20*, 36. [CrossRef]
30. Salem, M.A.; Lim, H.S.; Diong, K.S.; Alaghbari, K.A.; Zarakovitis, C.C.; Chien, S.F. Electromagnetic Field-Aware Radio Resource Management for 5G and Beyond: A Survey. *Computers* **2025**, *14*, 51. [CrossRef]
31. International Commission on Non-Ionizing Radiation Protection (ICNIRP). Guidelines for Limiting Exposure to Electromagnetic Fields (100 kHz to 300 GHz). *Health Phys.* **2020**, *118*, 483–524. [CrossRef]
32. Bailey, W.H.; Bodemann, R.; Bushberg, J.; Chou, C.-K.; Cleveland, R.; Faraone, A.; Foster, K.R.; Gettman, K.E.; Graf, K.; Harrington, T.; et al. Synopsis of IEEE Std C95.1^{TM}-2019 "IEEE Standard for Safety Levels With Respect to Human Exposure to Electric, Magnetic, and Electromagnetic Fields, 0 Hz to 300 GHz". *IEEE Access* **2019**, *7*, 171346–171356. [CrossRef]
33. Kodera, S.; Taguchi, K.; Diao, Y.; Kashiwa, T.; Hirata, A. Computation of Whole-Body Average SAR in Realistic Human Models From 1 to 100 GHz. *IEEE Trans. Microw. Theory Tech.* **2024**, *72*, 91–100. [CrossRef]

 https://doi.org/10.3390/s26020533

34. Chou, C.K. Controversy in Electromagnetic Safety. *Int. J. Environ. Res. Public Health* **2022**, *19*, 16942. [CrossRef] [PubMed]
35. Lin, J.C. Health and Safety Practices and Policies Concerning Human Exposure to RF/Microwave Radiation. *Front. Public Health* **2025**, *13*, 1619781. [CrossRef] [PubMed]
36. Foster, K.R.; Ziskin, M.C.; Balzano, Q. Time-Temperature Thresholds and Safety Factors for Thermal Hazards from Radiofrequency Energy above 6 GHz. *Health Phys.* **2021**, *121*, 234. [CrossRef]
37. International Commission on Non-Ionizing Radiation Protection (ICNIRP). Gaps in Knowledge Relevant to the "ICNIRP Guidelines for Limiting Exposure to Time-Varying Electric, Magnetic and Electromagnetic Fields (100 kHz TO 300 GHz)". *Health Phys.* **2025**, *128*, 190–202. [CrossRef]
38. Belyaev, I.; Dean, A.; Eger, H.; Hubmann, G.; Jandrisovits, R.; Kern, M.; Kundi, M.; Moshammer, H.; Lercher, P.; Müller, K.; et al. EUROPAEM EMF Guideline 2016 for the Prevention, Diagnosis and Treatment of EMF-Related Health Problems and Illnesses. *Rev. Environ. Health* **2016**, *31*, 363–397. [CrossRef]
39. Bioadmin. *The BioInitiative 2012 Report—Health Risks from EMF & RFR Exposure 2012*; BioInitiative Working Group: Columbia, SC, USA, 2012.
40. Bhatt, C.R.; Henderson, S.; Brzozek, C.; Benke, G. Instruments to Measure Environmental and Personal Radiofrequency-Electromagnetic Field Exposures: An Update. *Phys. Eng. Sci. Med.* **2022**, *45*, 687–704. [CrossRef]
41. Ramirez-Vazquez, R.; Escobar, I.; Arribas, E.; Vandenbosch, G.A.E. Systematic Review of Exposure Studies to Radiofrequency Electromagnetic Fields: Spot Measurements and Mixed Methodologies. *Appl. Sci.* **2024**, *14*, 11161. [CrossRef]
42. Ahsan Ashraf, M.; Celik, T. Evaluating Radiofrequency Electromagnetic Field Exposure in Confined Spaces: A Systematic Review of Recent Studies and Future Directions. *Radiat. Prot. Dosim.* **2024**, *200*, 598–616. [CrossRef]
43. Vogt-Ardatjew, R.; Dudareva, D.; Wiid, G.; Leferink, F. Low-Cost Measurement Devices for Education. *IEEE Electromagn. Compat. Mag.* **2023**, *12*, 100–104. [CrossRef]
44. Alali, R. Evaluation of Electromagnetic Field Levels Generated from Mobile Communication Stations near the University of Karbala. *Tech. Biochem.* **2025**, *13*, 93–111. [CrossRef]
45. Calabrò, E.; Magazù, S. Monitoring Electromagnetic Field Emitted by High Frequencies Home Utilities. *J. Electromagn. Anal. Appl.* **2010**, *2*, 571–579. [CrossRef]
46. Ibrani, M.; Mekuli, A. Experimental Assessment of Electric Field Levels Emitted by UHF TV Broadcasters. In Proceedings of the 2019 42nd International Convention on Information and Communication Technology, Electronics and Microelectronics (MIPRO), Opatija, Croatia, 20–24 May 2019; pp. 486–490. [CrossRef]
47. Matei, M.; Doru, A.; Radu, C. Analysis in Hospital Operating Rooms: Understanding Perturbations from Electrocautery. *IOP Conf. Ser. Mater. Sci. Eng.* **2024**, *1320*, 012016. [CrossRef]
48. Vasylieva, O.; Neyezhmakov, P. Metrological Traceability of the Results of Testing for Electromagnetic Compatibility in Accordance with the NATO Standards. *Ukr. Metrol. J.* **2023**, 7–15. [CrossRef]
49. Gaydos, D.; Nayeri, P.; Haupt, R. Experimental Demonstration of a Software-Defined-Radio Adaptive Beamformer. In Proceedings of the 2018 15th European Radar Conference (EuRAD), Madrid, Spain, 26–28 September 2018; pp. 561–564.
50. Mokhtari, M.; Popović, M. Low-Cost SDR-Based RF Transceiver for Microwave Breast Screening. In Proceedings of the 2024 IEEE MTT-S International Microwave Biomedical Conference (IMBioC), Montreal, QC, Canada, 11–13 June 2024; pp. 143–145.
51. Watne, Å.K.; Linden, J.; Willhelmsson, J.; Fridén, H.; Gustafsson, M.; Castell, N. Tackling Data Quality When Using Low-Cost Air Quality Sensors in Citizen Science Projects. *Front. Environ. Sci.* **2021**, *9*, 733634. [CrossRef]
52. Kotzagianni, M.; Hassani, A.; Morresi, N.; Udina, S.; Kyfonidis, C.; Roussos, A.; Casaccia, S.; Revel, G.M.; Noriega-Ortega, B. Calibration Strategies for Low-Cost Compact Field Sensors in Citizen Science Air Quality Measurements: Insights from SOCIO-BEE Project. In Proceedings of the 2023 8th International Conference on Smart and Sustainable Technologies (SpliTech), Split and Bol, Croatia, 20–23 June 2023; pp. 1–6.
53. Lin, Z.Y.; Du, Y.C.; Huang, W.; Tang, J.H.; Tang, C.; Huang, G.; Xu, Y.L.; Zhang, D.; Sun, Z.C. The Preliminary Long-Term Slow Drift Calibration Study in Low- Level RF System. *J. Phys. Conf. Ser.* **2019**, *1350*, 012161. [CrossRef]
54. Smith, K.R.; Edwards, P.; Evans, M.; Lee, J.D.; Shaw, M.; Squires, F.; Wilde, S.; Lewis, A. Clustering Approaches to Improve the Performance of Low Cost Air Pollution Sensors. *Faraday Discuss.* **2017**, *200*, 621–637. [CrossRef]
55. Delaine, F.; Lebental, B.; Rivano, H. Rendez-Vous Based Drift Diagnosis Algorithm for Sensor Networks Toward In Situ Calibration. *IEEE Trans. Autom. Sci. Eng.* **2023**, *20*, 1393–1407. [CrossRef]
56. Wilson, P.F.; Remley, K.A.; Young, W.F.; Gentile, C.A.; Ladbury, J.M.; Williams, D.F. A NIST Perspective on Metrology and EMC Challenges for 5G and Beyond. *IEEE Electromagn. Compat. Mag.* **2018**, *7*, 77–85. [CrossRef]
57. Adda, S.; Bottura, V.; D'Elia, S.; Pasquino, N.; Suman, R.; Vaccarono, M. Scalar Measurements of 5G Mm-Wave EMF Power in NLoS Propagation. In Proceedings of the 2024 IEEE International Symposium on Measurements & Networking (M&N), Rome, Italy, 2–5 July 2024; pp. 1–5.

58. Shastri, A.; Valecha, N.; Bashirov, E.; Tataria, H.; Lentmaier, M.; Tufvesson, F.; Rossi, M.; Casari, P. A Review of Millimeter Wave Device-Based Localization and Device-Free Sensing Technologies and Applications. *IEEE Commun. Surv. Tutor.* **2021**, *24*, 1708–1749. [CrossRef]
59. Wu, W.; Lee, J.; Lau, P.-K.; Kang, T.; Lau, K.K.; Yoo, S.-W.; Zhao, X.; Verma, A.; Lu, I.S.-C.; Yao, C.-W.; et al. A 14-Nm Low-Cost IF Transceiver IC with Low-Jitter LO and Flexible Calibration Architecture for 5G FR2 Mobile Applications. In Proceedings of the 2023 IEEE Radio Frequency Integrated Circuits Symposium (RFIC), San Diego, CA, USA, 11–13 June 2023; pp. 277–280.
60. Pawlak, R.; Krawiec, P.; Żurek, J. On Measuring Electromagnetic Fields in 5G Technology. *IEEE Access* **2019**, *7*, 29826–29835. [CrossRef]
61. Wang, Z. Evaluating the Efficacy of Machine Learning in Calibrating Low-Cost Sensors. *Appl. Comput. Eng.* **2024**, *71*, 30–38. [CrossRef]
62. Chu, H.-J.; Ali, M.; He, Y.-C. Spatial Calibration and PM2.5 Mapping of Low-Cost Air Quality Sensors. *Sci. Rep.* **2020**, *10*, 22079. [CrossRef]
63. Hirafuji, M.; Yoichi, H.; Miki, Y.; Kiura, T.; Tanaka, K.; Fukatsu, T.; Ninomiya, S. Deployment and Improvement of Field Servers in India. In Proceedings of the SICE Annual Conference 2010, Taipei, Taiwan, 18–21 August 2010; pp. 2925–2927.
64. Sitzia, D.; Castello, P.; Muscas, C.; Pegoraro, P.A.; Sulis, S.; Yildiz, S. Performance Evaluation of Particulate Matter Low-Cost Sensors Under Power Supply Variations. In Proceedings of the 2024 IEEE International Instrumentation and Measurement Technology Conference (I2MTC), Glasgow, Scotland, 20–23 May 2024; pp. 1–6.
65. Taylor, M.D. Low-Cost Air Quality Monitors: Modeling and Characterization of Sensor Drift in Optical Particle Counters. In Proceedings of the 2016 IEEE SENSORS, Orlando, FL, USA, 30 October–2 November 2016; pp. 1–3.
66. Hassani, A.; Castell, N.; Watne, Å.K.; Schneider, P. Citizen-Operated Mobile Low-Cost Sensors for Urban Pm2.5 Monitoring: Field Calibration, Uncertainty Estimation, and Application. *Sustain. Cities Soc.* **2023**, *95*, 104607. [CrossRef]
67. Yaqoob, I.; Kumar, V.; Chaudhry, S.A. Machine Learning Calibration of Low-Cost Sensor PM2.5 Data. In Proceedings of the 2024 IEEE International Symposium on Systems Engineering (ISSE), Perugia, Italy, 16–18 October 2024; pp. 1–8.
68. Balteanu, F. RF Circuit Techniques for Transition to 5G Advanced. *Int. J. Microw. Wirel. Technol.* **2025**, *17*, 215–233. [CrossRef]
69. Feng, Z.; Zheng, L.; Ren, B.; Liu, D.; Huang, J.; Xue, N. Feasibility of Low-Cost Particulate Matter Sensors for Long-Term Environmental Monitoring: Field Evaluation and Calibration. *Sci. Total Environ.* **2024**, *945*, 174089. [CrossRef]
70. Okafor, N.U.; Alghorani, Y.; Delaney, D. Improving Data Quality of Low-Cost IoT Sensors in Environmental Monitoring Networks Using Data Fusion and Machine Learning Approach. *ICT Express* **2020**, *6*, 220–228. [CrossRef]
71. Karahan, E.A.; Gupta, A.; Khankhoje, U.K.; Sengupta, K. Deep Learning Based Modeling and Inverse Design for Arbitrary Planar Antenna Structures at RF and Millimeter-Wave. In Proceedings of the 2022 IEEE International Symposium on Antennas and Propagation and USNC-URSI Radio Science Meeting (AP-S/URSI), Denver, CO, USA, 10–15 July 2022; pp. 499–500.
72. Ye, Q.; Li, H.; Lin, Z.; Zhu, Q.; Gao, Q.; Cao, B.; Liu, T.; Zhou, J. Data Correction-Based Model-Driven 3D Radiation Directional Pattern Construction for 5G Base Stations. In Proceedings of the 2024 IEEE 100th Vehicular Technology Conference (VTC2024-Fall), Washington, DC, USA, 7–10 October 2024; pp. 1–5.
73. Scientific-Based Exposure and Risk Assessment of Radiofrequency and Mm-Wave Systems from Children to Elderly (5G and Beyond) | SEAWave | Project | Fact Sheet | HORIZON. Available online: https://cordis.europa.eu/project/id/101057622 (accessed on 26 November 2025).
74. World Health Organization. *IARC to Coordinate Production of a Risk Assessment on 5G Exposures as Part of the EU-Funded SEAWave Project 2023*; World Health Organization: Geneva, Switzerland, 2023.
75. SEAWave. Available online: https://www.enea.it/en/enea-projects/seawave.html?utm_source=chatgpt.com (accessed on 26 November 2025).
76. comunicacio.isg. GOLIAT Study Explores 5G Radiation Exposure Across 10 European Countries. *GOLIAT*. 2025. Available online: https://projectgoliat.eu/news/goliat-study-explores-5g-radiation-exposure-across-10-european-countries/# (accessed on 5 September 2025).
77. 5G expOsure, Causal Effects, and Risk Perception Through Citizen Engagement. Available online: https://www.isglobal.org/-/5g-exposure-causal-effects-and-risk-perception-through-citizen-engagement (accessed on 26 November 2025).
78. Al-Jumaily, A.; Sali, A.; Riyadh, M.; Wali, S.Q.; Li, L.; Osman, A.F. Machine Learning Modeling for Radiofrequency Electromagnetic Fields (RF-EMF) Signals From mmWave 5G Signals. *IEEE Access* **2023**, *11*, 79648–79658. [CrossRef]
79. Eichstädt, S.; Werhahn, O. Metrology for Sensor Networks: Metrological Traceability and Measurement Uncertainties for Air Quality Monitoring. *TM—Tech. Mess.* **2024**, *91*, 419–429. [CrossRef]
80. Orobinskaya, V.N.; Mishina, T.N.; Mazurenko, A.P.; Mishin, V.V. Problems of Interpretability and Transparency of Decisions Made by AI. In Proceedings of the 2024 6th International Conference on Control Systems, Mathematical Modeling, Automation and Energy Efficiency (SUMMA), Lipetsk, Russia, 13–15 November 2024; pp. 667–671.
81. Manassas, A.; Delidimitriou, S.; Wiart, J.; Samaras, T. Predicting Electromagnetic Field Exposure Using Artificial Intelligence Methods. *IEEE Access* **2025**, *13*, 79832–79844. [CrossRef]

82. Salem, M.A.; Lim, H.S.; Chua, M.Y.; Chien, S.F.; Zarakovitis, C.C.; Ng, C.Y.; Rahman, N.Z.A. Investigation of EMF Exposure Level for Uplink and Downlink of 5G Network Using Ray Tracing Approach. Available online: https://ijtech.eng.ui.ac.id/article/view/5928 (accessed on 27 November 2025).
83. Wang, S.; Zhang, Y.; Liu, Y.; Liu, J.; Conil, E.; Jawad, O.; Samaras, T.; Ourak, L.; Wiart, J. Comprehensive Measurement-Based Assessment of Downlink RF-EMF Exposure in Urban Environments: Multi-Method Analysis and Intercomparison. *Bioelectromagnetics* **2025**, *46*, e70033. [CrossRef]
84. Silva, A.T.d.A.; Dias, C.F.; de Lima, E.R.; Fraidenraich, G.; de Morais Eldorado, G.I. A Pathway on 5G EMC Testing: A Tutorial. *IEEE Electromagn. Compat. Mag.* **2022**, *11*, 63–72. [CrossRef]

https://doi.org/10.3390/s26020533

MDPI AG
Grosspeteranlage 5
4052 Basel
Switzerland
Tel.: +41 61 683 77 34

Sensors Editorial Office
E-mail: sensors@mdpi.com
www.mdpi.com/journal/sensors

www.ingramcontent.com/pod-product-compliance
Lightning Source LLC
Chambersburg PA
CBHW040836250926
51850CB00007BA/1329
9783725867141